Springer Series on Polymer and Composite Materials

Series editor

Susheel Kalia, Army Cadet College Wing, Indian Military Academy, Dehradun, India

More information about this series at http://www.springer.com/series/13173

Swati Sharma · Ashok Kumar
Editors

Keratin as a Protein Biopolymer

Extraction from Waste Biomass and Applications

Springer

Editors
Swati Sharma
Faculty of Chemical and Natural Resource Engineering
Universiti Malaysia Pahang
Kuantan, Malaysia

Ashok Kumar
Department of Biotechnology and Bioinformatics
Jaypee University of Information Technology
Waknaghat, Solan, India

ISSN 2364-1878 ISSN 2364-1886 (electronic)
Springer Series on Polymer and Composite Materials
ISBN 978-3-030-02900-5 ISBN 978-3-030-02901-2 (eBook)
https://doi.org/10.1007/978-3-030-02901-2

Library of Congress Control Number: 2018960205

This Springer imprint is published by the registered company Springer Nature Switzerland AG
The registered company address is: Gewerbestrasse 11, 6330 Cham, Switzerland

Preface

This book was encouraged by the aspiration that we have to face the major challenge of environmental pollution. With increasing urbanization and industrialization, various kinds of waste products accumulated in the ecosystem which made the living habitat nuisance. Organic wastes from the household and food industry are also a major part of this problem. Poultry industry generated a huge amount of waste which is either dumped off or discarded in landfills and rivers which cause serious health concerns.

This book had been written to provide a framework for sustainable management of keratin-rich waste biomass from various sources. This book mainly emphasizes on the various techniques recently used for the extraction of valuable product like keratin from poultry waste biomass. Researchers from various countries have been trying to recycle the organic waste from different sources into value-added products. At present, millions of tons of keratin(s)-rich waste biomass have been yearly produced by food industry and that need to be faced with novel technological solutions able to afford valuable products. This book will be helpful to various groups in academia and industry to find the updated solution and methods for the conversion and application of keratin. As this text emphasizes, students must learn to identify various advanced techniques what skills are needed, and how effectively can be applied with appropriate procedures.

In essence, the purpose of this book is to provide a toolbox from which researchers, students, and industrialists who can plan a project in an academic or industrial context. Another major reason for writing the present book was our own research interest in the same topic. Through this book, we are collecting the information from various researchers engaged in the management of keratin-rich waste biomass from various sources. Generally, the students and early researchers tend to search the information on Internet for hours or days and left either with little valuable information or confusing results. Thus, in this book, we have chosen meticulous list of chapters to consider the need of quantifying the biological question of interest.

Chapter 1 described the basic structure function and sources of keratin from waste materials. Chapter 2 deals with production strategies of keratin used by various researchers which can directly provide table information about methods used. In this continuation, Chaps. 3 and 4 were dealing with the extraction and characterization and processing of keratin. Microbes also play an important role in the degradation of the insoluble keratin; Chap. 5 deals with the role of microorganism in the degradation of keratin-rich biomass. Chapters 6, 7, and 8 have summarized the application of extracted keratins in various fields such as biopolymer, biofilms, hydrogels, biofibers, and various other biotechnological aspects. We firmly hope that this book will benefit the students, academicians, and industrialists.

Kuantan, Malaysia Swati Sharma
Solan, India Ashok Kumar

Contents

1 **Keratin: An Introduction** 1
Swati Sharma, Arun Gupta and Ashok Kumar

2 **Keratin Production and Its Applications: Current and Future Perspective** 19
Anshuman Shah, Shaily Tyagi, Ram Naresh Bharagava, Dalel Belhaj, Ashok Kumar, Gaurav Saxena, Ganesh Dattatraya Saratale and Sikandar I. Mulla

3 **Extraction and Characterization of Keratin from Different Biomasses** 35
Claudia Vineis, Alessio Varesano, Greta Varchi and Annalisa Aluigi

4 **Keratin Processing** 77
Diego Omar Sanchez Ramirez, Riccardo Andrea Carletto and Francesca Truffa Giachet

5 **Degradation of Keratin Biomass by Different Microorganisms** 123
I. A. Adelere and A. Lateef

6 **Keratin as a Biopolymer** 163
Sarthak Saha, Muhammad Arshad, Muhammad Zubair and Aman Ullah

7 **Keratin-Based Biofilms, Hydrogels, and Biofibers** 187
James McLellan, Starla G. Thornhill, Spencer Shelton and Manish Kumar

8 **Keratin-Based Biotechnological Applications** 201
Kush Kumar Nayak, Piyush Parkhey and Bidyut Mazumdar

About the Editors

Dr. Swati Sharma has completed her PhD. from University Malaysia Pahang, Malaysia in 2018. She worked as a visiting researcher in college of life and environmental sciences at Konkuk University, Seoul South Korea. Dr. Sharma has completed her masters (M.Sc.) in 2013 from Dr. Yashwant Singh Parmar University of Horticulture and Forestry, Nauni Solan H. P. India. She has also worked as program co-coordinator at Himalayan action research center Dehradoon and Senior research fellow at India agricultural research institute in 2013–2014. Dr. Sharma has published her research papers in reputed international journals. Presently, Dr. Sharma's research is in the field of bioplastics, hydrogels, keratin nanofibres and nanoparticles, biodegradable polymers and polymers with antioxidant and anti-cancerous activities and sponges. Dr. Swati has published 14 research papers in various internationally reputed journals.

Dr. Ashok Kumar is working as an Assistant professor in Department of Biotechnology and Bioinformatics, Jaypee University of Information Technology, Waknaghat, Solan, Himachal Pradesh, India. He has completed his Ph.D. in Biotechnology from Department of Biotechnology, Himachal Pradesh University, Summer Hill, Shimla, India in 2013. He worked as a post-doctoral fellow in State Key laboratory of Agricultural Microbiology, Huazhong Agricultural University, Wuhan China for two years, 2014-2016. He also worked as Post Doctoral research associate at Konkuk University Seoul South Korea for 2016–2017. Dr. Ashok has keen interest in enzyme immobilization, biomaterials, biopolymers, nanobiotechnology, biocatalysis, waste management, biomass degradation, biofuel synthesis, gene cloning, characterization and biotransformation. Dr. Ashok has published 44 research papers in various internationally reputed journals and 11 book chapters.

Abbreviations

%	Percent
3D	Three-dimensional
4aPEONB	4-arm PEO norbornene
AAL	Allyl alcohol
Amim	1-allyl-3-methyl imidazolium
AMP	Allyl mercaptan
BABP	Bili acid-binding protein
BEA	2-Bromoethylamine
bFGF	Basic fibroblast growth factor
Bmim	1-butyl-3-methyl imidazolium
BOD	Biochemical oxygen demand
BSA	Bovine serum albumins
BSCA	Bromosuccinic acid
C_2C_{12}	Myofibroblast skin cell line
CAA	Monochloro acetic acid
CO_2	Carbon dioxide
COD	Chemical oxygen demand
DALS	Dialdehyde starch
DEP	Diethyl phosphate
DES	Deep eutectic solvents
DFP	Diisopropyl fluorophosphates
DFUs	Diabetic foot ulcers
DMF	Dimethylformamide
DMLS	Direct metal laser sintering
DMSO	Dimethyl sulfoxide
DO	Dissolved oxygen
DOX.HCl	Doxorubicin hydrochloride
DTT	Dithiothreitol
ECM	Extracellular matrix
EDTA	Ethylenediaminetetraacetic acid

EG	Ethylene glycol
EGDE	Ethylene glycol diglycidyl ether
Emim	1-ethyl-3-methyl imidazolium
FAL	Formaldehyde
FDA	Food and drug administration
FDM	Fused deposition modeling
FTIR-ATR	Fourier transform infrared spectroscopy-attenuated total reflection
GAL	Glutaraldehyde
GDE	Glycerol diglycidyl ether
GLX	Glyoxal
GNP	Genipin
HA	Hyaluronic acid
HFIP	Hexafluoroisopropanol
IAA	Iodoacetamide
IAC	Iodoacetic acid
IF	Intermediate filaments
ILs	Ionic liquids
ISO	International organization for standardization
KCN	Potassium cyanide
KH	Keratin hydrolysates
LMW	Low molecular weight
LS	Low sulfur content
MEC	Mercaptoethanol
MMP	Metalloproteinases
MSH	O-(2,4,6-trimethyl benzene sulfonyl) hydroxylamine
MW	Molecular weight
MWCO	Molecular weight cut-off
NEM	N-Ethylmaleimide
NFs	Nanofibers
nKer	Keratin nanoparticles
NPs	Nanoparticles
OSPW	Oil sand process affected water
OVAT	One variable at a time
PA6	Polyamide-6
PBS	Phosphate-buffered saline
PCA	Poly-e-caprolactone
PCL	Poly(ε-caprolactone)
PDA	Polydopamine
PDLA	Poly-*d*-lactic acid
PE	Polyethylene
PEG	Polyethylene glycol
PEGDE	PEO diglycidyl ether
PEI	Polyethylenimine
PEO (or PEG)	Polyethylene oxide (or Polyethylene glycol)

PEOMEM	PEO methyl ether methacrylate
PG	Propylene glycol
PGA	Polyglycolic acid
PGPR	Plant growth-promoting rhizobacteria
PHA	Polyhydroxyalkanoates
PHB	Polyhydroxybutyrate
PHBV	Poly(3-Hydroxybutyrate-co-3-hydroxyvalerate)
pKi	Isoelectric point
PLA (or PLLA)	Polylactic acid
PLGA	Poly(lactic-co-glycolic acid)
PLLA	Poly-*l*-lactic acid
PNPAMM	P(NIPAM)n-NH-maleimide
POSS	Polyhedral oligomeric silsesquioxanes
PP	Polypropylene
PPG	Polypropylene glycol
PS	Polystyrene
PVA	Polyvinyl alcohol
PVP	Polyvinylpyrrolidone
R′3P	Phosphines
RSM	Response surface methodology
SDS	Sodium dodecyl sulfate
SDS–PAGE	Sodium dodecyl sulfate–polyacrylamide gel electrophoresis
SE	Steam explosion
SEM	Scanning electron microscope
SFE	Steam flash explosion
SLA	Stereolithography
SLS	Selective laser sintering
TCEP	Tris(2-carboxyethyl)phosphine
TDI	Toluene diisocyanate
TFA	Trifluoroacetic acid
TFE	Trifluoroethanol
TGA	Thermogravimetric analysis
TLCK	Tosyl phenylalanyl chloromethyl ketone
UV	Ultraviolet
UVB	Ultraviolet B
VOCs	Volatile organic compounds

Chapter 1
Keratin: An Introduction

Swati Sharma, Arun Gupta and Ashok Kumar

Abstract What is keratin? And why to use the keratin? Well known that protein is a part of every cell in living organism's body which plays many different roles to keep living things alive and healthy. The importance of protein for the growth and repair of muscles, bones, skin, tendons, ligaments, hair, eyes, and other tissues is proven since a very long time. Proteins also exist in the form of enzymes and hormones needed for metabolism, digestion, and other important processes. *Natural proteins* are purified from *natural* sources. Keratin is among the most copious proteins found associated with the body of reptiles, birds, and mammals. It is a structural constituent of nail, wool, feathers, and hoofs which offers strength to body and muscles. Nowadays, the keratin-rich waste biomass produced from poultry and meat industry imposes serious threat to environment and living beings. We need to explore various techniques and methods for the extractions and use of keratin from waste biomass. From the industrial point of view, keratin is a useful product in the medical, pharmaceutical, cosmetic, and biotechnological industries. Materials obtained from keratin may be converted into porous foam of different sponges, shapes, coatings, mats, microfibers, gels, and materials of high molecular weight. In this chapter, we briefly describe the various sources, properties, and structures of keratin.

Keywords Protein · Waste · Biomass · Medical · Pharmaceutical · Application Extraction

---

S. Sharma · A. Gupta
Faculty of Chemical Engineering and Natural Resources,
Universiti Malaysia Pahang, 26300 Gambang, Kuantan, Malaysia

A. Kumar (✉)
Department of Biotechnology and Bioinformatics,
Jaypee University of Information Technology, Waknaghat, Solan 173234,
Himachal Pradesh, India
e-mail: ashok.nadda09@gmail.com

S. Sharma and A. Kumar (eds.), *Keratin as a Protein Biopolymer*, Springer Series on Polymer and Composite Materials, https://doi.org/10.1007/978-3-030-02901-2_1

1 Introduction

Growing need for sustainable and safe bio-based materials due to rising environmental concern has forced the use of available natural by-products as a substitution. By-products from different animal sources are recently being used for beneficial purposes such as drug delivery, medicines, cosmetics, and bioplastic.

Keratin is among the most abundant structural proteins (Coulombe and Omary 2002), and in animals together with collagen, it is the most important biopolymer (McKittrick et al. 2012). Keratinous materials, formed by specifically organized keratinized cells filled with mainly fibrous proteins (keratins), are natural polymeric composites that exhibit polypeptide chain structure, filament–matrix structure lamellar structure, and sandwich structure ranging from nanoscale to centimeter scale. Keratin is among the most copious proteins found associated with the body of reptiles, birds, and mammals. It is a structural constituent of nail, wool, feathers, and hoofs which offers strength to body and muscles (Reichl et al. 2011). They serve several functions, such as for predation and as armor, protection, and defense. Therefore, a thorough understanding of the relationships between the units that make up a functional keratinous material would expectantly provide useful knowledge in designing new materials. Keratin is chiefly found in epithelial cells in higher vertebrates (Korniłłowicz-Kowalska and Bohacz 2011). Keratins have high strength, stiffness, and insolubility in polar as well as nonpolar solvents. The stabilization is the result of intramolecular and intermolecular disulfide crosslinks, hydrogen bonding, and its crystallinity. These properties will differentiate it from other fibrous proteins like myofibrillar and collagen protein (Schrooyen et al. 2000). With increasing urbanization, food industries particularly the wool industry, slaughterhouse, and meat market produce million tons of keratin-containing biomass. The major producers including USA, Brazil, and China, report for more than 40 million tons annually. These proteins compose keratin by-products containing 15–18% nitrogen, 2–5% sulfur, 1.27% fat, 3.20% mineral elements, and 90% of proteins (Kunert 2000; Sangali and Brandelli 2000; Gessesse et al. 2003).

From past several years, keratin has been extracted from various sources such as horn, hoof, hair, beaks, shells, toenails, claws, fingernails, and feathers (Sharma and Gupta 2016; Sharma et al. 2017c). Keratin was used in medicine for the first time by Chinese herbalist Shi-Zen in sixteenth century. Hoffmeier, in 1905, first time extracted keratin from animal hoofs with the help of lime which further was used in making gels (Rouse and Van Dyke 2010). Keratins are cystine-rich proteins associated with intermediate filaments (IFs) which are cytoskeleton element having diameter of 8–10 nm (ARAI et al. 1983; Khosa et al. 2013). It is available in two forms α- and β-keratins. α-keratins are copiously found in the soft tissues such as sheep wool, hair, and skin. These are rich in cystine. β-keratins are present in hard tissue protein of nails, fish scales, bird feathers, and others. They are rich in glycine and alanine, poor in cystine, hydroxyproline, and proline (Gupta et al. 2012). Keratins are very much stable and insoluble in most of organic solvents. The presence of cystine in ample amount has made the keratin more susceptible to hydrolytic and oxidation reactions (Schrooyen et al. 2000; Barone et al. 2006b; Endo et al. 2008).

Nowadays, a large amount of keratin by-products are unused which is probable hazard to the environment (Cavello et al. 2012; Park et al. 2013). Keratin waste is classified into three categories in regulation (EC) 1774/2002 of the European Parliament and Council of 3rd October 2002 laying down health rules concerning animal by-products which are not intended for human consumption but does not spread diseases to humans or animals (Korniłłowicz-Kowalska and Bohacz 2011). In the form of solid biomass, keratin is less prone to enzymatic hydrolysis due to high cross-linking by hydrogen bonding, disulfide bonds, and hydrophobic interactions (Korniłłowicz-Kowalska and Bohacz 2011).

Thus, use of keratin as a biopolymer requires its extraction from the biomass. In recent years, several attempts have been made for the extraction of keratin using chemical, mechanical, and enzymatic methods (Korol 2012; Jeong et al. 2010; Chaudhari et al. 2013; Fang et al. 2013). There are number of ways for extraction of keratin from the waste biomass including acidic hydrolysis (Breinl and Baudisch 1907; Earland and Knight 1956), alkaline hydrolysis (Tsuda and Nomura 2014; Song et al. 2013; Poole et al. 2008), enzymatic hydrolysis (Eslahi et al. 2013), ionic liquid hydrolysis (Idris et al. 2014; Wang and Cao 2012), and alkaline–enzymatic hydrolysis (Yin et al. 2007). Acidic hydrolysis provides very severe conditions which can destroy some useful amino acids during hydrolysis. Conversely, enzymatic hydrolysis provides less species alteration but with a slower process and is more expensive (Staroń et al. 2014), which makes its commercial use more difficult. On the other hand, ionic liquids are too costly and protein recovery is very low (Cevasco and Chiappe 2014) to be used for industrial purpose. Hence, research into simple, cheap, environmentally sustainable, and industrial applicable method to extract keratin seems justifiable.

Keratin is a useful product in the medical, pharmaceutical, cosmetic, and biotechnological industry. Materials obtained from keratin may be converted into porous foam of different sponges, shapes, coatings, mats, microfibers, gels, and materials of high molecular weight. Keratin is attracting the attention of the researchers due to its abundance. Keratin biomaterial is applied in the development of wound healing gels, tissue engineering, drug delivery, trauma and medical devices, biomedical, and cosmetic applications. One of the impending applications of purified keratin is to produce biomaterials in regeneration and tissue repair (Alsarra 2009; Ramshaw et al. 2009; Natarajan et al. 2012; Ramadass et al. 2013; Kumaran et al. 2016).

These are the polymers formed by various amino acids capable of promoting intra- and intermolecular bonds, allowing the resultant materials to have a large variation in their functional properties (Gupta and Perumal 2013). Feather keratin is a potential source of abundant, inexpensive, eco-friendly, and commercial biomaterial (Poole et al. 2009; Shi et al. 2014). Keratin word first emerged around 1850 and it illustrates the material which constitutes hard tissues such as animal horns and hoofs. The name was taken from the word "kera" which means horn. Keratin is an insoluble, highly stable structure, small proteins, and uniform in size. Feather keratin has a molecular weight of about 10 kDa (Fraser et al. 1972; ARAI et al. 1983; Ullah et al. 2011; Kamarudin et al. 2017). It is composed of α-helix, β-sheet structures as discussed in further chapters. Also, the internal structure of every keratin has α-helices and β-sheets that support the protein. The elastic nature of keratin fiber is due to the interplay between α-helices and β-sheet configuration of the protein. Feathers consist of 50% of each fiber and quill by weight (Reddy and Yang 2007a). In a feather, fiber has a larger percentage of α-helix (41%) as compared to β-sheet (38%) and quill fraction consists of more β-sheet (50%) structure than α-helix (21%) (Barone et al. 2006a; Schmidt and Jayasundera 2004; Wallenberger and Weston 2003; Fraser et al. 1972). According to a previous study (Sun et al. 2009a), feather has 9.38% α-helix, 47.19% β-sheet, 32.25% β-turn, and 11.18% in random.

Keratin has about 7% cystine, which forms S–S bonds with other cystine molecules (ARAI et al. 1983) and forms cysteine by disulfide bridges. The presence of disulfide, hydrophobic, and hydrogen bond (Onifade et al. 1998; ARAI et al. 1983; Ullah et al. 2011; Cardamone 2010; Bulaj 2005) in keratin provides it strength, mechanical stability, rigidity, and resistance to degradation by proteolytic enzymes such as trypsin, pepsin, and papain (Yamamura et al. 2002; Agrahari and Wadhwa 2010; Paul et al. 2013) to keratin in the solid state. However, these cross-links are a hindrance to processing in the melt state (Barone et al. 2006b). The presence of reactive functional groups, especially peptide backbone, disulfide (–S–S) bridges, amino ($-NH_2$), and carboxylic acid (–COOH), makes it chemically reactive under favorable reaction conditions. During controlled reduction, protonation of keratin occurs. Thus, the keratin protein attains positive surface charge and becomes pseudo-cationic biopolymer.

Keratin is insoluble in polar and nonpolar solvents and has very low chemical reactivity. At low pH, high temperature and in presence of reducing agents the solubility of the keratin are increased. The biodegradability and nontoxic nature of keratin make it versatile biopolymer which can be modified and extended in various forms such as films, gel, beads, and nano/microparticles. The modified keratin has plenty of applications in food sciences, green chemistry, cosmetic industries, and pharmaceuticals.

2 Chemical Composition and Occurrence of Keratin

Keratin is the most important component of hair, wool, nails, hooves, claws, scales, horn, beaks, and feathers as shown in Table 1. These are least affected by chemical and physical environmental factors (Teresa and Justyna 2011). The keratin extracted is with 90–100 amino acids and 10.2–10.4 kDa molecular weight (Kamarudin et al. 2017; Barone and Schmidt 2005). Chemical structure of keratin showed α-helix, β-helix, or β-pleated sheet (Fraser et al. 1972; Lee and Baden 1975). Keratin has a high amount of cystine residues (7–15 mol% of amino acids) as compared to other which help to make intermolecular cross-links (Rouse and Van Dyke 2010). The amount of cysteine residues depends on the keratin source, which varies from 7% in feather to 15% in wool keratin (Fraser et al. 1972; ARAI et al. 1983).

Table 1 Distribution of α- and β-keratins

Types of keratin	Source organ
α-Keratin	Wool, quills, hair, horns, fingernails, hooves; stratum corneum
β-Keratin	Feathers, avian beaks and claws, reptilian claws and scales
	α- and β-keratin reptilia epidermis, pangolin scales

The high cystine content is the unusual characteristic of keratin which differentiates it from other structural proteins like elastin and collagen. Main amino acids present in keratin are cystine, proline, serine, and glycine (Fraser et al. 1972; Fraser and Parry 2003). In another study, γ-keratin was extracted which is nonstructural and associated with α- and β-sheets (Fraser et al. 1972; Hill et al. 2010). The sulfur content ratio plays an important role in keratin physical properties. Some researchers based upon sulfur content classified keratin as soft and hard forms (Rizvi and Khan 2008; Zoccola et al. 2009). Soft keratin has lower cystine content, week cross-linking, and smaller resistant to other chemicals found in the hair core and outer layer of epidermis (Fraser et al. 1972).

Hard keratin found in mammalian epidermal appendages, such as hairs, nails, horns, and in avian or reptilian tissues. In recent years, equine hoof (Douglas et al. 1996), bovine hoof, wool (Feughelman and Robinson 1971), and especially the sheep horn (Tombolato et al. 2010) are the attractive sources for keratin extraction.

3 Keratin Sources

Keratin biomass is derived from living organisms or from their body parts after death. The major livestock's of keratin includes sheepskins, goatskins, cattle hides, feathers, hairs, and buffalo hides as shown in Fig. 1. Skin and its appendages such as feathers, wool, nails, hooves, hair, scales, and stratum corneum are the richest sources of keratin (Kim 2007). It can be extracted from animal horns and hooves, wool, feathers, and human hairs (Fig. 1). Food industry produces million tons of keratin biomass. About 80% of human hair is formed of keratin only (Kaplin 1982; Wagner and Joekes 2005). It provides flexibility, strength, durability, and functionality to the hair in the form of different conformations (Velasco et al. 2009). Keratinous materials based on α- and β-keratins are discussed in Table 2.

Fig. 1 Main sources of keratin biomass

Table 2 Keratinous material based on the presence of α- and β-keratins

Keratinous materials based on α-keratin	Keratinous materials based on β-keratin	Keratinous materials based on α- and β-keratins
Stratum corneum	Feathers	Reptilian epidermis
Wool and hair	Beaks	Hard and soft epidermis of Testudines
Quills	Claws	Pangolin scales
Horns		
Hooves		
Nails		
Whale baleen		
Hagfish slime threads		
Whelk egg capsules		

4 Applications of the Keratin Protein

Keratin biomass is hydrolyzed by alkali, acid, or enzyme to extract keratin. The extracted keratin has various applications in various industries such as cosmetic, biomedical, and pharmaceutical industries. Furthermore, it does not have harmful effects and thus can be used for variety of cosmetics such as creams, shampoos, hair conditioners, and biomedical products. They have been used as a treatment of skin and human hair as reported before (Weigmann et al. 1990; Innoe 1992). Its existence in the hair cuticle and stratum corneum helps in preserving skin moisture while interacting with cosmetics. Its combination with other natural polymers such as chitosan, collagen, and silk fibroin was used as a component for cosmetic products (Sionkowska 2015). Keratin with high molecular weight is mostly used for skin care applications due to its individuality like film forming and hydrophilic. Keratin film or coating on skin provides smooth and soft sensation. Keratin-associated proteins from different sources were developed and applied as microscaffolds in medicine and cosmetics (Lipkowski et al. 2009). Proteins are useful ingredients for healthy skin and hair. There are studies which reported the role and efficacy of using protein in cosmetics; proteins can be obtained from simple and conjugated proteins (Secchi 2008). There are also studies which described the keratin derivatives and cationizing agent in the various cosmetics having specific functional groups (Matsunaga et al. 1983). Figure 2 shows the hair treatment cream formed using keratin extracted from chicken feathers. Other applications from different biomasses are discussed in Table 3.

Because the presence of high cross-linking by cystine formulation of micro- and nanoparticle from feathers is difficult, some researchers have prepared micro- and nanoparticles successfully from feather keratin. Keratin was converted into useful microparticles by treatment with ionic liquid, 1-butyl-3-methylimidazoliumchloride (Sun et al. 2009b). Treated feathers have low surface area but have higher ion sorption capacity than untreated feathers due to their hydrophilic nature. (Xu et al. 2014b) developed nanoparticles (50–130 nm) from feather keratin which showed good biocompatibility and stability essential for controlled drug release. Here, the chicken feather keratin nanoparticles were developed and used as a haemostatic agent which resulted in decrease in the bleeding time and blood loss in tail amputation and liver scratch rat models (Wang et al. 2016). In another study, quail feathers keratin incorporated into silver nanoparticles and formed the nanofibrous scaffold which gave 99.9% and 98% of the antibacterial activity against Gram-negative (*Escherichia coli*) and Gram-positive (*Staphylococcus aureus*) bacteria, respectively. Thus, it can be used for biomedical applications (Khajavi et al. 2016). As chitosan has good properties of biodegradation and biocompatibility, when they are mixed with keratin nanoparticles, they form scaffold. The biodegradation and protein adsorption of the scaffold had increased, and it was noncytotoxic to human osteoblastic cells. Thus, this scaffold can work as biomimetic substrate for bone tissue engineering applications (Saravanan et al. 2013). In one study, keratin nanopowder from chicken feather was produced by electrospraying which has a small particle size and has less crystallinity

Fig. 2 Hair treatment cream made by using keratin extracted from chicken feathers

than raw keratin (Rad et al. 2012). Water stable nanoparticles were developed from feather keratin. Figure 3 shows the keratin nanoparticles extracted from chicken feathers. Nanoparticles can be a good veterinary diagnostic, and it can penetrate

Table 3 Applications from different keratin biomasses

Sr. no.	Sources	Industrial application(s)	Reference(s)
1	Human hairs	Medicinal use	Zheng et al. (2005)
2	Hoof and horn's	Preparation of firefighting composition	Datta (1993)
3	Human hair and wool	To explore structural and biological properties of self-assembled keratins	Xu et al. (2014b)
4	Feathers	Development of protein fibers and 2D and 3D scaffolds for tissue engineering	Xu et al. (2014a), Rouse and Van Dyke (2010)
5	Chicken feathers	Keratin film for drug delivery system	Poole et al. (2009), Yin et al. (2013)
		Regenerated fibers	Xu et al. (2014a)
		Micro- and nanoparticles	Sun et al. (2009b)
		Graphene oxide and its derivative in biomaterials	Amieva et al. (2014)
		As a diet supplement for feeding ruminants	Coward-Kelly et al. (2006), Dalev (1994), Dalev et al. (1996), Dalev et al. (1997)
		Microporous material used as electrode material	Zhan and Wool (2011)
		Thermoplastic films	Reddy et al. (2013), Jin et al. (2011)
		Waste management using microorganisms for degradation	Vasileva-Tonkova et al. (2009), Syed et al. (2009), Grazziotin et al. (2006)
		Leather processing	Sastry et al. (1986), Sehgal et al. (1987), Karthikeyan et al. (2007)
		Handspun yarn	Reddy and Yang (2007b)
		Textile yarns	Reddy et al. (2014a, b), Yang and Reddy (2013)
		Keratinases in detergents formulation	Balakumar et al. (2013), Manivasagan et al. (2014), Rai et al. (2009)
		Flame retardant	Wang et al. (2014)
		Bio-composites or composite fabrication	Flores-Hernández et al. (2014), Spiridon et al. (2012), Huda and Yang (2008, 2009)
		Biofertilizer	Ichida et al. (2001), Kornillowicz-Kowalska and Bohacz (2010), Gurav and Jadhav (2013), Hadas and Kautsky (1994), Gousterova et al. (2012)

(continued)

Table 3 (continued)

Sr. no.	Sources	Industrial application(s)	Reference(s)
		Nanoparticle and microparticles for pharmaceutical application	Xu et al. (2014b), Sundaram et al. (2015), Yu et al. (2014), Sharma et al. (2017a, b)
		Keratin and graphene oxides within biomaterials	Amieva et al. (2014)
		Cement-bonded composites	Acda (2010)
		Keratin hydrogel	Wang et al. (2017), Barati et al. (2017), Priyaah et al. (2017)
		Tissue regenerative applications	Li et al. (2013), Saravanan et al. (2013), Kumar et al. (2017)
		Paper production	Tesfaye et al. (2017)
		Bioplastic	Sharma et al. (2018), Ramakrishnan et al. (2018)

easily into cells and organs due to their nano size. They are good for the drug delivery as compared to synthetic polymer and carbohydrates. Xu et al. (2014b) found that keratin nanoparticles have supportive function for the cell growth and stable in physiological environment for up to 7 days. The keratin nanoparticle shows anticancer properties. When keratin nanoparticles combine with chlorin e6, it resulted in greater cell death percentage (90%) as compared to free chlorin e6 on osteosarcoma (U2OS) and glioblastoma (U87) cells lines, and thus keratin nanoparticles are effective and promising delivery vehicles for photodynamic therapy applications (Aluigi et al. 2016). In one study, keratin-based drug-loaded nanoparticles were made and it was showing pH and glutathione dual-responsive behavior. Thus, from the results of the study conducted, it was concluded that keratin-based drug carriers had potential for drug delivery and cancer therapy in clinical medicine (Li et al. 2017).

Biocompatible composites were developed which can be used for high-performance dressing, to treat chronic ulcerous infected wounds by the combination of cellulose and keratin with the silver nanoparticles (Tran et al. 2016). Hair keratin nanoparticle had faster clotting time, and it significantly reduces blood loss and coagulation time which forms the viscosity gel on wound; hence, it has a great potential for haemostatic application (Luo et al. 2016). The soluble keratin would have applications in tissue regeneration, cell seeding, wound healing, and drug delivery (Yamini Satyawali 2013). Soluble keratin can be used to make 2D and 3D scaffolds, and protein fibers which further utilized for tissue engineering (Xu et al. 2014b). Due to self-congregation and polymerization property of keratin proteins has led to work as scaffolds for tissue engineering (Rouse and Van Dyke 2010).

When the partially oxidized keratin is added into water, it forms the hydrogel which can be used as an absorbent material, as a therapeutic for skin. The hydrogel can be used for the implantation. The keratin can be incorporated into films which are

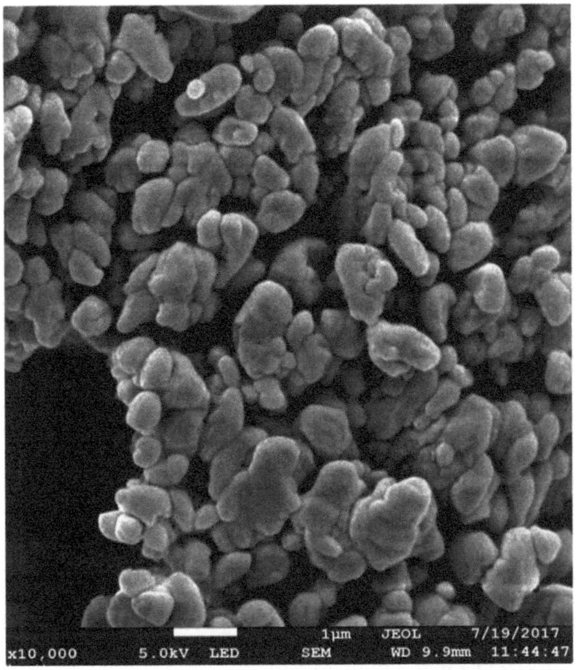

Fig. 3 Keratin nanoparticles extracted from chicken feathers

nonwoven, and these two materials are suitable for use in tissue engineering scaffolds (Blanchard et al. 2002). Keratin biomaterials have the potential to interact with cells and tissues but the composition, structure, and cell-instructive characteristics are not clear. Burnett et al. (2013) made keratin-based biomaterial, demonstrate self-assembly of cross-linked hydrogels, investigate a cell-specific interaction, and find the utilization in drug and cell delivery, tissue engineering, regenerative medicine, and trauma. The huge amount of waste biomass generated by animals as well as food industries can be used as raw material for the production of keratin at industrial level. The keratin waste biomass management by reconversion into industrially used products will save the ecosystem from large amount of sludge and boost up industries such as pharmaceutical and cosmetic industry economically.

5 Conclusion

To develop the strategies for efficient extraction of keratin from poultry biomass will prove to be very beneficial for sustainable management of huge waste. Researchers are working to develop various chemical, biological, and physical methods individually as well as in combined form for keratin extraction. The insoluble protein has many advantages in biomedical industry to develop products of pharmaceutical use, in tissue engineering as well as in agriculture industry too. Recent advances in the sustainable management of poultry waste biomass, extraction techniques used by various researchers, and numerous applications of the extractions have been discussed in the subsequent chapters of this book.

References

Acda MN (2010) Waste chicken feather as reinforcement in cement-bonded composites. Philippine J Sci 139(2):161–166

Agrahari S, Wadhwa N (2010) Degradation of chicken feather a poultry waste product by keratinolytic bacteria isolated from dumping site at Ghazipur poultry processing plant. Int J Poult Sci 9(5):482–489

Alsarra IA (2009) Chitosan topical gel formulation in the management of burn wounds. Int J Biol Macromol 45(1):16–21

Aluigi A, Sotgiu G, Ferroni C, Duchi S, Lucarelli E, Martini C, Posati T, Guerrini A, Ballestri M, Corticelli F (2016) Chlorin e6 keratin nanoparticles for photodynamic anticancer therapy. RSC Adv 6(40):33910–33918

Amieva EJ-C, Fuentes-Ramirez R, Martinez-Hernandez A, Millan-Chiu B, Lopez-Marin LM, Castaño V, Velasco-Santos C (2014) Graphene oxide and reduced graphene oxide modification with polypeptide chains from chicken feather keratin. J Alloy Compd 643:S137–S143

Arai KM, Takahashi R, Yokote Y, Akahane K (1983) Amino-acid sequence of feather keratin from fowl. Eur J Biochem 132(3):501–507

Balakumar S, Mahesh N, Arunkumar M, Sivakumar R, Hemambujavalli V (2013) Optimization of keratinase production by keratinolytic organisms under submerged fermentation. Optimization 5(3):1294–1300

Barati D, Kader S, Pajoum Shariati SR, Moeinzadeh S, Sawyer RH, Jabbari E (2017) Synthesis and characterization of photo-cross-linkable keratin hydrogels for stem cell encapsulation. Biomacromolecules 18(2):398–412

Barone JR, Dangaran K, Schmidt WF (2006a) Blends of cysteine-containing proteins. J Agric Food Chem 54(15):5393–5399

Barone JR, Schmidt WF (2005) Polyethylene reinforced with keratin fibers obtained from chicken feathers. Compos Sci Technol 65(2):173–181

Barone JR, Schmidt WF, Gregoire N (2006b) Extrusion of feather keratin. J Appl Polym Sci 100(2):1432–1442

Blanchard CR, Van Dyke ME, Timmons SF, Siller-Jackson AJ, Smith RA (2002) Non-woven keratin cell scaffold. Google Patents

Breinl F, Baudisch O (1907) The oxidative breaking up of keratin through treatment with hydrogen peroxide. Z Physiol Chem 52:158–169

Bulaj G (2005) Formation of disulfide bonds in proteins and peptides. Biotechnol Adv 23(1):87–92

Burnett LR, Rahmany MB, Richter JR, Aboushwareb TA, Eberli D, Ward CL, Orlando G, Hantgan RR, Van Dyke ME (2013) Hemostatic properties and the role of cell receptor recognition in human hair keratin protein hydrogels. Biomaterials 34(11):2632–2640

Cardamone JM (2010) Investigating the microstructure of keratin extracted from wool: Peptide sequence (MALDI-TOF/TOF) and protein conformation (FTIR). J Mol Struct 969(1):97–105

Cavello I, Cavalitto S, Hours R (2012) Biodegradation of a keratin waste and the concomitant production of detergent stable serine proteases from *Paecilomyces lilacinus*. Appl Biochem Biotechnol 167(5):945–958

Cevasco G, Chiappe C (2014) Are ionic liquids a proper solution to current environmental challenges? Green Chem 16(5):2375–2385

Chaudhari PN, Chincholkar SB, Chaudhari BL (2013) Biodegradation of feather keratin with a PEGylated protease of chryseobacterium gleum. Process Biochem 48(12):1952–1963. https://doi.org/10.1016/j.procbio.2013.09.011

Coulombe PA, Omary MB (2002) 'Hard' and 'soft' principles defining the structure, function and regulation of keratin intermediate filaments. Curr Opin Cell Biol 14(1):110–122

Coward-Kelly G, Agbogbo FK, Holtzapple MT (2006) Lime treatment of keratinous materials for the generation of highly digestible animal feed: 2. Animal hair. Bioresour Technol 97(11):1344–1352

Dalev P, Ivanov I, Liubomirova A (1997) Enzymic modification of feather keratin hydrolysates with lysine aimed at increasing the biological value. J Sci Food Agric 73(2):242–244

Dalev P, Ljubomirova A, Simeonova L, Ivanov I (1996) Protein hydrolysates from waste feathers for feed and their enrichment with lysine trough enzyme catalyzed covalent binding. In: Mededelingen-faculteit landbouwkundige en toegepaste biologische wetenschappen, vol 61, pp 1641–1644

Dalev PG (1994) Utilisation of waste feathers from poultry slaughter for production of a protein concentrate. Biores Technol 48(3):265–267

Datta M (1993) Role of keratin in fire fighting. J Ind Leath Technol Assoc 43:297–299

Douglas J, Mittal C, Thomason J, Jofriet J (1996) The modulus of elasticity of equine hoof wall: implications for the mechanical function of the hoof. J Exp Biol 199(8):1829–1836

Earland C, Knight C (1956) Studies on the structure of keratin II. The amino acid content of fractions isolated from oxidized wool. Biochimica et Biophysica Acta 22(3):405–411

Endo R, Kamei K, Iida I, Kawahara Y (2008) Dimensional stability of waterlogged wood treated with hydrolyzed feather keratin. J Archaeol Sci 35(5):1240–1246

Eslahi N, Dadashian F, Nejad NH (2013) An investigation on keratin extraction from wool and feather waste by enzymatic hydrolysis. Prep Biochem Biotechnol 43(7):624–648

Fang Z, Zhang J, Liu BH, Du GC, Chen J (2013) Biodegradation of wool waste and keratinase production in scale-up fermenter with different strategies by *Stenotrophomonas maltophilia* BBE11-1. Bioresource Technol 140:286–291. https://doi.org/10.1016/j.biortech.2013.04.091

Feughelman M, Robinson M (1971) Some mechanical properties of wool fibers in the "Hookean" region from zero to 100% relative humidity. Text Res J 41(6):469–474

Flores-Hernández CG, Colín-Cruz A, Velasco-Santos C, Castaño VM, Rivera-Armenta JL, Almendarez-Camarillo A, García-Casillas PE, Martínez-Hernández AL (2014) All green composites from fully renewable biopolymers: chitosan-starch reinforced with keratin from feathers. Polymers 6(3):686–705

Fraser R, MacRae T, Rogers GE (1972) Keratins: their composition, structure, and biosynthesis. Thomas, Charles C

Fraser RDB, Parry DAD (2003) Macrofibril assembly in trichocyte (hard alpha-) keratins. J Struct Biol 142(2):319–325. https://doi.org/10.1016/S1047-8477(03)00027-3

Gessesse A, Hatti-Kaul R, Gashe BA, Mattiasson B (2003) Novel alkaline proteases from alkaliphilic bacteria grown on chicken feather. Enzyme and Microbial Technology 32(5):519–524

Gousterova A, Nustorova M, Paskaleva D, Naydenov M, Neshev G, Vasileva-Tonkova E (2012) Assessment of feather hydrolysate from thermophilic actinomycetes for soil amendment and biological control application. Int J Environ Res 6(2):467–474

Grazziotin A, Pimentel FA, de Jong EV, Brandelli A (2006) Nutritional improvement of feather protein by treatment with microbial keratinase. Anim Feed Sci Tech 126(1–2):135–144. https://doi.org/10.1016/j.anifeedsci.2005.06.002

Gupta A, Kamarudin NB, Kee CYG, Yunus RBM (2012) Extraction of keratin protein from chicken feather. J Chem Chem Eng 6(8):732

Gupta A, Perumal R (2013) Process for extracting keratin. Google Patents

Gurav RG, Jadhav JP (2013) A novel source of biofertilizer from feather biomass for banana cultivation. Environ Sci Pollut R 20(7):4532–4539. https://doi.org/10.1007/s11356-012-1405-z

Hadas A, Kautsky L (1994) Feather meal, a semi-slow-release nitrogen fertilizer for organic farming. Fertil Res 38(2):165–170

Hill P, Brantley H, Van Dyke M (2010) Some properties of keratin biomaterials: kerateines. Biomaterials 31(4):585–593. https://doi.org/10.1016/j.biomaterials.2009.09.076
Huda S, Yang YQ (2008) Composites from ground chicken quill and polypropylene. Compos Sci Technol 68(3–4):790–798. https://doi.org/10.1016/j.compscitech.2007.08.015
Huda S, Yang YQ (2009) Feather fiber reinforced light-weight composites with good acoustic properties. J Polym Environ 17(2):131–142. https://doi.org/10.1007/s10924-009-0130-2
Ichida JM, Krizova L, LeFevre CA, Keener HM, Elwell DL, Burtt EH (2001) Bacterial inoculum enhances keratin degradation and biofilm formation in poultry compost. J Microbiol Meth 47(2):199–208. https://doi.org/10.1016/S0167-7012(01)00302-5
Idris A, Vijayaraghavan R, Patti A, MacFarlane D (2014) Distillable protic ionic liquids for keratin dissolution and recovery. ACS Sustain Chem Eng 2(7):1888–1894
Innoe T (1992) Hair cosmetic for protection of skins. Eur Pat Appl, EP 469
Jeong JH, Lee OM, Jeon YD, Kim JD, Lee NR, Lee CY, Son HJ (2010) Production of keratinolytic enzyme by a newly isolated feather-degrading Stenotrophomonas maltophilia that produces plant growth-promoting activity. Process Biochem 45(10):1738–1745. https://doi.org/10.1016/j.procbio.2010.07.020
Jin E, Reddy N, Zhu Z, Yang Y (2011) Graft polymerization of native chicken feathers for thermoplastic applications. J Agr Food Chem 59(5):1729–1738
Kamarudin NB, Sharma S, Gupta A, Kee CG, Chik SMSBT, Gupta R (2017) Statistical investigation of extraction parameters of keratin from chicken feather using Design-Expert. 3 Biotech 7(2):127
Kaplin I (1982) Effect of cosmetic treatments on the ultrastructure of hair. Cosmet Toiletries 97(8):22–26
Karthikeyan R, Balaji S, Sehgal P (2007) Industrial applications of keratins—a review. J Sci Ind Res 66(9):710
Khajavi R, Rahimi MK, Abbasipour M, Brendjchi AH (2016) Antibacterial nanofibrous scaffolds with lowered cytotoxicity using keratin extracted from quail feathers. J Bioact Compat Polym 31(1):60–71
Khosa MA, Wu J, Ullah A (2013) Chemical modification, characterization, and application of chicken feathers as novel biosorbents. RSC Adv 3(43):20800–20810
Kim J-D (2007) Purification and characterization of a keratinase from a feather-degrading fungus, *Aspergillus flavus* strain K-03. Mycobiology 35(4):219–225
Kornillowicz-Kowalska T, Bohacz J (2010) Dynamics of growth and succession of bacterial and fungal communities during composting of feather waste. Bioresource Technol 101(4):1268–1276. https://doi.org/10.1016/j.biortech.2009.09.053
Korniłłowicz-Kowalska T, Bohacz J (2011) Biodegradation of keratin waste: theory and practical aspects. Waste Manag 31(8):1689–1701
Korol J (2012) Polyethylene matrix composites reinforced with keratin fibers obtained from waste chicken feathers. J Biob Mater Bio 6(4):355–360. https://doi.org/10.1166/jbmb.2012.1237
Kumar SL, Anandhavelu S, Sivaraman J, Swathy M (2017) Modified extraction and characterization of keratin from Indian goat hoof: a biocompatible biomaterial for tissue regenerative applications. Integr Ferroelectr 184(1):41–49
Kumaran P, Gupta A, Sharma S (2016) Synthesis of wound-healing keratin hydrogels using chicken feathers proteins and its properties. 2016:8. https://doi.org/10.22159/ijpps.2017v9i2.15620
Kunert J (2000) Physiology of keratinophilic fungi. Revista Iberoamericana de Micologia 1:77–85
Lee L, Baden H (1975) Chemistry and composition of the keratins. Int J Dermatol 14(3):161–171
Li J-S, Li Y, Liu X, Zhang J, Zhang Y (2013) Strategy to introduce an hydroxyapatite–keratin nanocomposite into a fibrous membrane for bone tissue engineering. J Mater Chem B 1(4):432–437
Li Y, Zhi X, Lin J, You X, Yuan J (2017) Preparation and characterization of DOX loaded keratin nanoparticles for pH/GSH dual responsive release. Mater Sci Eng, C 73:189–197
Lipkowski AW, Gajkowska B, Grabowska A, Kurzepa K (2009) Keratin-associated protein micromaterials for medical and cosmetic applications. Polimery 54(5):386–388

Luo T, Hao S, Chen X, Wang J, Yang Q, Wang Y, Weng Y, Wei H, Zhou J, Wang B (2016) Development and assessment of kerateine nanoparticles for use as a hemostatic agent. Mater Sci Eng, C 63:352–358

Manivasagan P, Sivakumar K, Gnanam S, Venkatesan J, Kim S-K (2014) Production, biochemical characterization and detergents application of keratinase from the marine actinobacterium *Actinoalloteichus* sp. MA-32. J Surfactants Deterg 17 (4):669–682

Matsunaga K, Okumura T, Tsushima R (1983) Hair treatment cosmetics containing cationic keratin derivatives. Google Patents

McKittrick J, Chen P-Y, Bodde S, Yang W, Novitskaya E, Meyers M (2012) The structure, functions, and mechanical properties of keratin. JOM 64(4):449–468

Natarajan K, Xie Y, Baer MR, Ross DD (2012) Role of breast cancer resistance protein (BCRP/ABCG2) in cancer drug resistance. Biochem Pharmacol 83(8):1084–1103

Onifade A, Al-Sane N, Al-Musallam A, Al-Zarban S (1998) A review: potentials for biotechnological applications of keratin-degrading microorganisms and their enzymes for nutritional improvement of feathers and other keratins as livestock feed resources. Biores Technol 66(1):1–11

Park M, Kim B-S, Shin HK, Park S-J, Kim H-Y (2013) Preparation and characterization of keratin-based biocomposite hydrogels prepared by electron beam irradiation. Mater Sci Eng, C 33(8):5051–5057

Paul T, Halder SK, Das A, Bera S, Maity C, Mandal A, Das PS, Mohapatra PKD, Pati BR, Mondal KC (2013) Exploitation of chicken feather waste as a plant growth promoting agent using keratinase producing novel isolate Paenibacillus woosongensis TKB2. Biocatal Agric Biotechnol 2(1):50–57

Poole AJ, Church JS, Huson MG (2008) Environmentally sustainable fibers from regenerated protein. Biomacromolecules 10(1):1–8

Poole AJ, Church JS, Huson MG (2009) Environmentally sustainable fibers from regenerated protein. Biomacromolecules 10(1):1–8. https://doi.org/10.1021/bm8010648

Priyaah K, Gupta A, Sharma S (2017) Synthesis of wound-healing keratin hydrogels using chicken feathers proteins and its properties. Int J Pharm Pharm Sci 9(2):171–178

Rad ZP, Tavanai H, Moradi A (2012) Production of feather keratin nanopowder through electrospraying. J Aerosol Sci 51:49–56

Rai SK, Konwarh R, Mukherjee AK (2009) Purification, characterization and biotechnological application of an alkaline β-keratinase produced by *Bacillus subtilis* RM-01 in solid-state fermentation using chicken-feather as substrate. Biochem Eng J 45(3):218–225

Ramadass SK, Perumal S, Jabaris SL, Madhan B (2013) Preparation and evaluation of mesalamine collagen in situ rectal gel: a novel therapeutic approach for treating ulcerative colitis. Eur J Pharm Sci 48(1):104–110

Ramakrishnan N, Sharma S, Gupta A, Alashwal BY (2018) Keratin based bioplastic film from chicken feathers and its characterization. Int J Biol Macromol

Ramshaw JA, Peng YY, Glattauer V, Werkmeister JA (2009) Collagens as biomaterials. J Mater Sci - Mater Med 20(1):3

Reddy N, Chen L, Zhang Y, Yang Y (2014a) Reducing environmental pollution of the textile industry using keratin as alternative sizing agent to poly (vinyl alcohol). J Clean Prod 65:561–567

Reddy N, Jiang QR, Jin EQ, Shi Z, Hou XL, Yang YQ (2013) Bio-thermoplastics from grafted chicken feathers for potential biomedical applications. Colloid Surf B 110:51–58. https://doi.org/10.1016/j.colsurfb.2013.04.019

Reddy N, Shi Z, Temme L, Xu H, Xu L, Hou X, Yang Y (2014b) Development and characterization of thermoplastic films from sorghum distillers dried grains grafted with various methacrylates. J Agr Food Chem 62(11):2406–2411

Reddy N, Yang Y (2007a) Structure and properties of chicken feather barbs as natural protein fibers. J Polym Environ 15(2):81–87

Reddy N, Yang YQ (2007b) Structure and properties of chicken feather barbs as natural protein fibers. J Polym Environ 15(2):81–87. https://doi.org/10.1007/s10924-007-0054-7

Reichl S, Borrelli M, Geerling G (2011) Keratin films for ocular surface reconstruction. Biomaterials 32(13):3375–3386

Rizvi TZ, Khan MA (2008) Temperature-dependent dielectric properties of slightly hydrated horn keratin. Int J Biol Macromol 42(3):292–297. https://doi.org/10.1016/j.ijbiomac.2008.01.001
Rouse JG, Van Dyke ME (2010) A review of keratin-based biomaterials for biomedical applications. Materials 3(2):999–1014
Sangali S, Brandelli A (2000) Feather keratin hydrolysis by a Vibrio sp. strain kr2. J Appl Microbiol 89(5):735–743
Saravanan S, Sameera D, Moorthi A, Selvamurugan N (2013) Chitosan scaffolds containing chicken feather keratin nanoparticles for bone tissue engineering. Int J Biol Macromol 62:481–486
Sastry T, Sehgal P, Gupta K, Kumar M (1986) Solubilised keratins as a filler in the retanning of upper leathers. Leather Sci 33:345
Schmidt W, Jayasundera S (2003) In: Wallenberger F, Weston N (eds) Natural fibers plastics, and composites-recent advances. Kluwer Academic: USA
Schmidt WF, Jayasundera S. (2004) Microcrystalline Avian Keratin Protein Fibers. In: Wallenberger FT, Weston NE (eds) Natural Fibers, Plastics and Composites. Springer, Boston, MA
Schrooyen PM, Dijkstra PJ, Oberthür RC, Bantjes A, Feijen J (2000) Partially carboxymethylated feather keratins. 1. Properties in aqueous systems. J Agric Food Chem 48(9):4326–4334
Secchi G (2008) Role of protein in cosmetics. Clin Dermatol 26(4):321–325
Sehgal P, Sastry T, Kumar M (1987) Effect of keratin filler in retanning of nappa garment leathers. Leather Sci 34(1)
Sharma S, Gupta A (2016) Sustainable management of keratin waste biomass: applications and future perspectives. Braz Arch Biol Technol 59
Sharma S, Gupta A, Chik SMS, Kee CG, Mistry BM, Kim DH, Sharma G (2017a) Characterization of keratin microparticles from feather biomass with potent antioxidant and anticancer activities. Int J Biol Macromol 104:189–196
Sharma S, Gupta A, Chik SMSBT, Kee CYG, Poddar PK (2017b) Dissolution and characterization of biofunctional keratin particles extracted from chicken feathers. In: IOP conference series: materials science and engineering, vol 1. IOP Publishing, p 012013
Sharma S, Gupta A, Chik SMST, Kee CYG, Podder PK, Subramaniam M, Thuraisingam J (2017c) Study of different treatment methods on chicken feather biomass. IIUM Eng J 18(2):47–55
Sharma S, Gupta A, Kumar A, Kee CG, Kamyab H, Saufi SM (2018) An efficient conversion of waste feather keratin into ecofriendly bioplastic film. Clean Technol Environ Policy 1–11
Shi Z, Reddy N, Hou XL, Yang YQ (2014) Tensile properties of thermoplastic feather films grafted with different methacrylates. ACS Sustain Chem Eng 2(7):1849–1856. https://doi.org/10.1021/sc500201q
Sionkowska A (2015) The potential of polymers from natural sources as components of the blends for biomedical and cosmetic applications. Pure Appl Chem 87(11–12):1075–1084
Song N-B, Jo W-S, Song H-Y, Chung K-S, Won M, Song KB (2013) Effects of plasticizers and nano-clay content on the physical properties of chicken feather protein composite films. Food Hydrocolloids 31(2):340–345
Spiridon I, Paduraru OM, Rudowski M, Kozlowski M, Darie RN (2012) Assessment of changes due to accelerated weathering of low-density polyethylene/feather composites. Ind Eng Chem Res 51(21):7279–7286. https://doi.org/10.1021/ie300738d
Staroń P, Banach M, Kowalski Z, Staroń A (2014) Hydrolysis of keratin materials derived from poultry industry. In: Proceedings of ECOpole 8
Sun P, Liu Z-T, Liu Z-W (2009a) Particles from bird feather: a novel application of an ionic liquid and waste resource. J Hazard Mater 170(2):786–790
Sun P, Liu ZT, Liu ZW (2009b) Particles from bird feather: a novel application of an ionic liquid and waste resource. J Hazard Mater 170(2–3):786–790. https://doi.org/10.1016/j.jhazmat.2009.05.034
Sundaram M, Legadevi R, Banu NA, Gayathri V, Palanisammy A (2015) A study on anti bacterial activity of keratin nanoparticles from chicken feather waste against *Staphylococcus aureus* (Bovine Mastitis Bacteria) and its anti oxidant activity. Eur J Biotechnol Biosci 6:1–5

Syed DG, Lee JC, Li WJ, Kim CJ, Agasar D (2009) Production, characterization and application of keratinase from *Streptomyces gulbargensis*. Bioresource Technol 100(5):1868–1871. https://doi.org/10.1016/j.biortech.2008.09.047

Teresa KK, Justyna B (2011) Biodegradation of keratin waste: theory and practical aspects. Waste Manage 31(8):1689–1701. https://doi.org/10.1016/j.wasman.2011.03.024

Tesfaye T, Sithole B, Ramjugernath D, Chunilall V (2017) Valorisation of chicken feathers: application in paper production. J Clean Prod 164:1324–1331

Tombolato L, Novitskaya EE, Chen P-Y, Sheppard FA, McKittrick J (2010) Microstructure, elastic properties and deformation mechanisms of horn keratin. Acta Biomater 6(2):319–330

Tran CD, Prosenc F, Franko M, Benzi G (2016) One-pot synthesis of biocompatible silver nanoparticle composites from cellulose and keratin: characterization and antimicrobial activity. ACS Appl Mater Interfaces

Tsuda Y, Nomura Y (2014) Properties of alkaline-hydrolyzed waterfowl feather keratin. Anim Sci J 85(2):180–185

Ullah A, Vasanthan T, Bressler D, Elias AL, Wu J (2011) Bioplastics from feather quill. Biomacromolecules 12(10):3826–3832

Vasileva-Tonkova E, Gousterova A, Neshev G (2009) Ecologically safe method for improved feather wastes biodegradation. Int Biodeter Biodegr 63(8):1008–1012. https://doi.org/10.1016/j.ibiod.2009.07.003

Velasco MVR, Dias TCdS, Freitas AZd, Júnior NDV, Pinto CASdO, Kaneko TM, Baby AR (2009) Hair fiber characteristics and methods to evaluate hair physical and mechanical properties. Braz J Pharm Sci 45(1):153–162

Wagner RdCC, Joekes I (2005) Hair protein removal by sodium dodecyl sulfate. Colloids Surf, B 41(1):7–14

Wallenberger FT, Weston N (2003) Natural fibers, plastics and composites. Springer Science & Business Media,

Wang J, Hao S, Luo T, Cheng Z, Li W, Gao F, Guo T, Gong Y, Wang B (2017) Feather keratin hydrogel for wound repair: preparation, healing effect and biocompatibility evaluation. Colloids Surf, B 149:341–350

Wang J, Hao S, Luo T, Yang Q, Wang B (2016) Development of feather keratin nanoparticles and investigation of their hemostatic efficacy. Mater Sci Eng, C 68:768–773

Wang XY, Lu CQ, Chen CX (2014) Effect of chicken-feather protein-based flame retardant on flame retarding performance of cotton fabric. J Appl Polym Sci 131(15). https://doi.org/10.1002/app.40584

Wang Y-X, Cao X-J (2012) Extracting keratin from chicken feathers by using a hydrophobic ionic liquid. Process Biochem 47(5):896–899

Weigmann H, Kamath Y, Ruetsch S (1990) Characterization ofsurface deposits on human hair fibers. J Soc Cosmet Chem 41:379–390

Xu H, Cai S, Xu L, Yang Y (2014a) Water-stable three-dimensional ultrafine fibrous scaffolds from keratin for cartilage tissue engineering. Langmuir 30(28):8461–8470

Xu H, Shi Z, Reddy N, Yang Y (2014b) Intrinsically water-stable keratin nanoparticles and their in vivo biodistribution for targeted delivery. J Agric Food Chem 62(37):9145–9150

Yamamura S, Morita Y, Hasan Q, Yokoyama K, Tamiya E (2002) Keratin degradation: a cooperative action of two enzymes from *Stenotrophomonas* sp. Biochem Biophys Res Commun 294(5):1138–1143

Yamini Satyawali SS (2013) Enzymatic electrosynthesis: an overview on the progress in enzyme-electrodes for the production of electricity, fuels and chemicals. J Microb Biochem Technol. https://doi.org/10.4172/1948-5948.s6-007

Yang YQ, Reddy N (2013) Potential of using plant proteins and chicken feathers for cotton warp sizing. Cellulose 20(4):2163–2174. https://doi.org/10.1007/s10570-013-9956-9

Yin J, Rastogi S, Terry AE, Popescu C (2007) Self-organization of oligopeptides obtained on dissolution of feather keratins in superheated water. Biomacromolecules 8(3):800–806

Yin XC, Li FY, He YF, Wang Y, Wang RM (2013) Study on effective extraction of chicken feather keratins and their films for controlling drug release. Biomater Sci-UK 1(5):528–536. https://doi.org/10.1039/c3bm00158j

Yu D, Cai JY, Liu X, Church JS, Wang L (2014) Novel immobilization of a quaternary ammonium moiety on keratin fibers for medical applications. Int J Biol Macromol 70:236–240

Zhan M, Wool RP (2011) Mechanical properties of chicken feather fibers. Polym Compos 32(6):937–944

Zheng Y, Du X, Wang W, Boucher M, Parimoo S, Stenn K (2005) Organogenesis from dissociated cells: generation of mature cycling hair follicles from skin-derived cells. J Invest Dermatol 124(5):867–876

Zoccola M, Aluigi A, Tonin C (2009) Characterisation of keratin biomass from butchery and wool industry wastes. J Mol Struct 938(1–3):35–40. https://doi.org/10.1016/j.molstruc.2009.08.036

Chapter 2
Keratin Production and Its Applications: Current and Future Perspective

Anshuman Shah, Shaily Tyagi, Ram Naresh Bharagava, Dalel Belhaj, Ashok Kumar, Gaurav Saxena, Ganesh Dattatraya Saratale and Sikandar I. Mulla

Abstract Keratin is a global class of biological material, which represents a group of cysteine-rich filament-forming proteins. They serve as a shielding layer for the epidermal appendages like nails, claws, beak, hair, wool, horns, and feathers. These proteins are further subdivided into two different class based on their secondary structure: α-keratin and β-keratin. Keratin is insoluble in hot or cold water; this unique property helps to prevent their digestion by proteolytic enzymes. Additionally, their complex hierarchical-like filament-matrix structure at nanoscale and the polypeptide chains create a robust wall for protection against heat stress,

A. Shah and S. Tyagi—Equally contributors.

A. Shah · S. Tyagi
ICAR-National Research Centre on Plant Biotechnology, New Delhi 110012, India

R. N. Bharagava · G. Saxena
Laboratory of Bioremediation and Metagenomics Research (LBMR),
Department of Environmental Microbiology (DEM), Babasaheb Bhimrao Ambedkar University (A Central University), Vidya Vihar, Raebareli Road, Lucknow 226025, UP, India

D. Belhaj
Laboratory of Biodiversity and Aquatic Ecosystems Ecology and Planktonology,
Department of Life Sciences, University of Sfax-Tunisia, FSS, Street of Soukra Km 3.5, 3.5, BP 1171, 3000 Sfax, Tunisia

A. Kumar
Department of Biotechnology and Bioinformatics, Jaypee University of Information Technology, 173234 Waknaghat, Solan, Himachal Pradesh, India

G. D. Saratale
Department of Food Science and Biotechnology, Dongguk University-Seoul, Ilsandong-Gu, Goyang-Si, Gyeonggi-Do 10326, Republic of Korea

S. I. Mulla (✉)
Key Laboratory of Urban Pollutant Conversion, Institute of Urban Environment,
Chinese Academy of Sciences, Xiamen, China
e-mail: sikandar.mulla7@yahoo.com

S. I. Mulla
Department of Biochemistry, Karnatak University, Dharwad 580003, Karnataka, India

S. Sharma and A. Kumar (eds.), *Keratin as a Protein Biopolymer*, Springer Series on Polymer and Composite Materials, https://doi.org/10.1007/978-3-030-02901-2_2

pathogen invasions (particularly through skin), mechanical damage, etc. In this review, we are trying to attempt a linear focus in the direction of structure, function, extraction of keratin, and its industrial applications.

Keywords Keratin · Keratin structure · Keratin production · Keratin application

1 Introduction

Natural elements like carbon, nitrogen, oxygen, sulfate, and phosphorous and also some proteins produced by bacteria including green fluorescent protein, etc., have superabundance impact on living system, and most of them are beneficially using in life science and molecular engineering. Virtually, all the biological materials are robustly consisting of biopolymers and some minerals. Such combination represents excellent properties and functionalities in many biomolecules (Wang et al. 2016a). In the early 1800s, there was no specific name of protein and was considered in the category of "albuminoids". Around 1849, one novel word "keratin" was appeared in the literatures that represent different properties and attract the scientist's attention for understanding the novel behavior of this hygroscopic protein. Keratin belongs to a very diverse family of fibrous protein. Keratinous material like horns, wools, nails, feathers, and hooves are the global class of biological co-products, which represent a group of cysteine-rich filament-forming proteins. Moreover, keratins are the protein that forms a shielding layer for the epidermal appendages and thereby imparting an important role in protection. The spacious arrangement of disulfide overpass significantly influences the keratin properties, in particular broad chemical and mechanical resistance. Keratin is insoluble in hot or cold water; this unique property helps to prevent their digestion with proteolytic enzymes such as pepsin, papain or trypsin (Bohacz 2017; Staron et al. 2011; Tesfaye et al. 2017).

Epithelial cells of cytoplasm are the richest compartment for keratinous structural protein, in which they were assembled into a network of 10–12 nm wide intermediate filament (Kim and Coulombe 2007). In human genome, almost 54 conserved genes encode keratin protein and these genes are further subdivided into type A and type B subgroupings of intermediate filament (IF)-encoding genes (Table 1) (Schweizer et al. 2006). There are 28 classes of type A and 26 classes of type B genes and each type of gene encode for a single polypeptide chain. Type A proteins are smaller and more acidic while type B protein is larger and more alkali than type A family (Schweizer et al. 2006). Type A and type B proteins interact together and form a heterodimer that promotes polymerization of keratin (Steinert et al. 1976; Fuchs and Cleveland 1998; Kim and Coulombe 2007). The arrangement of keratin is mediated by type A as well as type B genes and thus maintains the integrity of epithelial cell lines, but, on the other hand, their improper arrangement and misfolding may lead to severe epithelial disease such as epidermolysis bullosa

Table 1 Subgrouping of IF-encoding genes

Protein family	Size (kDa)	Pi
Type A	40–64	~4.7–6.1
Type B	52–70	~5.4–8.4

simplex (Coulombe and Lee 2012). In this review article, we briefly discuss the structure, function, source, and industrial application of keratin, and finally few important priorities as conclusion.

2 Structure and Classification of Keratin

Keratin is a global class of biological material, which represents a group of cysteine-rich filament-forming proteins. Keratin is the second most essential and abundant biopolymer that confronts in animals (Sharma and Gupta 2016; Sharma et al. 2016, 2017a, b, c, 2018). They serve as a shielding layer for the epidermal appendages like nails, claws, beak, hair, wool, horns, and feathers (Shavandi et al. 2017a, b). Their complex hierarchical structure like filament-matrix composition at nanolevel and polypeptide chains creates a robust wall for protection against heat stress, pathogen invasions (particularly through skin), mechanical damage, etc. (Kadir et al. 2017).

Keratins and keratinous materials are subdivided into two different classes of secondary structures of protein: alpha (α)-keratin and beta (β)-keratin (Shavandi et al. 2017b). On the basis of X-ray diffraction pattern, these proteins can also be classified as α-pattern of keratin, β-pattern of keratin, amorphous, and feather pattern of keratin (Fraser et al. 1972; Astbury and Woods 1934; Fraser and Parry 2011). The amorphous pattern elucidates the amorphous matrix component in α-keratinous tissue (Staron et al. 2011), while the feather pattern shows similarity to β-pattern (Fraser et al. 1972).

2.1 α- and β-Keratins

Keratin is made up of long polymerized chain of amino acids, and molecular weight of keratin in feather is about 10,500 Da (Staron et al. 2011). These α-helix and β-plated sheet structures arise due to the continuous folding of the backbone of the polypeptide chain due to hydrogen bonding between carboxyl and amino groups of the peptide chain. The structural configuration of α- and β-keratins is represented in Fig. 1 along with 3D structure of keratin protein.

In α-helical structure, polypeptide chain forms a hydrogen bond with the maximum possibilities of their interaction between carbonyl group of one amino acid and amino group of another amino acid. The diameter of α-helix structure is 7–10 nm, and their molecular mass is in the range of 40–68 kDa (Wang et al. 2016b, Alibardil et al.

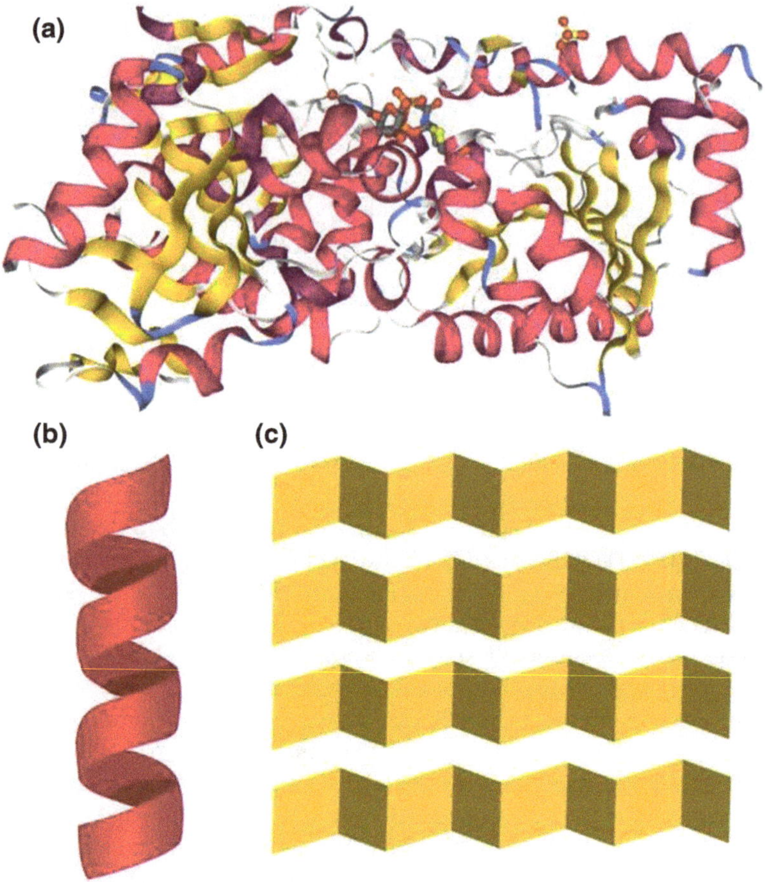

Fig. 1 **a** Schematic representation of 3D structure of keratin protein, **b** α-helix is formed by twisting of polypeptide chains into screw manner, and **c** β-sheet structure is the result of maximum extension and stretching of polypeptide chains

2006). α-Keratin is a most abundantly found in mammals. Such type of structure is organized as coiled coil and thereby participating in compact structural arrangement to elongate polypeptide chain (Crick 1952, 1953). On the other hand, β-structure carries intermolecular hydrogen bonding between carbonyl and amino group. The diameter of β-structure is about 3–4 nm and molecular mass is about 10–22 kDa (Table 2). These types of structure are most abundantly found in reptilian tissues, such as feathers, claws, and beaks of birds, etc. (Chen et al. 2012).

Table 2 Structure-based comparison and distribution of α- and β-keratins

Characterization	α-Keratin	β-Keratin
Bonding	Intramolecular hydrogen bonding	Intermolecular hydrogen bonding
Diameter (nm)	7–10	3–4
Structural features	Intermediate filament matrix	Amorphous matrix
Molecular mass (kDa)	40–68	10–22
Distribution	Wool, horns, hairs, nails, and hooves	Avian beaks, feathers, claws, reptilian epidermis, and claws
Stiffness	High	Lower than α-keratin

3 Sources of Keratin

Sources of keratin include wools, nails, hairs, hooves, feathers, reptilian scales, and avian (Fig. 2). Reptilian scales, feathers, and avian display almost similar keratin genes (Gregg et al. 1984), and it has been also proved that these feathers and scales are β-sheet based on transmission electron microscope (Filshie et al. 1964) and X-ray diffraction patterns (Astbury and Marwick 1932; Fraser and Parry 2008).

3.1 Feather, Beaks, Wool, and Hair

Since very early days, humans are very fascinated with understanding the concept of how birds can fly, and how birds (like duck) can protect their wings during swimming or rainy condition. You have been already noticed that the bird's wing does not get sloppy, easily during rain or rainy season. So, this is mainly considered keratin (β-form) protein may involve for providing mechanical support to the feather, and thereby prevents feather to get sloppy. To further understand this concept, Filshie and Gogers (1962) analyze feather under transmission electron micrographs, and their study reveals the filament-matrix structure with the β-keratin filaments. This structure was about 3 nm in diameter and embedded in an amorphous matrix.

The keratinous material is also present in rhamphotheca (outer surface layer) of bird beaks, and it allows the beaks to participate in a variety of functions, like feeding, social interactions, fighting, and grooming (Luester 2006). Usually, bird beaks fall into two categories: (1) long thin and (2) short/thick except in toucan beak, as it is both thick and long (Wang et al. 2016b). Wool and hair is the remarkable example of the hard-keratinous material. Wool is the excellent animal fiber, containing 82% keratinous protein. It is widely used in textile application and is scientifically important because of its mechanical and structural behavior (Feughelman 1997).

Hair is another widely used and studied fiber. Study on hair and wool revels that they both have common features except wool has large diameter. In many cases, hair plays an important role in terms of protection against dust and pathogens (hairs in nostril).

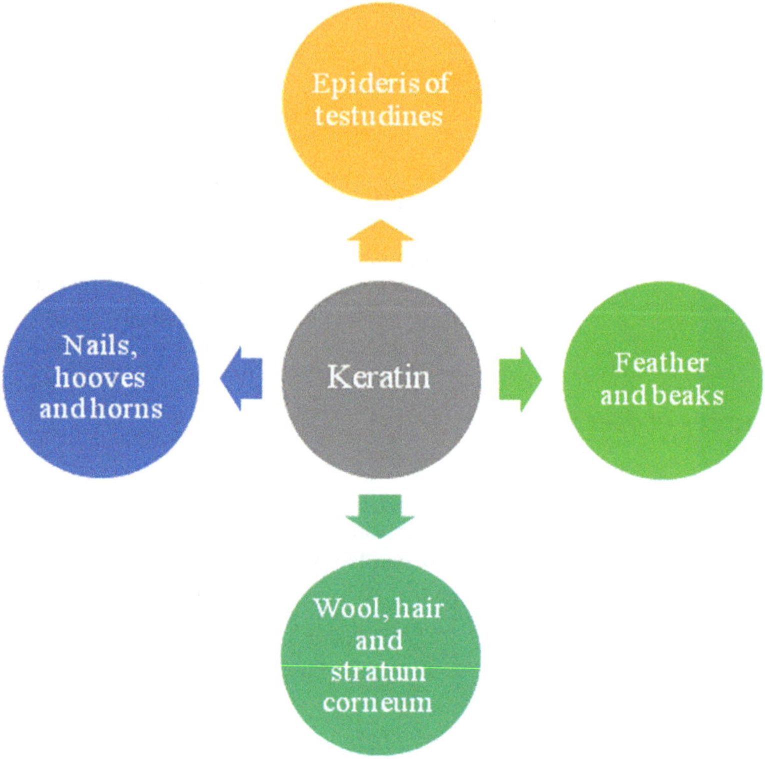

Fig. 2 Different sources of keratin

3.2 *Stratum Corneum, Horns, and Hooves*

Light and skin interaction regulates the optical visibility of the skin to human eye. Skin can absorb and reflect light in a diverse manner (Lee et al. 2018). Skin is also responsible for regulating the body's temperature, and it is actively participated in defense mechanism by acting as a barrier for pathogen(s). Mammalian skin is covered by outermost layer of stratum corneum, and the thickness of this layer is about 20–40 μm (Tomlinson et al. 2004). This layer act as defense layer from external attack (predators) and serve as diffusion barrier (Meyers et al. 2008).

Stratum corneum is composed of keratinocytes; these cells are embedded in an intracellular matrix that is highly rich in lipid content. Their cytoplasmic content is arranged in web-like pattern and robustly integrates at cell-to-cell junctions. Such interaction(s) in a diversely integrated manner maximize the mechanical support (Coulombe and Wong 2004).

Bovid animals like cattle, waterbuck, sheep, buffalo, and gazelle carry keratinous material in the form of horn. Horns act as a weapon, sometime(s) as shield to protect from predator and combat with other animals for their defense (Trim et al. 2011).

It may also involve for regulating body temperature (Geist 1966). These are highly tough, and can show resistance against opposite force, but cannot grow, once it gets broken (Tombolato et al. 2010). Hooves are the toughest keratinous materials and are still being an attractive topic for the research. Hoof wall is composed of flattened, keratinocytes. Hooves have a complex structure, and these complexities allow the hoof wall to absorb enough energy as the crack grows (Cook et al. 1964).

3.3 Epidermis of Testudines and Nails

The group "Testudines" usually represents the turtles, tortoises, and terrapins. Recent studies on Testudines indicate turtles are rich in both α- and β-keratins (Dalla Valle et al. 2013). The β-keratin/α-keratin ratio is higher in the carapace in hard-shelled turtle. This ratio provides an important factor for further study in the background of keratin.

Nails are the important source of α-keratin. In primates and few mammals, it covers the ends of the fingers and toes (as their ends have soft skin, so indirectly these proteins participate in protection). In cats, these protein forms curved claws and show α-type pattern during the analysis under X-ray diffraction pattern (Bear and Rugo 1951). Fingernails are one of the important features in primates (Hamrick 1998), as they can use their nails during fighting, scratching, and opening of some object.

4 Various Methods Used for the Keratin Production and Degradation

Poultry, meat, and leather processing industries generated every year large amount of keratin-containing wastes. Globally, every year, poultry processing industries generate around 8.5 million feather waste (Fellahi et al. 2014). Currently, these keratins-containing waster products are buried, dumped, and used as land filing, which generates problem in storage, ash disposal, and emission control (Agrahari and Wadhwa 2010). Here, we are trying to focus on some important methods for keratin production and degradation (Fig. 3).

4.1 Hydrolysis of α- and β-Keratins by Bacterial Cultures

The static extension of poultry industry facilitates a huge amount of wastes. Recycling of such waste(s) does not affect the environmental pollutant problems, but it also improves the rate of keratin production for various purposes. Poultry meat

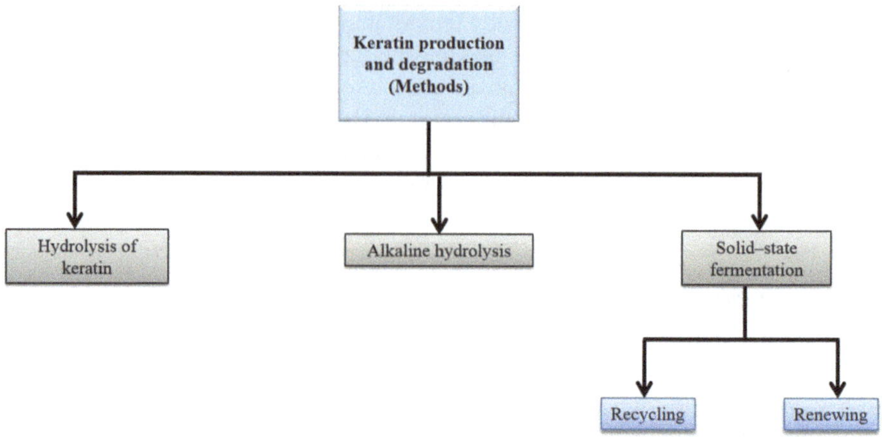

Fig. 3 Methods of keratin production

industries provide gigantic waste such as bones, feathers, and viscera. Bioconversion of such wastes into the useful material using a specific enzyme(s) is a fascinating paradigm. Presently, enzymes are widely used at broad spectrum for the conversion and production of poultry waste into feeding and fertilizers.

Keratin proteins separated from feather by several chemical modifications. Poultry (including birds) feathers contain around 90% keratin along with some important amino acids like cysteine, threonine, and arginine (Sinkiewicz et al. 2017; Tiwary and Gupta 2012).

Initially, protease (keratinases, EC 3.4.99.11)-producing microorganisms were isolated, screened and most potent protease-producing microorganisms were identified. The microbial enzyme (protease) attacks insoluble keratinous protein substrates and thereby releasing free amino ($-NH_2$) group containing molecules (Abdel-Naby et al. 2017; Bagewadi et al. 2018). In order to check the hydrolysis property of keratin components, different kinds of substrates like chicken feather, sheep wool, pea pods and more were used as substrate. Generally, after hydrolyzing, these substrates were further proceed by following chopping, washing, and then overnight incubation at 60 °C. Finally, freely released of amino acids were analyzed at 0–3 days of interval (Fellahi et al. 2014). The brief summary of keratin production is summarized in Fig. 4.

4.2 *Production of Keratin Using Alkaline Hydrolysis*

Keratin production and their extraction can also be done by alkaline hydrolysis. Extraction and production, these two processes mainly depend on two different complementary factors (1) chemicals, as it is important for extracting spe-

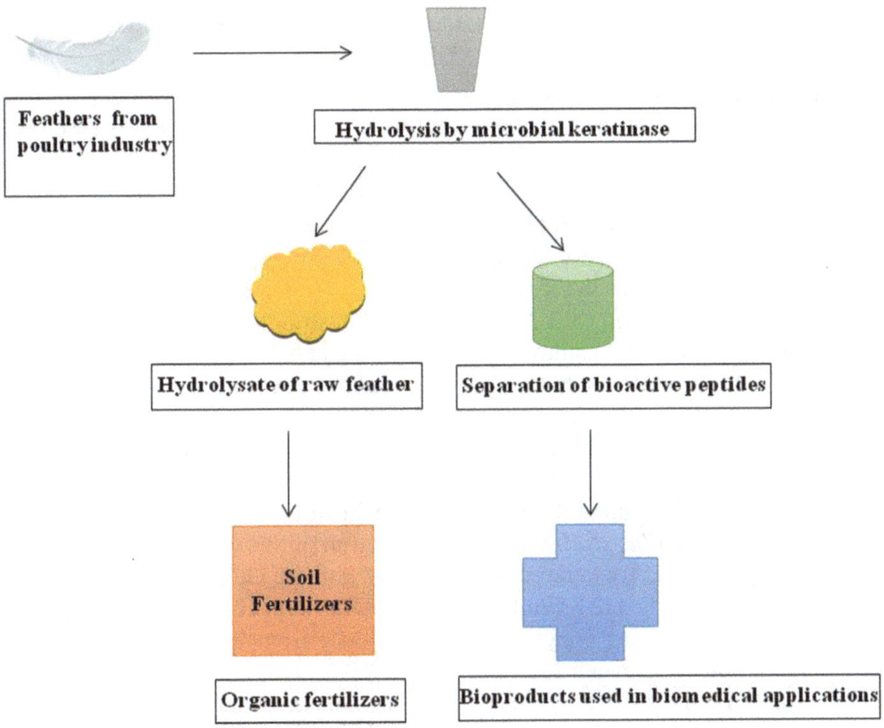

Fig. 4 Microbial keratinase activities for the conversion of wastes from poultry industry to useful by-products. Following hydrolysis, the hydrolysates can be used as organic fertilizers, and these hydrolysates can also be used in biomedical applications

cific cellular proteins from the cell, tissue or organ, and (2) optimal condition(s), as it is important, because, it provides all the ideal support for protein stability. For example, keratin from bird's feather can be obtained by alkaline hydrolysis method. During the process chemicals like ammonium sulfide or sodium sulfide can reduce disulfide group to thiol group. And also, by using β-mercaptoethanol, free –SH group can be partially modified. The process was carried out for 90 min at 45–80 °C. The obtained keratin molecular weight was approximately between 1 kDa and 11 kDa (U.S Patent 2007). Recently, Sinkiewicz et al. (2017) demonstrated the influence of thermochemical treatments using different reducing agents (2-mercaptoethanol, dithiothreitol, sodium m-bisulphite, and sodium bisulphite, as well as sodium hydroxide) on the production of keratin (in chicken feathers extract). 2-Mercaptoethanol and sodium bisulphite treatment yields about 84% and 82% of keratin. Additionally, sodium hydroxide (2.5%) was further improved the production of keratin up to 94% (Sinkiewicz et al. 2017).

4.3 Solid-State Fermentation (SSF) for Keratin Degradation

Keratinases are the enzymes having proteolytic activity for the keratin degradation (Bagewadi et al. 2018). Consumption and utilization of poultry animals as the food product tremendously increase the consequence of biowaste, and it is very difficult to degrade such wastes within the short period of time. Keratin and their keratinous component, carries numerous amino acids, and their decomposition in the form of feather does not only cause environmental problem but it also leads to loss of several important amino acids that can be used in making nutrient-rich food products.

Solid-state fermentation (SSF) is a booming industrial process in biotechnology and microbiology field; it is mainly used for recycling or renewing biological waste(s) at large scale. In this process, microorganism grows in aqueous-free environment or sometimes at very low water content (Saratale et al. 2014). Submerged fermentation is the analogue of solid-state fermentation in which all the natural conditions are optimized and they mimic the natural environment (Bagewadi et al. 2018; Soccol et al. 2017). Chitturi and Lakshmi (2016) demonstrate an attractive method for keratinase production using *Bacillus strains*. The beneficial trait of this research was indicated as an improved method for keratinase production, and the yield was increased from 283–367 KU/mL to 415–442 KU/mL (Chitturi and Lakshmi 2016). Not only bacterial cultures, but the keratinase enzyme can also produce from fungal culture(s) under optimal conditions (Bagewadi et al. 2018), as it has a great influence on the feather degradation. SSF is eco-friendly process, significantly contributing to minimize the bio-pollutants and to maximize the area for food industries (Bagewadi et al. 2018).

5 Industrial and Clinical Applications of Keratins

It is very difficult to degrade keratins and their disposal may lead to serious harmful impact on the environment. Therefore, research should be proceeding by focusing on keratin waste utilization (Bagewadi et al. 2018; Bohacz 2017; Karthikeyan et al. 2007). India has broad livestock populations, which display annually: 82 million goatskins, 30 million sheepskins, and 28 million buffalo hides (Karthikeyan et al. 2007). Around 8–10% of chicken weight is due to their feather and approximately millions of tons of feather are generated annually (Karthikeyan et al. 2007). Besides this, a large quantity of horns, hairs, hooves, and feathers are wasted every year (Onifade et al. 1998). From the past few years, these wastes are utilizing at industrial level to extract some important and insoluble (keratin) proteins. To get rid of this serious threat, there are various applications that can utilize these biowastes to convert them into beneficial form. Polyvinyl alcohol fibers that contain keratin were found to be an extensive industrial application, for example; they can be used as absorbents for toxic/hazardous substances like heavy metals ions as well as formaldehyde gas (Rouse and Van Dyke 2010).

5.1 Alkali and Hydrothermal Treatment

Hydrothermal treatment depends on high steam pressure of 10–15 psi with or without high temperature (80–140 °C) in the presence of acidic solutions (hydrochloric acid, sulfuric acid and more) (Eggum 1970) or alkali solution (sodium hydroxide, potassium hydroxide, sodium carbonate and more) (Gousterova et al. 2003; Papadopoulos et al. 1985). Combining acid or alkali treatment at the boiling temperature for about 2-3 h could open disulfide bridges of keratin and forms water-soluble polypeptides and oligopeptides or sometimes few amino acids. The crucial point of this process is the degradation of amino acids and polypeptide chains at high temperature that affects the nutritional improvement values.

5.2 Use of Keratin in Leather Industries

Keratin is a biopolymer having multifunctional properties, and these properties make (turn on the pathways of) keratinous material utilization at industrial level. Systematic optimization of the keratinous component form biological waste products, participate in recovery of natural resources and aids in maintaining pollution free environment (Niculescu et al. 2016).

Wool is the keratinous material having well-defined structure and mechanical properties (Wang et al. 2016b). Wool fiber contains approximately 82% of keratinous protein and high percentage of cysteine, around 17% along with small number of polysaccharides and lipid wax (Lewis and Rippon 2013). All these properties made keratin useful in leather industries (2007).

5.3 Keratin in Drug Delivery System

Drug delivery is a dynamic process which involves the process of administration of drugs, their absorption, and interaction with the target site. For many drug applications, drugs are designed in such a way that it should be able to provide maximum desired therapeutic effect or drug efficacy with less toxicity (Nelson et al. 2002; Parveen and Sahoo 2008). Drug delivery should be maintained in a controlled manner; if the condition for drug delivery is not maintained at high level, then it may cause serious complications, and even it may lead to the death of the person. Moreover, a diverse family of polymers have been identified in drug delivery purposes. Drug delivery already has been achieved by using synthetic and natural polymers. Synthetic polymers including polyphosphazenes, polyorthoesters, polyanhydrides, polyesters, and polyphosphoesters have found in broad application (Nair and Laurencin 2006), while natural polymers including collagen, cellulose, gelatin, and alginates are broadly used because of their biocompatibility and biodegradability (Malafaya et al. 2007; Mano et al. 2007).

Over the last few years, keratin is widely used for the drug delivery system (DDS) in a controlled manner because of their stability, less reactivity, structural, and mechanical properties. Being natural polymer, keratins have advantage as it is rich in reactive chemical groups like amide, carboxyl, hydroxyl, as well as sulfhydryl, and hence, these properties enhance their activity to interact with biologically active molecules. One simplest way of incorporating drugs into keratin biomaterials is to dissolve or mix before processing into the keratin solution. Such protocols were designed for the synthesis of drug loaded keratin films (Fujii et al. 2004; Vasconcelos et al. 2010). For example, Li and co-workers developed a nano-based PEG-induced keratin method for drug delivery to treat cancer (Li et al. 2012).

5.4 Keratin in Surgery and Repairing

Keratins have complex bio-protectant and strong structural properties. They show similar properties to collagen like insolubility, high tensile strength, etc., and these properties illustrate the importance of keratin in surgery.

Partial and peripheral burns are the consequence of patient management with respect to multiple dressing. It is important to protect the skin from infection, wound, trauma, and external injuries to minimize the possibilities of major (supreme) damages in skin and thereby, promoting rapid epithelial cells proliferation (Loan et al. 2016). Skin is the outermost layer of the body that protects every living organism from serious threats and it also regulates homeostasis during unfavorable conditions. Loan and co-group reported that "keragel and keramatrix" (keratin-based products) can be used during damaging of dermis and epidermis layers (Loan et al. 2016).

Skin injury and their repairing possibilities can be enhanced by stimulating migration activity of keratinocytes at injured, damaged, and wounded site(s). These proliferation and migration rate of keratinocytes significantly promote basal membrane's protein expression including types IV and VII collagens (Tang et al. 2012) and show improvement in clinical trials such as skin surgery, leg ulcers (Hammond et al. 2010), and in epidermolysis bullosa (Than et al. 2013).

6 Conclusion and Important Aspects

Keratin depicts toughest and significant biological material; act as protective integument in mammals, primates, and in Testudines. Keratin has compact structural and robust mechanical properties which attracts life science researchers for the proper understanding of physical, chemical, and biological properties of keratin. For example, keratinous materials provide the toughest source for protection (nails over the skin, horn, and hooves in horse), feeding and interaction (outermost layer of beaks).

Research on keratin is directed toward the advancement of numerous biomaterials having keratin for practice in application of biomedical field. For example, in recent days, hair damage (caused by UV radiation from sunlight, pollution, and nutrient deficient food supplements) can be repaired by synthetic treatment of keratin. Although such treatment is temporary, still it is widely used in metropolitan area to minimize hair loss. Additionally, keratinous materials have diverse hierarchal structures and functions that can be useful for development or architecture of new structure for the human benefit. In spite of the fact that remarkable research has been conducted on keratin, still some area of research is unclear and insufficient. Here, we are now trying to put some important factors that may illustrate some beneficial points in keratin research:

- Although a significant amount of details is available for keratin, α- and β-keratinous material are not well understood, including their biosynthesis, molecular assembly, gene sequence, mechanical, and structural nature. This indicates and opens a new door in research for diverse investigation of keratinous material.
- Keratin has high mechanical and compact biological properties, this knowledge we can apply on synthesizing and assembly of new products with low molecular weight and high mechanical strength.
- To predict the novel structure and their function, it is essential to understand the morphological and structural configuration relationship.
- More studies required to understand the importance of keratin-based nanomaterials that may be useful for drug delivery and providing therapeutic response.

Acknowledgements The authors are very thankful to the Mr. Chaitanya Sharma and Ms. Arshia Gupta for their valuable suggestions during designing of 3D structure of keratin protein.

References

Abdel-Naby MA, El-Refai HA, Ibrahim MHA (2017) Structural characterization, catalytic, kinetic and thermodynamic properties of keratinase from *Bacillus pumilus* FH9. Int J Biol Macromol 105:973–980

Agrahari S, Wadhwa N (2010) Degradation of chicken feather a poultry waste product by keratinolytic bacteria isolated from dumping site at Ghazipur poultry processing plant. Int J Poult Sci 9:482–489

Alibardi L, Dalla Valle L, Toffolo V, Toni M (2006) Scale keratin in lizard epidermis reveals amino acid regions homologous with avian and mammalian epidermal proteins. Anat Rec A Discov Mol Cell Evol Biol 288:734–752

Astbury WT, Marwick TC (1932) X-ray interpretation of the molecular structure of feather keratin. Nature 130:309–310

Astbury WT, Woods HJ (1934) X-ray studies of the structure of hair, wool, and related fibres. II. The molecular structure and elastic properties of hair keratin. Trans R Soc (London) 232:333–394

Bagewadi ZK, Mulla SI, Ninnekar HZ (2018) Response surface methodology-based optimization of keratinase production from *Trichoderma harzianum* isolate HZN12 using chicken feather waste and its application in dehairing of hide. J Environ Chem Eng 6:4828–4839

Bear RS, Rugo HJ (1951) The results of X-ray diffraction studies on keratin fibers. Ann N Y Acad Sci 53:627–648

Bohacz J (2017) Biodegradation of feather waste keratin by a keratinolytic soil fungus of the genus *Chrysosporium* and statistical optimization of feather mass loss. World J Microbiol Biotechnol 33:13
Chen PY, McKittrick J, Meyers MA (2012) Biological materials: functional adaptations and bioinspired designs. Prog Mater Sci 57:1492–1704
Chitturi MKCh, Lakshmi VV (2016) Development of semi-solid state fermentation of Keratinase and optimization of process by cheaper and alternative agricultural wastes. Eur J Biotechnol Biosci 4:1–4
Cook J, Gordon JE, Evans CC, Marsh DM (1964) A mechanism for the control of crack propagation in all-brittle systems. Proc R Soc Lond A 282:508–520
Coulombe PA, Wong P (2004) Cytoplasmic intermediate filaments revealed as dynamic and multipurpose scaffolds. Nat Cell Biol 6:699–706
Coulombe PA, Lee CH (2012) Defining keratin protein function in skin epithelia: epidermolysis bullosa simplex and its aftermath. J Invest Dermatol 132:763–775
Crick F (1952) Is [α]-keratin a coiled coil? Nature 170:882–883
Crick FH (1953) The packing of α-helices: simple coiled-coils. Acta Cryst 6:689–697
Dalla Valle L, Michieli F, Benato F, Skobo T, Alibardi L (2013) Molecular characterization of alpha-keratins in comparison to associated beta-proteins in soft-shelled and hard-shelled turtles produced during the process of epidermal differentiation. J Exp Zool Part B Mol Dev Evol 320:428–441
Eggum BO (1970) Evaluation of protein quality of feather meal under different treatments. Acta Agricul Scand 20:230–234
Fellahi S, GAD MH, Zaghlou TI, Feuk-Lagerstedt E, Taherzadeh MJ (2014) A *Bacillus* strain able to hydrolyze α- and β-keratin. J Bioprocess Biotech 4:181
Feughelman M (1997) Mechanical properties and structure of alpha-keratin fibers: wool, human hair and related fibers. University of New South Wales Press
Filshie BK, Gogers GE (1962) An electron microscope study of the fine structure of feather keratin. J Cell Biol 13:1–12
Filshie BK, Fraser RD, MacRae TP, Rogers GE (1964) X-ray-diffraction and electron-microscope observations on soluble derivatives of feather keratin. Biochemical J 92(1):18–2
Fraser RD, Parry DA (2008) Molecular packing in the feather keratin filament. J Struct Biol 162:1–13
Fraser RD, Parry DA (2011) The structural basis of the filament-matrix texture in the avian/reptilian group of hard β-keratins. J Struct Biol 173:391–405
Fraser RDB, MacRae TP, Rogers GE (1972) Keratins: their composition, structure and biosynthesis. Thomas
Fuchs E, Cleveland DW (1998) A structural scaffolding of intermediate filaments in health and disease. Science 279:514–519
Fujii T, Ogiwara D, Arimoto M (2004) Convenient procedures for human hair protein films and properties of alkaline phosphatase incorporated in the film. Biol Pharm Bull 27:89–93
Geist V (1966) The evolution of horn-like organs. Behavior 27:175–214
Gousterova A, Nustorova M, Goshev I, Christov P, Braikova D, Tishinov K, Haertle T, Nedkov P (2003) Alkaline hydrolysate of waste sheep wool aimed as fertilizer. Biotechnol Biotechnol Equip 17:140–145
Gregg K, Wilton SD, Parry DA, Rogers GE (1984) A comparison of genomic coding sequences for feather and scale keratins: structural and evolutionary implications. EMBO J 3:175–178
Hammond C, Than M, Walker J (2010) From the laboratory to the leg: patient's and nurses perceptions of product application using three different dressing formats. Wound Pract Res 18:189–195
Hamrick MW (1998) Functional and adaptive significance of primate pads and claws: evidence from New World anthropoids. Am J Phys Anthropol 106:113–127
Kadir M, Wang X, Zhu B, Liu J, Harland D, Popescu C (2017) The structure of the "amorphous" matrix of keratins. J Struct Biol 198:116–123
Karthikeyan R, Balaji S, Sehgal PK (2007) Industrial applications of keratins—a review. J Sci Ind Res 66:710–715

Kim S, Coulombe PA (2007) Intermediate filament scaffolds fulfill mechanical, organizational, and signaling functions in the cytoplasm. Genes Dev 21:1581–1597
Lee SH, Jun SH, Yeom J, Park SG, Lee CK, Kang NG (2018) Optical clearing agent reduces scattering of light by the stratum corneum and modulates the physical properties of coenocytes via hydration. Skin Res Technol. https://doi.org/10.1111/srt.12439
Lewis DM, Rippon JA (2013) The coloration of wool and other keratin fibres. https://doi.org/10.1002/9781118625118
Li QM, Zhu LJ, Liu RG, Huang D, Jin X, Che N, Li Z, Qu XZ, Kang HL, Huang Y (2012) Biological stimuli responsive drug carriers based on keratin for triggerable drug delivery. J Mater Chem 22:19964–19973
Loan F, Cassidy S, Marsh C, Simcock J (2016) Keratin-based products for effective wound care management in superficial and partial thickness burns injuries. Burns 42:541–547
Luester AU (2006) Manual of Pirrot behavior. Wiley Blackwell Publishing
Malafaya PB, Silva GA, Reis RL (2007) Natural-origin polymers as carriers and scaffolds for biomolecules and cell delivery in tissue engineering applications. Adv Drug Deliv Rev 59:207–233
Mano JF, Silva GA, Azevedo HS, Malafaya PB, Sousa RA, Silva SS, Boesel LF, Oliveira JM, Santos TC, Marques AP, Neves NM, Reis RL (2007) Natural origin biodegradable systems in tissue engineering and regenerative medicine: present status and some moving trends. J R Soc Interface 4:999–1030
Meyers MA, Chen P-Y, Lin AY-M, Seki Y (2008) Biological materials: structure and mechanical properties. Prog Mater Sci 53:1–206
Nair L, Laurencin C (2006) Polymers as biomaterials for tissue engineering and controlled drug delivery. Adv Biochem Eng Biotechnol 102:47–90
Nelson JL, Roeder BL, Carmen JC, Roloff F, Pitt WG (2002) Ultrasonically activated chemotherapeutic drug delivery in a rat model. Cancer Res 62:7280–7283
Niculescu M-D, Berechet MD, Gaidau C, Ignat M, Radu M (2016) Study on obtaining keratin extracts from leather industry by-products, ICAMS (2016). In: 6th international conference on advanced materials and systems. http://icams.ro/icamsresurse/2016/proceedings/IV_Industrial_Ecology_07.pdf
Onifade AA, Al-Sane NA, Al-Musallam AA, Al-Zarban S (1998) A review: potentials for biotechnological applications of keratin degrading microorganisms and their enzymes for nutritional improvement of feathers and other keratins as livestock feed resources. Biores Technol 66:1–11
Papadopoulos MC, El Boushy AR, Ketelaars EH (1985) Effect of different processing conditions on amino acid digestibility of feather meal determined by chicken assay. Poult Sci 64:1729–1741
Parveen S, Sahoo SK (2008) Polymeric nanoparticles for cancer therapy. J Drug Target 16:108–123
Rouse JG, Van Dyke ME (2010) A review of keratin-based biomaterials for biomedical applications. Materials 3:999–1014
Saratale GD, Kshirsagar SD, Sampange VT, Saratale RG, Oh SE, Govindwar SP, Oh MK (2014) Cellulolytic enzymes production by utilizing agricultural wastes under Solid State Fermentation and its application for Biohydrogen production. Appl Biochem Biotechnol 174:2801–2817
Schweizer J, Bowden PE, Coulombe PA, Langbein L, Lane EB, Magin TM, Maltais L, Omary MB, Parry DA, Rogers MA, Wright MW (2006) New consensus nomenclature for mammalian keratins. J Cell Biol 174:169–174
Sharma S, Gupta A (2016) Sustainable management of keratin waste biomass: applications and future perspectives. Braz Arch Biol Technol 59
Sharma S, Gupta A, Chik S, Kee CG, Mistry BM, Kim DH, Sharma G (2017a) Characterization of keratin microparticles from feather biomass with potent antioxidant and anticancer activities. Int J Biol Macromol. https://doi.org/10.1016/j.ijbiomac.2017.06.015
Sharma S, Gupta A, Chik SMSBT, Kee CYG, Poddar PK (2017b) Dissolution and characterization of biofunctional keratin particles extracted from chicken feathers. In: IOP conference series: materials science and engineering, vol 1. IOP Publishing, p 012013

Sharma S, Gupta A, Chik SMST, Kee CYG, Podder PK, Subramaniam M, Thuraisingam J (2017c) Study of different treatment methods on chicken feather biomass. IIUM Eng J 18(2):47–55
Sharma S, Gupta A, Kumar A, Kee CG, Kamyab H, Saufi SM (2018) An efficient conversion of waste feather keratin into ecofriendly bioplastic film. Clean Technol Environ Policy 1–11
Sharma S, Gupta A, Saufi SM, Chik T, Chua GK, Pradeep Kumar P, Jayshree T, Malini S (2016) Extraction and characterization of keratin from chicken feather waste biomass: a study
Shavandi A, A Bekhit AE-D, Carne A, Bekhit A (2017a) Evaluation of keratin extraction from wool by chemical methods for bio-polymer application. J Bioact Compat Polym 32:163-177
Shavandi A, Silva TH, Bekhit AA, Bekhit AEA (2017b) Keratin: dissolution, extraction and biomedical application. Biomater Sci 5:1699–1735
Sinkiewicz I, Sliwinska A, Staroszczyk H, Kołodziejska I (2017) Alternative methods of preparation of soluble keratin from chicken feathers. Waste Biomass Valor 8:1043–1048
Soccol CR, da Costa ESF, Letti LAJ, Karp SG, Woiciechowski AL, de Souza Vandenberghe LP (2017) Recent developments and innovations in solid state fermentation. Biotechnol Res Innov 1:52–71
Staron P, Banach M, Kowalski Z (2011) Keratin-Origins, properties, application. Chemik 65:1019–1026
Steinert PM, Idler WW, Zimmerman S (1976) Self-assembly of bovine epidermal keratin filaments in vitro. J Mol Biol 108:547–567
Tang L, Sierra JO, Kelly R, Kirsner RS, Li J (2012) Wool-derived keratin stimulates human keratinocyte migration and types IV and VII collagen expression. Exp Dermatol 21:458–460
Tesfaye T, Sithole B, Ramjugernath D, Chunilall V (2017) Valorisation of chicken feathers: characterisation of chemical properties. Waste Manag 68:626–635
Than MP, Smith RA, Cassidy S, Kelly R, Marsh C, Maderal A, Kirsner RS (2013) Use of a keratin-based hydrogel in the management of recessive dystrophic epidermolysis bullosa. J Dermatol Treat 24:290–291
Tiwary E, Gupta R (2012) Rapid conversion of chicken feather to feather meal using dimeric keratinase from *Bacillus licheniformis* ER-15. J Bioprocess Biotech 2:123
Tombolato L, Novitskaya EE, Chen PY, Sheppard FA, McKittrick J (2010) Microstructure, elastic properties and deformation mechanisms of horn keratin. Acta Biomater 6:319–330
Tomlinson DJ, Mülling CH, Fakler TM (2004) Invited review: formation of keratins in the bovine claw: roles of hormones, minerals, and vitamins in functional claw integrity. J Dairy Sci 87:797–809
Trim MW, Horstemeyer MF, Rhee H, El Kadiri H, Williams LN, Liao J, Walters KB, McKittrick J, Park SJ (2011) The effects of water and microstructure on the mechanical properties of bighorn sheep (*Ovis canadensis*) horn keratin. Acta Biomater 7:1228–1240
U.S Patent No. 7,169,896 (2007) Keratin-based products and methods for their productions. USA
Vasconcelos A, Pêgo AP, Henriques L, Lamghari M, Cavaco-Paulo A (2010) Protein matrices for improved wound healing: elastase inhibition by a synthetic peptide model. Biomacromolecules 11:2213–2220
Wang J, Hao S, Luo T, Yang Q, Wang B (2016a) Development of feather keratin nanoparticles and investigation of their hemostatic efficacy. Mater Sci Eng C 68:768–773
Wang B, Yang W, McKittrick J, Meyers MA (2016b) Keratin: structure, mechanical properties, occurrence in biological organisms, and efforts at bioinspiration. Prog Mater Sci 76:229–318

Chapter 3
Extraction and Characterization of Keratin from Different Biomasses

Claudia Vineis, Alessio Varesano, Greta Varchi and Annalisa Aluigi

Abstract Keratin is a high-sulphur content protein, highly abundant in nature since it is the major component of feathers, hair, wool, horns and nails. In recent years, keratin-based materials have received great consideration due to its unique features in terms of ability to absorb heavy metals and other toxic compounds, thus resulting particularly useful for water and air purification. Moreover, due to its intrinsic efficacy in promoting cells growth, along with its ability to encapsulate both hydrophobic and hydrophilic drugs, keratin has been increasingly studied for the preparation of a wide range of bio-medical devices, especially in the field of tissue engineering and controlled drug delivery. Extraction of keratin from low-cost biomasses deriving from food industry by-products (especially slaughterhouse, dairy and poultry industry) is a challenging process hampered by the presence of a high content of disulphide bonds that bestow the protein with high resistance to chemical, enzymatic and thermal treatments. Thus, the large-scale use of keratin strongly depends on the development of cost-effective and time-efficient extraction methods. This chapter gives an overview on the availability of different keratinous biomasses and examines the various extraction methods proposed in the literature, underlining their advantages and limitations. Moreover, a detailed comparison between the chemical–physical properties of keratins obtained from different biomasses is here reported.

Keywords Keratin wastes · Wool · Feather · Hair · Reduction · Oxidation Sulphitolysis · Alkaline extraction · Microwave · Steam explosion Superheated water · Ionic liquids · Enzymatic treatments

---

C. Vineis · A. Varesano
Institute for Macromolecular Studies (ISMAC), Consiglio Nazionale delle Ricerche, C.so G. Pella 16, 13900 Biella, Italy

G. Varchi · A. Aluigi (✉)
Institute for Organic Synthesis and Photoreactivity (ISOF), Consiglio Nazionale delle Ricerche, via P. Gobetti 101, 40129 Bologna, Italy
e-mail: annalisa.aluigi@isof.cnr.it

G. Varchi · A. Aluigi
Kerline Srl, via P. Gobetti 101, 40129 Bologna, Italy

S. Sharma and A. Kumar (eds.), *Keratin as a Protein Biopolymer*, Springer Series on Polymer and Composite Materials, https://doi.org/10.1007/978-3-030-02901-2_3

1 Introduction

In recent years, we witnessed an increasing interest in the recovery and reuse of biomass especially as a valuable alternative to traditional plastic materials for manufacturing high-quality, cost-competitive and biodegradable consumer goods. This natural source for bioplastic materials not only would reduce the dependence of our economy on fossil sources, but also would help to diminish environmental pollution by reducing the net carbon dioxide emissions. Among natural sources, protein-based materials are widely considered for biotechnological applications (Altman et al. 2003; Goo et al. 2003).

In this view, keratin is one of the most promising protein to be exploited for the design of advanced biomaterials, due to some relevant key features: (a) keratin is an efficient adsorbent of heavy metals and volatile organic compounds (VOCs), thus useful for water and air purification (Aluigi et al. 2009); (b) its intrinsic bioactive properties, such as the ability to promote cells adhesion and proliferation, allow a wide range of medical applications (Shavandi et al. 2017a); (c) due to its flame retardant (Sanchez-Olivares et al. 2017) and gas barrier properties, keratin represents a very promising biomaterial for the development of compostable food packaging (Ramirez et al. 2017; Ramakrishnan et al. 2018). Keratin is a high-sulphur-content protein and it exists in two different structural forms with different X-ray diffraction pattern: α-keratins, mostly containing, α-helix structures, and β-keratins, essentially made of β-sheet structures (Wang et al. 2016a).

In particular, α-keratins are made of two sub-filaments: the acid keratin (Type I) with molecular weights of 40 and 50 kDa and the neutral or basic keratin (Type II) with molecular weights of about 55–65 kDa (Heid et al. 1986). α-keratins are found in mammalian body parts and tissues such as wool, hair, nails, horn and hooves (Wang et al. 2016a). Instead, the β-keratins are characteristic of nails, claws, shells and beaks of birds and they are made of 10 kDa units. Up to 95% by weight of wool is made of pure keratin; in particular, the wool cortex (85% of weight fibre) is composed of intermediate-filament proteins (IFPs), characterized by a predominantly α-helical structure embedded in a matrix of high-sulphur keratins made of β-sheet and amorphous structures (Cardamone 2010).

Horn and hooves are composed of a core of bone (the distal phalanx in the hoof) surrounded by a thick keratin covering are expressed in the hoof tissue (Pollitt 2004). Feathers, which constitute nearly the 10% of birds body weight, are composed of proteins (90–92%), water (6–8%) and lipids (1–3%). Among that 92% of protein, keratin represents the major constituent, conferring to feathers their specific organization and structural properties.

Currently, the disposal of keratinous wastes represents an issue; indeed, more than 40 million tonnes per year are produced by the constantly increasing food industry (meat, slaughter house and dairy products), whose production and market is led by United States, Brazil and China. Keratin biomasses disposal represents a problem to be faced because their burning for fuel is inefficient (keratin is self-extinguishing) and polluting (because of the high sulphur content). Moreover, the continuous accumulation in the ecosystem gives rise to landscape degradation, flies and rodents, as well as pollution of soil and groundwater (Zhang et al. 2014; Dou et al. 2015). Based on these considerations, it is evident that setting up efficient processing techniques for extracting keratin from biomasses would represent an important advancement for an efficient exploitation of this biomaterial.

However, keratin extraction can be achieved only by cleavage of disulphide covalent bonds and hydrogen inter-chain bonds. Available extracting methods can be classified into denaturing and hydrolyzing procedures. Denaturing methods allow obtaining keratin with its native molecular weight distribution and amino acid composition; indeed, this protein is a biopolymer which can be processed into several kinds of materials such as films (Posati et al. 2016), sponges (Posati et al. 2016) and nanofiber membranes (Aluigi et al. 2017).

On the other hand, hydrolyzing methods produce a powder made of a mixture of low molecular weight proteins and polypeptides both having negligible sulphur content; this powder is generally employed as filler for other polymer materials (Mazotto et al. 2017). In the following sections, the various methods used for the keratin extraction will be reviewed, with detailed reference to their impact on protein structure and properties.

2 Keratinous Biomasses Availability

Keratins are cysteine-rich proteins associated with intermediate filaments (IFs) which are cytoskeleton element with diameter of 8–10 nm (Bragulla and Homberger 2009). This protein is mainly found in two forms α and β-keratin. The α-keratins are rich in cysteine and contain a lower number of hydroxyproline and proline amino acids: these are largely found in soft tissues, such as wool, skin and hair (Gillespie 2007). β-keratins are rich in alanine and glycine, but poor in cysteine, hydroxyproline and proline (Wang et al. 2016a): these proteins are mostly found in hard tissue, such as bird feathers, fish scales, nails and others (Gupta et al. 2011). Keratins are highly chemically stable and insoluble compounds in most of the organic solvents. Some researchers classified keratin as soft and hard forms based on their sulphur content (Rizvi and Khan 2008; Sharma and Gupta 2016). Soft keratin has lower cystine content, weak cross-linking and is found in hairs core and outer layer of epidermis (Tsuda and Nomura 2014). Hard keratin has higher cystine content, higher cross-linking degree and it is found in mammalian epidermal appendages, such as horns, hairs, nails, and in avian or reptilian tissues (McKittrick et al. 2012). The EC regulation (CE 1069/2009) lays down health rules as regards animal by-products not intended for animal consumption, including keratinous wastes (http://eur-lex.europa.eu/LexUriServ/LexUriServ.do?uri=OJ:L:2009:300:0001:0033:EN:PDF).

According to the EC regulation, animal by-products are divided into three categories based on their risk analysis: (i) Category I: animal by-products in which the presence of risks for humans are possible or confirmed, such as BSE, prohibited chemicals and/or substances, and/or environment pollutants; (ii) Category II: animal by-products containing pathogens and/or chemicals (drugs) above the allowed limit; (iii) Category III: animal by-products originated from healthy and non-contaminated animals. Depending on the source, keratin biomasses can potentially enter all these categories but in most cases they are considered as category II or III.

At present, millions of tonnes of biomasses congaing keratin(s) are yearly produced by food industry and, at a lower extent, by textiles companies; therefore, their safe disposal poses serious environmental issues that need to be faced with novel technological solutions able to afford valuable products (Jayathilakan et al. 2012; Tesfaye et al. 2017). Moreover, due to outbreaks of prion diseases in the past, the application of keratin meals in livestock feeding has been lately subject to severe legal restrictions in most European countries, thus the development of novel conversion methods of keratin-based biomasses has become critical.

The major producers of keratin biomasses include United States, Brazil and China that together account for more than 40 million tonnes per year of keratin-based wastes (Sharma and Gupta 2016). Data regarding the total availability of keratin biomasses in terms of type of waste is very difficult to be determined due to inefficient and inconsistent data availability (Gupta et al. 2011; Taskin and Kurbanoglu 2011; Gupta 2014; Gonzalo et al. 2016; Oladele et al. 2017). However, as reported in Fig. 1a, the greater amount of biomasses are produced by poultry and pork meat industries, which lead the worldwide market of meat production (Fig. 1b).

As an instance, a slaughtered pig delivers about 0.9 kg bristles, which consist primarily of keratin (90% or more); thus, if considering the world production of pork meat, the available amount of keratin-rich biomasses only from this source are in the range of millions of tonnes (Gonzalo et al. 2016; Łaba et al. 2017). Globally chicken feathers are the most common keratin-based waste product, with high amounts produced in poultry slaughterhouses. The development of the poultry industry in the world has led to the generation of more than 4 million tonnes per year of waste feathers. Consumption of poultry meat grows year by year, which results in an increase in its production. Feathers consist of 90% of keratin and they make up 5–7% of the

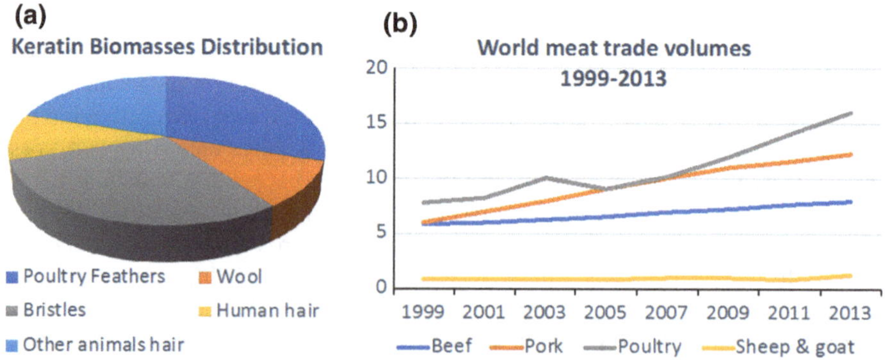

Fig. 1 **a** Worldwide availability distribution of keratin-based biomasses. **b** World meat production trade volumes 1999–2013

total weight of adult chickens, thus representing the main source of keratin (Stroń et al. 2014). These organic wastes can be utilized as a natural source to extract the keratin to be used in commercial applications (Tesfaye et al. 2017). Management of waste wool is a problem related to sheep farming and butchery (Zoccola et al. 2015). The wool obtained from sheep breeds is categorized into carpet grade (<50 μm fibre diameter) and coarse type of wool (>50 μm fibre diameter) (Holkar et al. 2016). Coarse wool is not suitable for making carpets and apparel and therefore has very low market value. Indeed, the cost of wool production is not even able to compensate the cost of shearing, sheep maintenance, land renting, etc. Hence, sheep wool of coarse grade has been considered as a worthless byproduct of sheep farming, which is or illegitimately thrown over the land for burning. Due to environmental pollution and risks of spreading contagious disease such as anthrax (Gousterova et al. 2005), valorization technologies of coarse wool for the production of more valuable products based on keratin are constantly sought.

3 Extraction Methods Based on Protein Denaturation

Keratin extraction based on 'denaturating' methods employs chemicals that can disrupt protein's supermolecular structures (e.g. secondary, tertiary, etc.) without degrading it in terms of molecular weights. These methods allow to obtain a keratin powder with a well-defined molecular weight distribution that is characteristic of the starting keratinous biomass. Generally, extracting solutions are made of buffers containing denaturing agents, surfactants and chemicals that can break down disulphide covalent bonds.

The most used denaturing agents are guanidine or urea and their function is to swell the biomasses structure by breaking the hydrogen bonds between polypeptides chains (Katoh et al. 2004). On the other hand, surfactants help to increase the extraction yield by increasing biomasses wettability and by increasing keratin solutions stability through the weakening of the hydrophobic interactions between the protein chains (Yamauchi et al. 1996). Instead, based on the kind of reaction used to break the disulphide bonds, the denaturing extraction methods are classified into reductive or oxidative methods, sulphites- or phosphines-mediated extraction.

3.1 Reductive Methods

Reductive extraction methods produce a class of proteins known as kerateines; most common reducing agents employed are thioglicolic acid, 2-mercaptoethanol and dithiothreitol (Shavandi et al. 2017b). The general procedure for extracting keratin from wool by reductive method includes washing the wool with water, drying and degreasing it with Soxlet extractor using a mixture of hexane and dichloromethane (1:1) or petroluem ether. Clean wool is mixed with the extracting solution made

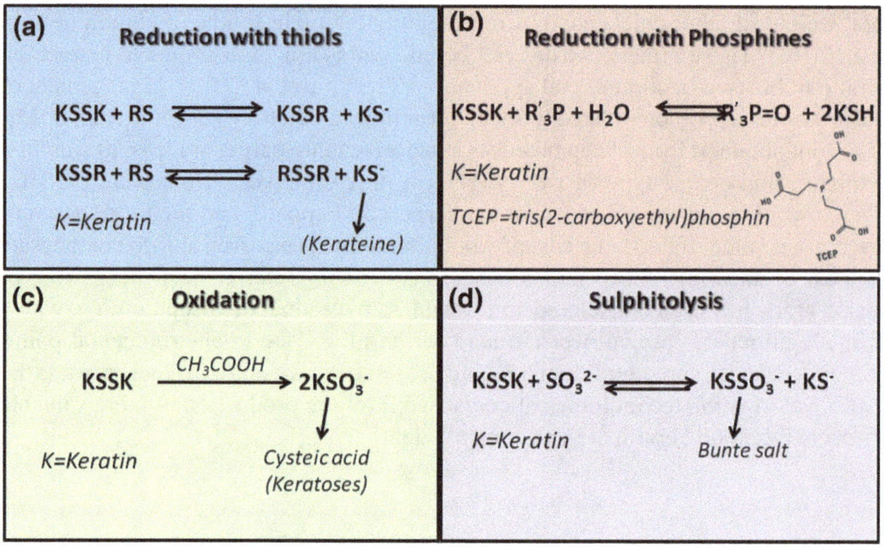

Fig. 2 Schematic diagrams of the keratin extraction by **a** reduction with thiols, **b** reduction with phosphines, **c** oxidation with peracetic (or performic) acid and **d** sulphitolysis

of urea (7–8 M), sodium dodecyl sulphate (SDS) and a reducing agent, and the resulting mixture is stirred at 50 °C for a desired time. The resulting mixture is then filtered (5 micron cut-off) and the filtrate is dialyzed against distilled water using a cellulose tube membrane (MWCO: 12–14 kDa). The obtained aqueous solution is then freeze-dried in order to obtain keratin powder (Yamauchi et al. 1996).

The reaction mechanism consists of two reversible nucleophilic displacement reactions (Fig. 2a). Under basic conditions, the thiol anion is formed and attacks the disulfide bridge favouring a double nucleophilic substitution. Thiol groups of the reduced wool proteins are easily oxidized to disulfide groups when they are exposed to air: for this reason, thiols are usually blocked by alkylation immediately after the reduction. The yield of protein extraction is strictly correlated with the reaction conditions (Table 1). Most of the extraction processes based on reduction are carried out using 2-mercaptoethanol (MEC) or thioglycolic acid at pH > 5, because keratin undergoes decomposition at alkaline pH (>11). This was confirmed by Patterson et al. (1951), who reported a significant hydrolysis of peptide bonds at pH higher than 11. Nevertheless, by using thioglycolic acid, a great protein extraction yield (47%) was reported by Savige (1960), at pH 2 and using moderate temperatures (0–60 °C). However, the authors also observed a high amount of lanthionine formation that is probably related to the processing temperature.

Nakamura et al. (2002) developed a rapid and convenient extraction procedure of keratin from human hair, using thiourea as denaturing agent. As shown in Table 1, while thiourea alone is less efficient than urea, their combination (Shindai method)

Table 1 Keratin extraction yield by reduction

Keratin source	Extraction solution	Liquor ratio (solid:liquid) (w/w)	Temperature (°C)	Time (h)	Yield (%)
Wool (Savige 1960)	Thioglycolic acid; pH = 2	–	50–60	–	47
Wool (Shavandi et al. 2017b)	Urea = 8 M; Thioglycolic acid = 0.2 M; pH = 11	–	50	3	70
Wool (Yamauchi et al. 1996)	Urea = 7 M; MEC = 1 M; SDS (g)/wool (g) = 0.6 g; pH = 7	1:20	50	24	60
Wool (Gillespie 2007)	Urea = 8 M; DTT = 0.2 M EDTA = 3 mM; 0.2 M Tris–HCl pH 9	–	20	16	60
Wool (Zoccola et al. 2009)	Urea = 8 M; DTT = 0.14 M; 0.05 M Tris–HCl pH 8.6	1:50	25	4	50
Wool (Wang et al. 2016b)	Urea = 8 M; l-cysteine = 0.165 M; NaOH up to pH 10.5	1:10	75	12	72
Wool (Wang et al. 2012)	TCEP/$Na_2S_2O_5$	–	80	5	80
Hair (Nakamura et al. 2002)	Urea = 8 M; MEC = 0.7 M (5% v/v); 0.25 M Tris–HCl pH 8.5	1:25	50	48	27
Hair (Nakamura et al. 2002)	Thiourea = 2.4 M; MEC = 0.7 M (5% v/v); 0.25 M Tris–HCL pH 8.5	1:25	50	48	19
Hair (Nakamura et al. 2002)	Urea = 5 M; Thiourea = 2.4 M; MEC = 0.7 M (5% v/v); 0.25 M Tris–HCL pH 8.5	1:25	50	48	61
Hair (Nakamura et al. 2002)	Urea = 5 M; Thiourea = 2.4 M; DTT = 5%; 0.25 M Tris–HCL pH 8.5	1:25	50	48	67
Feathers (Schrooyen et al. 2000)	Urea = 8 M; MEC = 0.125 M; 0.25 M Tris–HCL pH 7–9	1:25	40	0.5	75
Feathers (Ma et al. 2016)	Urea = 8 M; l-cysteine = 10% vs feather; NaOH up to pH 10.5	1:17	70	12	60

(continued)

Table 1 (continued)

Keratin source	Extraction solution	Liquor ratio (solid:liquid) (w/w)	Temperature (°C)	Time (h)	Yield (%)
Feathers (Sharma et al. 2018)	Na_2S [500 mM]	–	50	6	80
Horn-hoof (Zoccola et al. 2009)	Urea = 8 M; DTT = 0.14 M; 0.05 M Tris–HCL pH 8.6	1:50	25	4	3
Bovine hoof (Kakkar et al. 2014)	Urea = 7 M; MEC = 0.7 M (5% v/v); SDS(g)/Bovine Hoof(g) = 0.6; pH = 7	1:30	60	12	44

results in an increased extraction yield (61%). Similar results were obtained when the mercaptoethanol was substituted with dithiothreitol (DTT).

Reduction methods can be applied also to other keratin sources; Schrooyen et al. (2001) optimized the reductive keratin extraction from feathers, demonstrating that by increasing the pH as well as the urea amount, mercaptoethanol provides higher extraction yields. A keratin extraction yield up to 75% from feathers was achieved by using mercaptoethanol (0.125 mM) and urea>6 M at pH ranging from 7 to 9 pH for 30 min at 40 °C. Zoccola et al. (2009) applied the reductive extraction method using DTT to horn-hoof wastes. Although the horn-hoof contains less cystine than wool, the keratin extraction yield from horn-hoof by reduction is significantly lower than that of wool. This is probably due to the higher crystallinity and higher hardness of horn-hoof compared to wool. On the other hand, Kakkar et al. (2014) obtained a considerably higher extraction yield (44%) from bovine hoof by using MEC at 60 °C for 12 h.

The use of sodium dodecyl sulphate (SDS) as surfactant has different effects on the different biomasses. As an instance, Yamauchi et al. found that SDS accelerated keratin extraction from wool, while increasing the process yield (Fig. 3). Besides, SDS prevents the aggregation of keratin polypeptide chains during dialysis, when the reducing and denaturing agents are removed (Yamauchi et al. 1996). On the other hand, Schrooyen et al. (2001) proved that the addition of SDS to the extraction solution did not have significant effects on the yield of keratin extraction from feathers. However, the authors demonstrated that adding SDS to the keratin-extracted solution just before dialysis aided solution stabilization by preventing aggregation and intramolecular disulphide bonds formation by cisteyne oxidation.

Overall, the use of thiols for extracting keratin displays several advantages in terms of yield and stability of the solution; however, it has some drawbacks, such as reagents toxicity and difficult handling. In this view, Bomou et al. (2016) recently proposed l-cysteine ($HOOCCH(NH_2)CH_2SH$) as reducing agent for extracting keratin from feathers. In another work, Wang et al. (2016b) used the α-amino acid L-cysteine to extract keratin from wool. Both authors obtained undamaged keratins with an extraction yield of 60% for feather and 72% for wool.

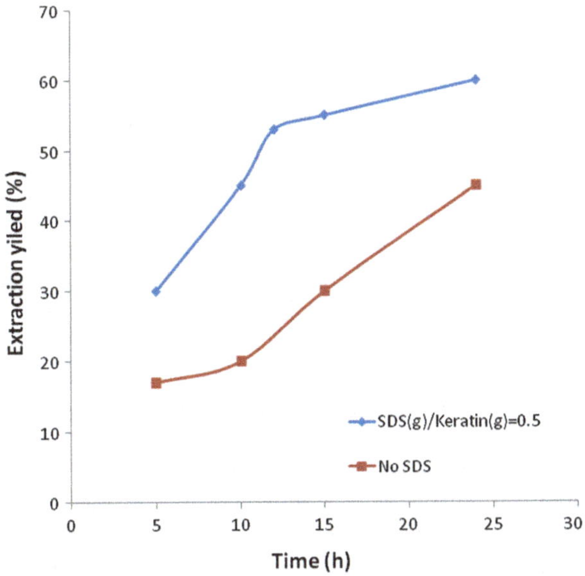

Fig. 3 Effect of SDS on the keratin extraction yield through reduction. Image reproduced according to references (Schrooyen et al. 2001; Shavandi et al. 2017a)

Phosphines (R′3P) are also a very efficient class of reducing agents for extracting keratin. Although the reaction of phosphines with disulphides in the presence of water to form thiols (Fig. 2b) is known since 1935, this procedure has been little investigated due to the unpleasant odour of phophines and their insolubility in water (Burns et al. 1991). Moreover, the tris(2-carboxyethyl) phosphine (TCEP), having the structure reported in Fig. 2b, is an odourless, crystalline, air-stable and hydro soluble solid, which rapidly reacts with disulphides at room temperature. In a work by Wang and co-workers (2012), TCEP was used to dissolve wool fibres, resulting in a almost complete wool dissolution and a keratin yield of 80%.

In a work of Sharma et al. (2018), keratin was extracted from chicken feathers, using sodium sulphide (Na_2S) as reducing agent, under alkaline conditions. The maximum extraction yield of 80.2% wt was obtained by using a concentration of sodium sulphide of 500 mM and an incubation time of 6 h at 50 °C. The obtained keratins are not damaged and were succesfully used for the preparation of bioplastic with excellent mechanical properties (Ramakrishnan et al. 2018).

3.2 Oxidative Methods

Keratin extraction from sheep wool by oxidation was earlier demonstrated in 1950 (Alexander and Earland 1950) and lately applied to human hair (Widra 1966). The oxidation treatment is generally carried out with performic or peracetic acid and results in a breakage of disulphide covalent bonds into cysteic acid residues (Fig. 2c). These oxidized keratins are commonly referred as keratoses, which are characterized by a non-covalent cross-linked structure. de Guzman et al. (2011) extracted keratose from human hairs using 2% peracetic acid for 10 h at 37 °C on a 150 rpm orbital shaker. Treated hairs were washed with water to eliminate any oxidant traces, suspended in 100 mM Tris base for 2 h, strained through a 500 μm sieve, and the liquid recovered. The protein solution was then adjusted to pH 7.4, centrifuged, filtered and dialyzed. Characterization of the isolated keratin pointed out that these keratoses miss the capacity to form disulfide cross-links due to oxidation of free thiol groups.

Keratins extracted by oxidation with performic or peracetic acid can be separated into three fractions exploiting their different solubility at different pH (Earland and Knight 1955). In particular, α-keratoses, deriving from cortex and having a crystalline structure, are isolated by solubilization in ammonina followed by precipitation at pH 4; on the other hand, β-keratoses, which derive from the cell membrane complex and from the cuticle, are insoluble in ammonia. Finally, γ-keratoses are soluble in ammonia but do not precipitate in acidic conditions (Earland and Knight 1955). It has been reported that the ratio between $\alpha/\beta/\gamma$ keratoses is of 60/10/30% for wool and of 31/18/35% for feathers barbs (Earland and Knight 1955; Corfield 1963).

3.3 *Methods Based on Sulphitolysis*

The sulphitolysis is the reaction of cleavage of disulphide bonds by sulphite ions (sulphite: SO_3^{2-}; bisulphite: HSO_3^{-} and disulphite: $S_2O_5^{2-}$), which results in hte formation of a thiol and an S-sulphonate anion (Bunte salt), as shown in Fig. 2d. Soudium sulphite (SO_3^{2-}), bisulphite (HSO_3^{-}) and disulphite ($S_2O_5^{2-}$) co-exist at the equilibrium in aqueous solution and can all be involved in the sulphitolysis reaction. The sulphitolysis reaction of cystine has been extensively studied by Cecil and co-workers (Cecil and Mcphee 1959). Sulphite ions (SO_3^{2-}) react with cystine more rapidly than bisulphites (HSO_3^{-}), and their concentration increases with increasing the pH, thus higher rates of sulphitolysis are achieved by increasing the pH. However, cystine sulphitolysis is a reversible reaction and, at pH>9, its rate decreases due to the repulsion between the protein carboxylic anions and sulphite ions and for the same reason sulphitolysis usually does not reach completion. However, in the presence of urea and SDS all disulphide bonds can be broken, thus improving the extraction yield; in general, it has been observed that the extent of reaction completion strictly depends on the composition of the extraction solution, on the process parameters, as well as on the kind of starting keratin biomasses (Table 2).

As concerns wool, the degree of sulphytolysis completion increases with increasing the SDS/keratin ratio and with decreasing the liquor ratio (up to 1:15). Instead, Aluigi et al. observed that by increasing the treatment temperature (up to 100 °C), the keratin extraction yield decreased to 22% due to partial protein degradation (Tonin et al. 2007). Moreover, the extraction yield collapses to 14% when pigmented wool

Table 2 Results about keratin extraction by sulphitolysis

Keratin source	Extraction solution	Liquor ratio (solid:liquid) (w/w)	Temperature (°C)	Time (h)	Yield (%)
Wool (Tonin et al. 2007)	Urea = 8 M; Sodium metabisulphite = 0.5 M; NaOH up to pH 6.5	1:30	100	5	22
Wool (Tonin et al. 2007)	Urea = 8 M; Sodium metabisulphite = 0.5 M; NaOH up to pH 6.5; SDS (g)/wool(g) = 0.3	1:30	65	5	30
Wool (Tonin et al. 2007)	Urea = 8 M; Sodium metabisulphite = 0.5 M; NaOH up to pH 6.5; SDS (g)/wool (g) = 0.6	1:30	65	5	33
Wool (Aluigi et al. 2014)	Urea = 8 M; Sodium metabisulphite = 0.5 M; NaOH up to pH 6.5; SDS (g)/wool (g) = 0.6	1:20	65	2	38
Wool (Shavandi et al. 2017a)	Urea = 8 M; Sodium metabisulphite = 0.5 M; NaOH up to pH 6.5; SDS (g)/wool (g) = 0.6	1:15	60	5	41
Wool (Zeng and Lu 2014)	Urea = 8 M; Bisulphite = 0.5 M; LiBr = 0.1 M; SDS = 0.02; NaOH pH12	–	90	4	50
Brown Alpaca (Aluigi et al. 2014)	Urea = 8 M; Sodium metabisulphite = 0.5 M; NaOH up to pH 6.5; SDS (g)/wool (g) = 0.6	1:20	65	2	14
Feathers (Sinkiewicz et al. 2017)	Urea = 8 M; Sodium metabisulphite = 0.5 M; SDS = 0.2 M	1:25	50	2	63
Feathers (Sinkiewicz et al. 2017)	Urea = 8 M; Sodium bisulphite = 0.5 M; SDS = 0.2 M	1:25	50	2	82
Feathers (Sinkiewicz et al. 2017)	Urea = 8 M; Sodium metabisulphite = 0.2 M; SDS (g)/feather (g) = 0.6 M; NaOH up tp pH 6.5	1:36	65	5	88

fibres (e.g. brown alpaca) were used; indeed, the egg-shaped pigments (melanins) interfere with the filtration steps of the extraction procedure, causing a significant loss of materials (Aluigi et al. 2014). In a work of Zeng and Qi, the sulphitolysis reaction was carried out with sodium bisulphite and SDS, in the presence of LiBr. The metal salt swell the wool fibres by opening the hydrogen bonds and favouring wool dissolution: under optimized process conditions ($NaHSO_3$ 0.5 M, LiBr 0.1 M, SDS 0.02 M at 90 °C for 4 h) an extraction yield of 50.2% was achieved (Zeng and Lu 2014).

Sinkiewicz et al. (2017) applied the sulphitolysis reaction to feathers; the authors showed that bisulphite and metabisulphite ions, at a concentration of 0.5 M, provide a keratin extraction yield of 82% and 63%, when the process was carried out at 50 °C for 2 h. Instead, Ayutthaya et al. extracted keratin from feather using metabisulphite ions at a higher temperature of 65 °C and for longer reaction time (5 h), demonstrating that the highest yield (88%) was achieved using a metabilphite concentration of 0.2 M (Isarankura Na Ayutthaya et al. 2015).

4 Comparison of Keratins Obtained from Different Biomasses

The extraction methods based on reduction with thiols or sulphitolysis give rise to a well and reproducible electrophoretic separation pattern, thus these methods were used to study the physic–chemical characteristics (molecular weight distributions, iso-electric point, conformational structure) of the keratins deriving from different biomasses.

Keratins from mammals have similar electrophoretic separation pattern, as well as similar amino acid composition and protein conformational structure. In a work of Zoccola et al. (2009) wool keratin was compared with that of horn-hoof. The authors showed that, as respect to horn–hoof keratin, wool keratin has a higher amount of cystine and a lower amount of amino acids that favour α-helix formation (e.g. glutamic acid, aspartic acid, leucine, lysine, and arginine) (Fig. 4a). Instead, feather keratin shows the lowest amount of cystine compared to mammalian keratins.

In Fig. 4b, the molecular weight distribution of keratins extracted from different biomasses by sulphitolysis, is reported. As shown, mammalian keratin (wool, horn-hoof and pig) shows two major fraction bands between 40 and 60 kDa, related to the low sulphur content intermediate filaments that have an α-helix structure (Shavandi et al. 2017a). The minor bands between 30 and 6 kDa derive from the amorphous and high-sulphur inter-filament proteins (Aluigi et al. 2014). The electrophoretic pattern of horn-hoof shows different molecular weight bands between 40 and 60 kDa. Zoccola et al. (2009) attributed these bands to the different kinds of tissue according to their origin and function. On the other hand, keratins from feathers show lower molecular weights between 30 and 10 kDa, associated to β-keratins typical of reptiles and birds (Bragulla and Homberger 2009).

(a) Amino-acid composition

AA/Sample	Wool (mol%)	Horn hoof (mol%)	Pig hair	Feather
CYA	0.2	0	6.05	5.42
ASP	9.3	8.7		8.41
SER	11.7	7.2	11.30	9.33
GLU	15.6	16.3	12.55	#
GLY	7.3	4.0	9.25	7.16
HIS	0.5	1.3	0.65	0.50
ARG	5.9	10.8	7.65	4.96
THR	6.8	5.1	6.95	4.04
ALA	5.7	3.7	4.90	3.31
PRO	3.1	4.4	7.15	10.03
LANT	0.4	#		#
1/CYS	9.5	7.1	10.75	5.07
TYR	2.5	4.5	3.85	4.15
VAL	5.5	5.3	4.85	6.06
MET	0.4	1.0	0.65	0.41
LYS	4.0	4.6	2.60	1.12
ILE	2.9	4.1	3.15	3.51
LEU	7.1	8.7	6.95	6.16
PHE	1.8	3.2	2.30	3.18

(b) Electrophoresis SDS-PAGE

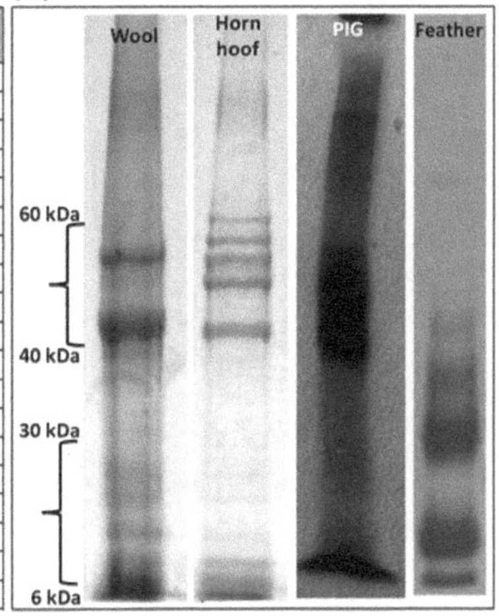

Fig. 4 Amino acid composition and molecular weight distribution of keratin extracted from different biomasses by sulphitolysis

5 Extraction Methods Based on Protein Degradation

To better exploit keratin-based biomasses in a green and economic manner, several methods based on the use of green chemicals or strong chemical–physical conditions have been developed. Under these conditions, keratin extraction from biomasses occurs through a complete damage of cystine and a partial degradation of the proteins into oligopeptides with molecular weights lower than 14 kDa. With these methods, keratin-based materials (feathers or wool) are reduced to a yellowish slurry that is separated by filtration into a solid fraction made of damaged fibres and a liquid fraction, from which it is possible to recover a powder made of proteins and oligopeptides by purification (through precipitation or dialysis) followed by freeze-drying (Fig. 5). The key methods used to solubilize keratin biomasses by partial protein degradation are: (a) the chemical treatment with alkali or ionic liquids and (b) the physical methods based on steam explosion, microwave and superheated water. Also, combinations of chemical and physical methods have been investigated.

5.1 Alkaline Extraction

It has been known that keratin degrades in alkali at pH > 9 since (Blackburn and Lee 1956) hot and strong alkali solutions produce an irreversible hydrolysis of keratin (Shavandi et al. 2017b). Indeed, under these conditions, the primary chain of keratin is damaged by breaking amide bonds and its structure is disrupted to free amino

Fig. 5 Photographic image of solid fraction and powder from liquid fraction obtained from the yellow slurry after steam explosion treatment

acids. Usually, alkaline extraction requires high amounts of alkaline chemicals and consequently, high amounts of acids to neutralize the products. Moreover, as a keratinous material is put in contact with an alkali solution, loss of sulphur (Harris 1935; Goddard 1934) occurs and degradation of the cystine residues takes place. Cystine is a major amino acid in keratinous materials, e.g. ~7% in feathers and 11–17% in wool (Fraser and Parry 2008; Arai et al. 1983), which also has important properties, therefore, its preservation during protein extraction process should be considered. On the other hand, it has been reported that cystine is very sensitive to the presence of alkali and decomposes producing aminoacrylic acid and pyruvic acid (Clifford 1933) (Fig. 6a). The breakdown of disulphide bonds can lead to the formation of sulphur odour during the alkali extraction. A most important step in the explication of the reaction involving cystine and cysteine residues was the isolation of lanthionine in keratin hydrolysates by alkali dissolution. Blackburn and Lee (1956) studied the problem by carrying out hydrolysis of wool by boiling at different pH. They detected increasing lanthionine amount as the severity of the alkaline conditions increased, while pristine wool has a negligible content of lanthionine. Moreover, an equal amount of meso- and dl-lanthionine was found, therefore the authors stated that lanthionine was produced during alkaline treatments. Figure 6a depicts cystine decomposition into lanthionine residues under alkaline conditions reported by these authors.

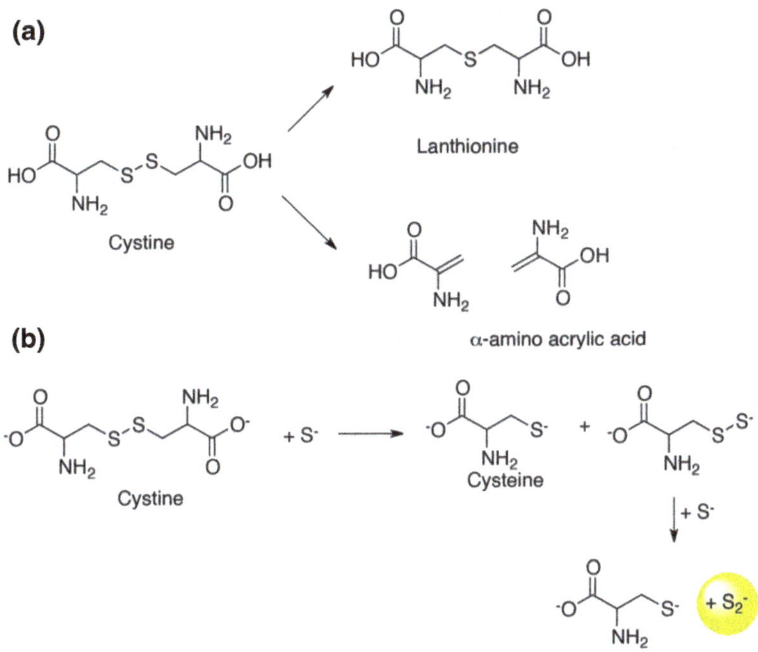

Fig. 6 **a** Chemical reaction of cystine degradation to α-amino-acrylic acid and lanthionine under alkaline condition. **b** Degradation of cystine to cysteine and sulphur

On the other hand, the solubility tests of keratin in alkali can provide useful information of the extent of the change in its chemical structures brought about by certain chemical and/or physical agents, such as acids, alkalis, oxidation, reduction, cross-linking, heat and light. Acids, oxidation, reduction, dry heating and exposure to light cause an increase in the alkali solubility of keratin; whereas, the contact with alkalis and a treatment with cross-linking agents causes decrease of solubility. An International Standard (Wool: 'Determination of solubility in alkali', ISO 3072:1975) specifies a method for measuring the solubility of wool in alkali in order to assess the severity of treatments.

Harris (1935) reported that the addition of sodium sulphide to NaOH solution speeds up and increases the solubility of the wool fibres (>50% of the wool mass was solubilized) as well as their sulphur content as compared to NaOH solution alone. These observations led to conclude that disulphide sulphur is more sensitive to alkaline treatment compared to the –SH groups, and so cysteine is more stable to alkali than cystine (Fig. 6b).

Shavandi et al. (2017a) characterized and compared the keratin extracted from wool using five extraction methods, namely, alkali hydrolysis, sulphitolysis, reduction, oxidation and extraction with ionic liquid. Alkali hydrolysis was carried out with 2%NaOH (wt.) at 80 °C until complete solubilisation of the fibres in 3 h. The solution was then neutralized with HCl, centrifuged, dialyzed and lyophilized. As compared to the other procedures, this method has the lowest keratin extraction yield (~25%) calculated by the ratio between the weight after lyophilisation and the initial weight of the wool, probably due to extensive hydrolysis of the protein which would lead to a great amount of protein lost during dialysis. SDS–PAGE revealed that the typical bands of wool disappeared in the alkali extract and the resulting protein fraction was close to 3.5 kDa, meaning that alkali extraction is unable to preserve the molecular weight of wool keratin. However, this method is relatively fast and cheap in comparison with other procedures.

Cardamone (2010) investigated the microstructure of keratin extracted from wool by two hydrolysis methods: reduction and alkaline hydrolysis. Microstructures of the extracts were characterized by MALDI-TOF/TOF spectroscopy to detect the peptide sequences and by FTIR spectroscopy to identify the protein conformation. Results from the authors show that both methodologies afford keratins with similar amino acid composition and comparable Amides I, II, and secondary structure. SDS–PAGE electrophoresis confirmed that keratin obtained by reduction had protein homologs of molecular weight around 40–60 kDa, while keratin isolated by alkaline hydrolysis contained peptide fragments of approximately 6–8 kDa. FTIR spectroscopy analysis of keratins secondary structure allowed to determine the content of α-helix/β-sheet/disordered regions, which highlighted an increase of β-sheet and disordered structures as compared to pristine wool [(25.7/51.8/22.5%) vs. (36.7/50.2/13.1%)].

Tsuda and Nomura (2014) compared the properties (amino acid composition and molecular weight) of alkaline hydrolysates from feathers and wool produced with a 10 g/L NaOH solution at 120 °C in 10 min. The hydrolysates were neutralized with HCl and dialyzed against water. Amino acid analysis showed that the hydrolysate from feathers contained more hydrophobic amino acids than the hydrolysate from wool, e.g. 44.3% versus 32.0%, (Fig. 7). Moreover, the glycine/proline content was higher for keratin from feathers than those from wool, also indicating a higher number of β-sheet structure in feathers' keratin as compared to wool. Gel permeation chromatography indicated that both hydrolysates had peptides with molecular weights lower than pristine feathers and wool.

Grewther et al. (1970) treated feathers powder with 0.1 M NaOH at 90 °C for 15 min (1 g of powder in 40 mL of liquid) achieving an extraction yield of 90%. Alkali-attacked feather keratin afforiding free amino acids (half of the constituent residues) and water-soluble protein fraction. The presence of amino acids in the extracts was assessed by paper and column chromatography. However, the chromatographic composition of the amino acids in the extracts was considerably different from that of amino acids in the starting feathers, suggesting that some amino acid residues which constituted feather keratin were degraded to some extent during extraction. On the other hand, the remaining protein fraction was richer in methionine, lysine and glutamic acid, and was poor in threonine, serine, cystine and arginine as compared to the pristine feather powder.

In another work, Yasutoyo (1970) reported the treatment of chicken feathers powder with Schweitzer's reagent (i.e. $[Cu(NH3)4(H_2O)_2]^{2+}$) produced by precipitating copper(II) hydroxide from an aqueous solution of copper sulphate using NaOH, then dissolving the precipitate in a solution of ammonia. The extraction was performed

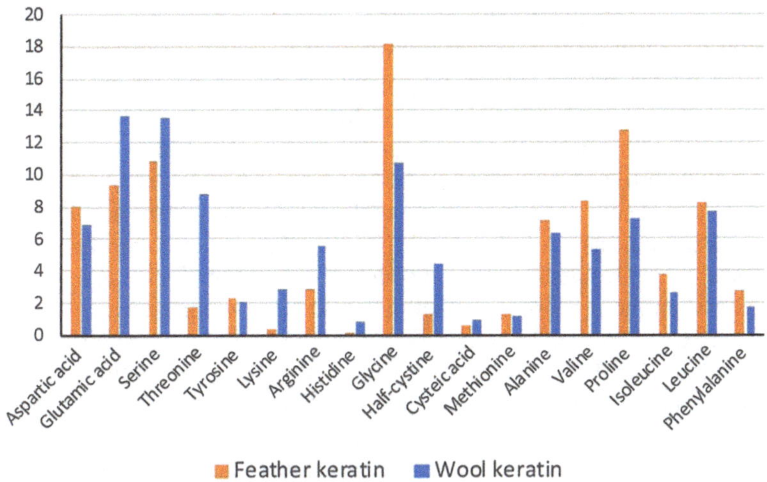

Fig. 7 Amino acid distribution of keratin extracted from feathers and wool with alkali hydrolysis. Image reproduced according to data of reference (Tsuda and Nomura 2014)

in the presence of air at room temperature for 24 h. The solid fraction was separated by filtration, while the liquid fraction was acidified in order to precipitate the protein fractions. Copper was included in the precipitate. The maximum recovery of the protein was obtained at pH 7. Further acidification of the extract decreased protein recovery because proteins dissolved in the acid. Protein precipitates were dissolved in minimum amount of phosphate buffer at pH 7.5 and dialyzed to remove free copper. The authors could not explain the mechanism of keratin dissolution by Schweitzer's reagent, even if the binding of cupric ion to proteins is a well-known reaction exploited in the biuret test. In fact, the ultimate extraction product obtained by treatment with this reagent is a keratin–copper complex since copper was not eliminated during dialysis. Interestingly, the authors proved that cystine residues in the feather powder were converted to cysteic acid by Schweitzer's reagent in the presence of atmospheric oxygen, and despite the high pH, no lanthionine was formed.

Feathers were also dissolved by Song et al. (2013) for 24 h at room temperature with 1 M NaOH solution. The extracted solution was centrifuged for 1 h and the filtrate was neutralized through the addition of 1 M HCl and then dialyzed. SDS–PAGE analysis showed a single band at about 5.8 kDa. Almost the same procedure was followed by Esparza et al. (2017) on feather powder. The final recovery of keratin was 54.5% by weight after freeze-drying. Soluble keratin obtained by alkaline extraction did not show cytotoxic effect on fibroblasts in vitro. In vivo tests of films containing feather keratin extracted by a controlled alkaline method (Thilagar et al. 2009) were carried out on dogs, indicating that the film grafts facilitate early formation of sebaceous hair follicle. Therefore, the authors conclude that alkaline hydrolysate keratin from poultry feathers can be used in wound healing patches for dogs.

Staroń and co-workers (2014) reported a study on the effect of process parameters on the extraction of keratin from poultry feathers for producing digestible proteins. The authors analysed the degree of keratin degradation in order to optimize the process and achieve the highest yield with the lowest degradation. Alkaline hydrolysis was performed with calcium hydroxide and was carried out in two stages: at first, the material was properly sterilized in agreement with regulatory laws, e.g. 133 °C for 20 min at an absolute steam pressure of three bars. During the second stage, the reactor temperature was lowered to 90–110 °C and the pressure was taken to ambient value. The obtained results show that the highest amount of digestible protein (89%) is obtained when the process is carried out at 90 °C in the presence of 2 g of H_2O and 0.75 g of Ca(OH)2 for 240 min.

In addition, alkaline conditions were used in combination with other reagents/conditions for extracting keratin from feathers; as an instance, Ma et al. (2016) reported a combination of NaOH, urea and L-cysteine which afforded keratin powder in approximately 60% yield. Alkali extraction of human hair has been studied in toxicological analysis of hairs for drugs detection (Sporkert and Pragst 2000). The proposed procedure employs a solution of 4% NaOH and a large amount of Na_2SO_4 (50%). An alkali treatment was also used to detect mercury and methylmercury in hair (Sarmani et al. 1997) using 10 M NaOH solution at 90 °C for 15 min.

The effect of NaOH concentration in the dissolution of human hair using sodium sulphite and sodium dodecyl sulphate at 80 °C was studied (Shui-qing et al. 2017). NaOH concentration was in the range from 0 to 10 g/L. NaOH detaches the cuticle layer of hairs so that sodium sulphite could enter the hair structure more easily. The highest extraction yield was reached at 6 g/L of NaOH (75.3%); a further increase in NaOH concentration likely lead to protein degradation. Alkaline conditions (pH ~ 10) were also used to extract keratin with thioglycolic acid as reducing agent (Aboushwareb et al. 2008). For the production of keratin-based haemostatic gel, small amounts of NaOH were periodically added during dialysis in order to maintain the pH of the keratin solution higher than 7.

Borrelli et al. (2015) used the Shindai method, previously described by Nakamura et al. (2002) or the extraction of keratin from human hair to be used in the production of keratin films for ocular surface reconstruction. The authors compared the effects of the dialysis against water and against a 0.25 M NaOH solution (alkaline dialysis), demonstrating that the alkaline dialysis prevents the formation of disulphide bonds and consequently particle aggregation. The alkaline dialysate, which was yellow-brown at the beginning due to hair melanin, was clear at the end. On the other hand, SDS–PAGE analysis showed that alkaline dialysis produced the loss of two bands at 40 and 60 kDa and a decrease in intensity for the band related to higher molecular weight proteins. Therefore, a hydrolytic degradation occurred during alkaline dialysis. As a result, ~40% of proteins had molecular weight smaller than 5 kDa and thus lost during dialysis. However, an improvement in transparency was reported.

A two-step method of alkali and reduction combination was used to extract keratin from human hair (Xu et al. 2015). Hairs were first treated with 0.1 M NaOH for 1 h, and then were treated with 0.75 M Na_2SO_3, 8 M urea, 0.02 M sodium dodecyl sulphate at 80 °C for 5 h. The dissolution yield was ~ 55%. Molecular weight of the dissolved keratin was between 25 and 37 kDa (evaluated by SDS–PAGE).

Recently Holkar and co-workers (2016) reported the effect of wool pretreatment with acoustic cavitation on keratin alkaline hydrolysis and compared it with a conventional pretreatment with steam. Different amounts of wool were mixed with an alkali solution containing KOH:NaOH in a 14:1 ratio and with concentrations ranging from 0.5 to 3% (w/v). Solutions were then exposed to ultrasonic irradiation for 30 min using a 20 kHz sonicator (40% amplitude of 750 W). The authors demonstrated that both processes (acoustic cavitation and steam) produced disulfide bonds cleavage at similar extent; however, acoustic-assisted alkaline hydrolysis represents an environmentally friendly and less energy demanding process thus representing a promising procedure for keratin extraction. Moreover, the authors demonstrated the effectiveness of the extracted keratins as fertilizers, both on number of seeds produced over time and the average height of plants. Results related to the alkaline extraction of keratin are summarized in Table 3.

5.2 Enzymatic and Microbial Methods

Keratin-rich materials can be degraded and hydrolyzed to peptides usable for biotechnological or food applications, using microbial and enzymatic treatments. Main sources of keratin are wool and feathers, providing a number of valuable nutrients such as carbon, nitrogen and sulphur for microorganisms. Keratin is highly resistant to proteolysis by enzymatic microorganisms; therefore, feathers and other agricultural by-products are commonly subjected to thermochemical treatment at industrial level to produce so-called 'feather meal' having a poor protein quality. However, hydrolysates generated from enzymatic hydrolysis of keratin can be used as flavouring agent because of the high amount of cysteine which can assist in the production of meat-like flavoring when heated with reducing sugars (Lasekan et al. 2013). Enzymes have many advantages over chemical or thermal treatments, so they are commonly used in various industrial and biotechnological processes. Most enzymes are produced via microbial pathways and almost all industrial enzymes are used for hydrolysis reactions. The enzymatic hydrolysis of keratin has low energy consumption, mild treatment conditions compared to chemical methods and it is environmentally safe.

Keratinases are microbial proteases produced by some microorganisms capable to hydrolyze keratin; they have several potential applications such as wool cleaning, finishing treatment of fabrics, the obstructions treatments in sewage systems and the removal of hair from leather industry, which allows avoiding the use of strong chemicals that affect the mechanical properties of the skin. The hydrolysis of keratin by microbial enzymes is a green method for the valorization of recalcitrant keratin by-products from different production streams (Brandelli 2008). Several feather-degrading bacteria, Streptomyces and *Bacillus*, have been isolated from soils and poultry wastes (Brandelli 2008). As largely reported in the literature, *Bacillus* strains

Table 3 Results about alkaline extraction of keratin

Keratin source	Extraction conditions	Notes
Wool (Blackburn and Lee 1956)	Boiling under reflux; buffer (pH from 7 to 10)/wool = 150/1 (w/w)	• Complete wool dissolution was achieved • Amino acids degradation (threonine, serine, and cystine) • Direct measurement of lanthionine
Wool (Harris 1935)	Different concentrations of Na_2CO_3/NaOH solutions; wool/solution = 1 g/50 mL at 55 °C for 30’	• Loss of sulphur occurs and degradation of the cystine residues takes place
Wool (Shavandi et al. 2017a)	Wool (10 g) treated with 2 wt% NaOH at 80 °C for 3 h. Solution was neutralized with HCl to pH 7.0, centrifuged, dialyzed and lyophilized	• Keratin extraction yield: 25 (%) • Average MW ~3.5 KDa; no typical bands of wool keratin were detected
Wool (Cardamone 2010)	0.5 N NaOH (pH 13.9); weight of bath/weight of wool = 20/1; heating at 62–65 °C. for 3 h	• Peptide fragments of ~6–8 kDa • α-helix/β-sheet/disordered regions = 25.7/51.8/22.5% (hydrolysed keratin); 58.2/37.9/3.9% (pristine wool)
Wool (Holkar et al. 2016)	Wool mixed with an alkali solution of KOH:NaOH (14:1) at concentrations ranging from 0.5 to 3% (w/v). Solutions were then exposed to ultrasonic irradiation for 30 min using a 20 kHz sonicator (40% amplitude of 750 W)	The authors demonstrated that both processes (acoustic cavitation and steam) produced disulfide bonds cleavage at similar extent extraction yields up to 100%
Wool Feathers (Tsuda and Nomura 2014b)	NaOH solution (10 g/L) at 120 °C for 10 min Wool/bath = 1/10	• Hydrophobic amino acids (feathers hydrolysates) = 35.4% Hydrophilic amino acids (feathers hydrolysates) = 44.3% • Hydrophobic amino acids (wool hydrolysates) = 32.0% Hydrophilic amino acids (wool hydrolysates) = 53.5% • Feathers hydrolysates: high glycine and proline content; more β-sheet

(continued)

Table 3 (continued)

Keratin source	Extraction conditions	Notes
Chicken feathers (Grewther et al. 1970)	0.2 N Na_2CO_3 (100 mL/gr of feathers) 1 h at 90 °C or 0.1 N NaOH (different volumes/gr of feathers) for 30′ at 90 °C	• Extraction yield 40% (0.2 N Na_2CO_3) • Extraction yield from 90% (40 mL of 0.1 N NaOH/g of feathers)
Chicken feathers (Yasutoyo 1970)	• Schweitzer's reagent $[Cu(NH_3)_4(H_2O)_2]^{2+}$ • Feather powder/reagent = 1.0 g/50 mL at room temperature for 24 h in presence of air	• The product a keratin–copper complex • Cystine residues were converted to cysteic acid and no lanthionine was formed • Low content of aromatic amino acids
Chicken feathers (Song et al. 2013)	1 N NaOH (300 mL) for 24 h at room temperature. After centrifugation the solution was neutralized with 1 N HCl	SDS–PAGE profile showed a single major band 5.8 kDa
Chicken feathers (Esparza et al. 2017)	1 M NaOH (feathers/bath = 1 g/25 mL) Soluble feather keratin was recovered by filtration (53 μm metallic mesh). The filtrate was neutralized with HCl and centrifuged	• Keratin recovery: 54.5% (by weight after freeze-drying) • Nitrogen content of feather keratin: 4.9% • Feather keratin was not toxic for fibroblasts (conc range 0.4-1.6 mg/mL)
Chicken feathers (Staroń et al. 2014)	Two steps extraction: (1) steam (130 °C; 20 min, 3 bars); (2) $Ca(OH)_2$ (water: 2–5 g; *T*: 90–110 °C; time: 60–240 min)	Highest amount of digestible protein: 89%
Chicken feathers (Ma et al. 2016)	8 M urea and l-cysteine (10% w/w), 50%NaOH to pH 10.5 (feathers/bath = 17/1) 12 h at 70 °C.	Yield of freeze-dried keratin powder: 60%
Duck feathers (Zhang et al. 2015)	Combination of steam explosion (1.4–2.0 MPa for 0.5–5 min) and alkali: 0.4%NaOH (solvent/material = 20/1 (v/w), for 1 h at 25 °C	• Steam explosion pretreatment promotes feathers dissolution • Extraction rate: 65.78% • Keratin yield: 42.78% (1.6 MPa, 1 min) • Most of the protein backbone, were maintained • Fragmentation of macromolecular chains and a loss of ordered structures

(continued)

Table 3 (continued)

Keratin source	Extraction conditions	Notes
Poultry feathers (Mokrejs et al. 2011)	Two-stage alkaline-enzymatic hydrolysis: (1) 0.1 or 0.3% KOH water solution (feathers/solution= 1/50) at 70 °C for 24 h (2) Adding of proteolytic enzyme (Savinase Ultra 16 L)	Hydrolysis efficiency (expressed as percentage of degraded feathers): 24%: 0.1% KOH at 70 °C for 8 h; 5% enzyme dose 90.8%: 0.3% KOH at 70 °C for 8 h; 5% enzyme dose
Human hair (Sarmani et al. 1997)	10 M NaOH solution at 90 °C for 15 min (hair/solution= 0.1 g/0.2 mL)	• Alkali treatment was used to detect mercury and methylmercury in hair
Human hair (Sporkert and Pragst 2000)	4%NaOH and a large amount of Na_2SO_4 (50%)	Alkali extraction was studied in toxicological analysis of hairs for drugs detection
Human hair (Shui-qing et al. 2017)	NaOH 6 g/L. NaOH	Extraction yield: 75.3% FTIR analysis: the molecular structure of keratin was transformed from α-helix to β-sheet
Human hair (Aboushwareb et al. 2008)	1 M thioglycolic acid titrated to pH 10.2 with NaOH	• During the dialysis, the pH of the keratin solution was kept at 7.0 by periodically adding 0.1 M NaOH • Characteristic bands were detected by SDS–PAGE at 64 and 43 KDa • Lower molecular weight gamma-keratins at 10 and 16 KDa • Higher Threonine and Arginine content • Tryptophan is destroyed • Asparagine and Glutamine side chains are hydrolysed to form Aspartic and Glutamic Acid
Human hair (Borrelli et al. 2015)	Shindai method Comparison between dialysis against water and against a 0.25 M NaOH solution (alkaline dialysis)	• SDS–PAGE analysis showed that alkaline dialysis produced the loss of two bands at 40 and 60 kDa • 40% of proteins had molecular weight smaller than 5 kDa
Human hair (Xu et al. 2015)	0.1 M NaOH for 1 h followed by 5 h treatment with 0.75 M Na_2SO_3, 8 M urea, 0.02 M sodium dodecyl sulphate at 80 °C	Molecular weight of the dissolved keratin was between 25 and 37 kDa

have been recognized as the principal keratinolytic microorganism for keratin hydrolysis. Bacteria have been broadly investigated for keratin hydrolysis, especially for degrading materials containing β-keratin, such as feathers. On the contrary, the enzymatic hydrolysis of α-keratin (e.g. wool) by enzymes is reported in a little number of papers (Shavandi et al. 2017b).

Bacillus licheniformis and *Bacillus pumilus* produce subtilisins which have already been recognized as keratinases. Rajput et al. (2012) engineered and characterized enzymes from *B. licheniformis* and *B. pumilus* by pro-sequence swapping. One of the resulting keratinases became more thermally stable. A previously selected strain of *Bacillus cereus* B5esz (Łaba et al. 2017) was applied to keratin biodegradation at 30 °C in order to gathered amino acids. The medium used contains $MgSO_4$, KH_2PO_4, $FeSO_4$, $CaCl_2$, yeast extract and keratin (i.e. chicken feathers or pig bristle). pH value was adjusted to 7.1 before autoclaving. The following variants of the procedure were tested: a thermochemical pretreatment followed by an enzymatic degradation, an enzymatic degradation of untreated feathers in the presence of sulphite, and a thermochemical pretreatment followed by microbial degradation. The degree of bristle biodegradation was evaluated taking into account the pretreatment being linked to sulphite concentration. In general, hydrolysates amino acid composition from enzymatic degradations of bristle before a thermochemical treatment was substantially different from those produced with acidic or alkaline hydrolysis (Łaba et al. 2017).

Several feather-derived products constituted by β-keratin (namely, feather meal, white feathers and melanized feathers) were efficiently hydrolyzed by *Bacillus subtilis* FTC02PR1. On the contrary, human hair, which is composed of α-keratin, were poorly degraded (Ferrareze et al. 2016). Keratin degradations were carried out in a mineral medium containing NaCl, K_2HPO_4 and KH_2PO_4. The degradation was quantified according to the Folin phenol reagent method on soluble protein. A keratinolytic protease was identified and characterized as a serine proteinase with about 30 kDa molecular weight. The enzymatic activity was measured using azocasein and azokeratin. The highest activity was at 60 °C with a pH range from 6 to11. Moreover, the effect of many different chemicals and known protease inhibitors (i.e. phenylmethylsulfonyl fluoride and ethylenediaminetetraacetic acid) were tested. The enzyme is sensitive to the inhibitors, 2-mercaptoethanol, $ZnSO_4$ and $LaCl_3$.

Three new strains of *Bacillus* bacteria were isolated from poultry industrial residues. The bacteria were identified as *B. subtilis* 1271, *B. licheniformis* 1269 and *B. cereus* 1268 (Mazotto et al. 2011).The *Bacillus* spp. were incubated on phosphate buffered medium with feather as source of nitrogen and carbon for 7 days at 28 °C. Supernatant solutions after cultivation were used to analyse keratinolytic activity that was evaluated at pH ranging from 3 to 11 and temperatures from 10 to 80 °C. Results obtained under these conditions suggested that the extracellular enzymes were a serine peptidase. The best pH was in the range 7–11 and the best temperature in the range 40–70 °C.

Another strain of *Bacillus* (i.e. *Bacillus subtilis* AMR) was used to treat feathers to produce keratin peptides as an ingredient in hair care products (Villa et al. 2013). The hydrolysis was carried out on feathers at pH 8 and 28 °C with small amount of yeast and extracted for 5 days. The authors report that feathers were almost completely degraded (90–95%) to oligopeptides with molecular mass in the range 800–1079 Da. Commercial keratin hydrolysates with molecular mass in the range 900–1400 Da were used for comparison.

A potential bacterial strain, identified as *Bacillus cereus* H10, showing keratinolytic action was isolated from porcine hair (Huang 2007). Two types of extracellular enzymes with molecular mass of 50.0 and 44.8 kDa were isolated from H10 (Chen et al. 2011). Keratinolytic activity of the enzymes was determined by using azokeratin with optimal pH and temperature of 7.57 and 59 °C, respectively. Moreover, the enzymes showed an extraordinary thermal stability at temperatures above 80 °C, which could minimize risks of microbial contamination in practical applications. The proteolytic and keratinolytic activities were partially inhibited by ethylenediaminetetraacetic acid and 4-(2-aminoethyl) benzenesulfonyl floride, while Zn^{2+} completely inhibited of keratinolysis. On the contrary, keratinolytic activity was enhanced by the presence of sodium dodecyl sulphate and Fe^{3+}. The enzymes displayed low hydrolysis yield for feather meal and porcine hair, but high digestion (72%) was assessed for porcine hair meal, as well as for soybean meal and casein.

A strain of *Stenotrophomonas maltophilia* (BBE11-1) was isolated and cultivated for keratinase production. After purification, three enzymes were separated from the fermentation broth. Two enzymes demonstrated keratinolytic ability and one enzyme was able to cleave disulfide bonds. The effective biodegradation of feather and wool observed in *S. maltophilia* BBE11-1 is probably due to a synergic action all the three enzymes. An alkaline pH range (7–11) and mild temperatures (40–50 °C) were optimal conditions for the enzymes (Wang et al. 2007).

A preliminary study on *Aeromonas hydrophila* K12, *Chryseobacterium indologenes* A22 and *Serratia marcescens* P3 indicated that during growth in feather meal each strain produces a metalloprotease (Bach et al. 2011). Strains were tested as potential keratinolitic microorganisms suitable for both α (i.e. human hair, wool) and β-keratin (i.e. feather, feather meal) (Bach et al. 2015). Degradation was evaluated by weight differences of the keratin substrates after digestion. Human hair and wool were digested for 360 h, while feather and feather meal for 48 h. Because feather meal is composed by thermo-mechanically processed feathers, its degradation resulted more efficient than feather for K12 and P3. Interestingly, A22 showed slightly higher degradation of feather than feather meal. A22 and K12 were also tested on degradation of human hair and wool. Small amounts of soluble protein were produced by strains on both substrates. The degradation ranged from 13 to 24%.

A novel strain of *Serratia* (sp. HPC 1383) was used to hydrolyze feathers (Khardenavis et al. 2009). The isolated bacterium showed 98% identity with six already known *Serratia* strains, including *S. marcescens*. HPC 1383 exhibited the highest keratinolytic activity on feather meal in the initial phase of growth. Hydrolytic activity of *Chryseobacterium* sp. RBT was tested on silk, human hair, wool and chicken feathers (Gurav and Jadhav 2013). The bacterium showed a high degradation efficacy towards chicken feathers (98.6% by weight difference), while the degradation of a-keratin substrates was lower, in particular, degradations of wool, silk and hair were 34.6, 23.1 and 19.2%, respectively. The effect of several parameters (temperature, pH, carbon and nitrogen sources) was also considered. The optimal conditions for hydrolysis were at pH 8.6 and temperature 50 °C. Moreover, molasses enhanced the keratinolytic activity by two-fold.

Six non-pathogenic fungal strains were isolated from alkaline soils (*Acremonium murorum*, *Aspergillus sidowii*, *Cladosporium cladosporoides*, *Neurospora tetrasperma*, *Purpureocillium lilacinum* and *Westerdikella dispersa*). The fungi were grown on feather meal agar as well as in solid state and submerged cultures, in order to test their ability to produce keratinolytic enzymes. All the fungi grew on feather meal agar, but only three of them were able to hydrolyze keratin. In particular, *P. lilacinum* showed the highest proteolytic and keratinolytic activities (Cavello 2013). The related enzyme is believed to belong to the class of serine proteases because phenylmethylsulfonyl fluoride caused a strong inhibition.

The fungus *Paecilomyces lilacinus* PL-HN-16 was used to degrade feathers showing a high efficacy in the keratin degradation (Wang and Liao 2014). After a 3 days treatment, the barbules of feathers fall off from feather, which were softened and partly degraded with a degradation efficiency of 54% by weight difference; after 4 days, feathers were completely degraded. *P. lilacinus* PL-HN-16 is active in the 6–9 pH range and the highest activity was observed at pH 7. The optimal temperature was 29 °C. The enzymatic activity was completely inhibited by phenylmethanesulfonyl fluoride.

In addition to microorganisms, some researchers studied only enzymes with keratinolytic action. Ladjama and co-workers (Habbeche et al. 2014) purified and characterized an extracellular thermostable keratinase produced by the thermophilic actinomycete *Actinomadura keratinilytica* Cpt29, which exhibited a high production of keratinolytic enzyme when grown in chicken feather meal. The purified keratinase had a molecular mass of about 29 kDa and the amino acidic sequence showed a strong homology with proteases produced by *Streptomyces*. This keratinase probably belongs to the serine protease family because it was completely inhibited by phenylmethanesulfonyl fluoride and diiodopropyl fluorophosphates. The optimum pH was at 10 and highest keratinase activity is at 70 °C. Moreover, it is inactive in acidic pH conditions and by Hg^{2+}, Ni^{2+} and Cd^{2+}, but it is stable in the presence of several non-ionic and anionic surfactants and bleaching agents (such as hydrogen peroxide and sodium perborate).

In another work, an enzymatic digestion was used in combination with an alkaline hydrolysis on feathers (Mokrejs et al. 2011). The enzyme used was Savinase Ultra 16 L. The two-stage strategy was selected to enhance enzymatic action in breaking of disulfide and peptide bonds. The first stage consisted in a 24 h alkaline treatment under mild conditions using a diluted aqueous solution of KOH at 70 °C. The second stage was the enzymatic treatment at pH 9. The study evaluated three main parameters: the duration of the second-stage of hydrolysis, the ratio between enzyme and feathers and the temperature of second-stage hydrolysis. At the best conditions investigated, the amount of the total hydrolyzed feathers was 90.8%.

Savinase 16.0LEX was used by Eslahi et al. (2013a) to hydrolyze wool and feathers with the addition of a reducing agent (sodium hydrogen sulfite). Hydrogen sulfite breaks down the disulfide bonds and facilitates the enzymatic digestion. The extraction yield was evaluated by the weight loss of the solid fractions. The effects of three factors (namely the enzyme concentration, the reducing agent concentration, and the type of substrate) were considered to design the experiments. The extraction yield was in the range 21–26%. The extracted proteins had lower molecular mass than those of the untreated fibres, hydrolysates from wool had a main molecular mass within 11–28 kDa, while proteins from feather have a molecular mass of 10 kDa. On the other hand, the resulting solid fractions were composed of fibrils that still showed features of the pristine secondary structures. In a further work (Eslahi et al. 2015), the extract produced from feathers was applied as textile finishing to wool fabrics in order to prevent felting shrinkage and to increase tensile strength and bending length.

Enzymatic treatments have been also proposed to produce keratin nanoparticles. Savinase has used to treat wool (Eslahi et al. 2013b) to produce nanoparticles. The treatment condition with the highest enzymatic efficiency was achieved using an anionic surfactant together with sodium bisulfite as reducing agent. The mean size of hydrolyzed wool particles decreased with ultrasonic treatment. Microorganisms were also proposed to produce methane from wool textiles (Kabir et al. 2013) and feathers (Forgács et al. 2013).

5.3 Dissolution in Ionic Liquids

Ionic liquids (ILs) have attracted tremendous interest in the last decades due to their high chemical variability and their interesting and useful chemical, physical and biological properties. ILs are a group of salts made of ions and their combination, and they are characterized by relatively low melting temperature (below 100 °C) (Fernández et al. 2008). They are considered the 'solvent of the future' since they are eco-friendly and safe solvents due to their non-volatility, non-flammability, chemical and thermal stability, high solvation and easy recycling (Welton 1999).

For the reasons, ILs have been widely investigated as solvents for biopolymers including wool (Zhang et al. 2017b), silk (DeFrates et al. 2017), cellulose (Feng and Chen 2008), chitin and chitosan (Silva et al. 2017). The great property of the ILs is that several anions and cations can be combined in order to obtain the optimal solvent for the dissolution and regeneration of a desired polymer.

As concerns keratinous biomasses, ionic liquids have been used to dissolve especially wool and poultry feather. For the dissolution, keratinous materials are immersed in the liquid ionic at a desired temperature and for a desired period of time, and keratin is usually regenerated using water as coagulation bath. However, the high price of ILs is a strong limitation for their uses in the keratin extraction process at industrial level (Brennecke and Maginn 2001). Recycling and reuse of ILs is the best strategy to address the prices issue. Indeed, due to their low vapour pressure, they can be recovered from the water-based regeneration bath by distillation and reused for a further extraction process (Mai et al. 2014). Figure 8 reports structures of cations and anions of ionic liquids used for keratinous biomasses dissolution.

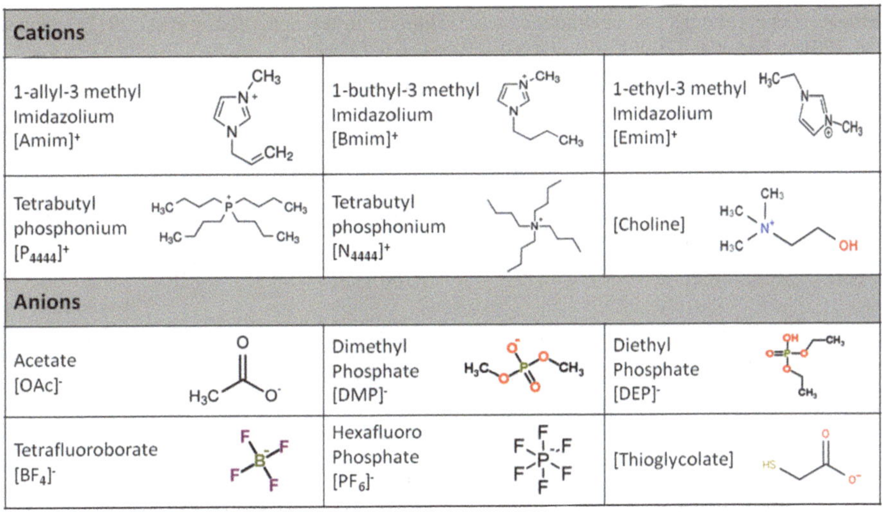

Fig. 8 Cation and anion structures of ionic liquids used for keratinous biomasses dissolution (Cl^- and Br^- are not reported)

The mechanism hypothesized for ILs action on wool foresaw that ILs first disrupt the outer lipid layer of wool fibres, mainly consisting of 18-methyleicosanoic acid along with other fatty acids are bounded through thioester linkages (Ghosh et al. 2014). The high polarity of ionic liquids makes them highly effective in the dissolution of polymers. Both ions are involved in the dissolution process, even if the anion is primarily responsible for the breakdown of inter- and intramolecular protein bonds. The dissolution occurs at higher temperatures (>100 °C); indeed, upon heating, the ion pairs in the ionic liquid dissociate into anion and cation. It is hypothesized that free anions associate with keratin hydroxyl protons, while cations complex the hydroxyl oxygens and this complex association disrupts the hydrogen bonds (Swatloski et al. 2002). Instead, the disulphide covalent bonds are reduced to H_2S (Zhang et al. 2017a).

Imidazolium-based ILs show higher extraction ability, giving higher extraction yields and $Bmim^+$ showed the best results; whereas, quaternary ammonium and phosphonium showed weak dissolution ability (Table 4) (Zhang et al. 2017a). As concern the anions, Cl^- showed the best performances in dissolving wool, allowing higher extraction yields (Zhang et al. 2017a). No dissolution was observed for non-coordinating anions such as BF_4^- and PF_6^- (Xie et al. 2005). However, at least 65% of the disulphide bonds in wool keratin should be cleaved to achieve the dissolution in ILs (Zhang et al. 2017a), while lower amount of S–S breakage strongly reduce the recovery of keratin. However, breaking down of cystine combines with the hydrolysis of peptide bonds, therefore an increase in the cystine (S–S) loss (>70%) results in lower yields of regenerated keratin (Ghosh et al. 2014; Zhang et al. 2017a).

Table 4 Results about keratin extraction by ionic liquids. References: Zhang et al. (2017a), Ghosh et al. (2014), Xie et al. (2005), Ji et al. (2014), Idris et al. (2013)

Keratin source	ILs based solvent	Liquor ratio (solid:liquid) (w/w)	Temperature (°C)	Time (h)	Yield (%)	Molecular weight (MW/kDa)
Wool	[Am223233m][Cl]	1:10	120	0.5	73.7	#
Wool	[Bmim][Cl]	1:6	120	0.5	57	20–30
Wool	[Bmim][Cl]	1:6	150	0.5	35	20–30
Wool	[Bmim][Cl]	1:6	180	0.5	18	20–30 and 10
Wool	[Bmim][Cl]	1:10	120	0.5	78.5	#
Wool	[Emim][Cl]	1:10	120	0.5	73.8	#
Wool	[Emim][Br]	1:10	120	0.5	# (PD)	#
Wool	[Bmim][OAc]	1:10	120	0.5	16.8	#
Wool	[Emim][OAc]	1:10	120	0.5	15.8	#
Wool	[Emim][DEP]	1:10	120	0.5	70.2	#
Wool	[Emim][DEP]	1:10	100	2	#ND	#
Wool	[Emim][DEP]	1:10	140	2	11.36	#
Wool	[Emim][DEP]	1:10	120	1	40.62	#
Wool	[Emim][DEP]	1:10	120	2	30.63	#
Wool	[Emim][DEP]	1:10	120	3	20.56	#
Feather	[Bmim][Cl]	1:20	80	2	2.5	20–30 and 10
Feather	[Bmim][Cl]	1:20	130	2	59	20–30 and 10
Feather	[Amim][Cl]	1:20	130	2	57	20–30 and 10
Feather	[Choline][Thioglycolate]	1:20	130	2	51	10
Feather	[Bmim][Cl] +10%Na_2SO	1:20	80	2	70	#
Feather	[Bmim][Cl] +10%Na_2SO	1:40	80	2	62	#
Feather	[Bmim][Cl] +10%Na_2SO	1:20	80	1	58	#
Feather	[Bmim][Cl] +10%Na_2SO	1:20	90	1	72	#

The highest extraction yields obtained with ILs is 78% wt and it was achieved by using a liquor ratio (wool:ionic liquid = 1:10), a temperature of 120 °C and a time treatment of 30 min. However, the yield was reduced to 57% when the liquor ratio was restricted to 1:6. It has been observed that both temperature and retention time have a strong effect on the solubility of wool in ILs. A relatively high temperature (>100 °C) reduces the solution viscosity and provides higher ions mobility, thus promoting the dissolution process (Ghosh et al. 2014; Zhang et al. 2017a). However, a temperature higher than 120 °C dramatically reduced the cystine content, as well as the molecular weight of the regenerated keratins. The protein was hydrolyzed in a mixture of peptide fragments with heterogeneous molecular masses. This resulted in a significant reduction of the extraction yield. As concern the time effect, it was observed that an increase in the dissolution time reduced the disulphide bonds content, as well as the extraction yield (Zhang et al. 2017a).

In the extraction process with ILs, the biopolymer is usually precipitated in water, therefore, the recycled ILs should be recovered after water removal. Indeed, it has been reported that water interacts with ILs through strong hydrogen bonds, affecting their viscosity, conductivity, surface tension and thus their dissolving power (Migliorati et al. 2011). The effect of water content of [Emim]DEP on its capability to dissolve keratin has been studied by Zhang et al. (2017b): by increasing water content, the [Emim]DEP viscosity, density and its glass transition temperature were decreased; as a consequence, its ability to dissolve keratin was compromised. Moreover, the amount of disulphide bonds on regenerated keratin increased with increasing the water content in ILs, confirming that the dissolving ability of [Emim]DEP was reduced when rising the water content. However, when the mole ratio water/IL ranges from 0.4 to 0.8, the time to complete the wool dissolution was not sharply increased; therefore it is acceptable and economic to recycling [Emim]DEP by maintaining water mole ration between 0.4 and 0.8 (Zhang et al. 2017b). Imidazolium-based ILs showed the best dissolving properties also for feathers and, among those, the chlorine-containing ones were reported for having the best performances. In particular, by using the [Bmim]Cl in a liquor ratio of 1:20 the keratin extraction yield was only of 2.5% wt when the process was carried out at 80 °C and for 2 h (Idris et al. 2013); on the other hand. The yield increased up to 60% when the temperature rose to 130 °C for 10 h. Comparable results were obtained with the [Amim]Cl (Ji et al. 2014).

In order to overcome the issues related to the disulphide bonds of cystine in keratin, sulphites salts, such as Na_2SO_3 were used to promote the disulphide bridges breakdown. Sodium sulphite (Na_2SO_3) can unfold the protein by breaking the disulphide covalent bonds through the equation of Fig. 2d. The extraction yield of keratin from feathers reached the 75.1% when working with an extraction solution at a liquor ratio of 1/20, a temperature of 90 °C and a time treatment of 1 h, when a 10% of sodium sulphite is added to a ILs solution containing the 20% of water (with respect IL). The 20% of water is advantageous to promote the dissolution of sodium sulphite. The yield decreased to 69.3%, when the temperature was increased to 100 °C due to keratin hydrolysis into polypeptides of lower molecular weights (Ji et al. 2014).

Another strategy applied to promote keratin dissolution is the use of new ionic liquids containing thio groups able to reduce the S–S bridges, such as [choline][thioglycolate] and [bis-(2-ethylhexyl)ammonium][thioglycolate]. However, no significant increase in feathers solubility was achieved with [choline][thioglycolate] as compared to [Amim]Cl and [Bmim]Cl. Feathers dissolution in the thioglycolate based IL was slightly more rapid in the first 10 min since the initial S–S breaking accelerates the process; however, this difference becomes negligible at a longer time. The highest extraction yield (51%) obtained with thioglycolate based ILs was observed when using a liquor ratio of 1:20, at 130 °C for 10 h. On the other hand, no dissolution was observed in [bis-(2-ethylhexyl)ammonium][thioglycolate], probably because of the longer cationic chain that reduces the solvating properties of the ionic liquid (Idris et al. 2013).

In the study of Ghosh et al. (2014) wool keratin extracted with [Bmim]Cl at 120 °C resulted partially degraded since the protein fractions at 60 and 45 kDa disappeared while new fractions at lower molecular weights of 20 and 30 kDa appeared. With increasing the dissolution temperatures (150 and 180 °C), the regenerated keratin powder was enriched with polypeptide fractions with molecular weights below 10 kDa. Compared to wool keratins, keratin from feathers has lower molecular weights (10–30 kDa), even though, the breakdown of protein chains into smaller fractions was observed also for feather treated with ILs.

FTIR and XRD analyses carried out on keratin extracted with ILs and regenerated in water, revealed that the α-helix and β-sheet crystalline structures of the starting materials tend to convert into unfolded and disordered structures. The reduction in these crystalline structures increased when increasing the treatment temperature and this match with the increase of the degradation extent and of the S–S covalent bonds reduction. The transition of α-crystallites into disordered/amorphous structures was also confirmed by DSC thermograms, in which the endothermic peak at about 230 °C, related to the α-helix denaturation of the starting materials (wool or feather) is shifted to lower temperatures in the keratin extracted with ILs and regenerated in water. In particular for wool, the shift decreased to 215 °C and 200 °C for the extraction temperatures of 120–150 °C, respectively (Ghosh et al. 2014). On the other hand, a recovery in β-sheet structures was observed when the protein was regenerated in a methanol bath (Xie et al. 2005).

5.4 Other Green Methods

To obtain a sustainable conversion of low-cost keratinous biomasses into valuable products like proteins or polypeptides, other green methods were proposed, essentially based on the treatment of keratinous biomasses under strong chemical–physical conditions (e.g. high pressures or high temperatures). Among those, keratin extraction assisted by steam explosion, microwave and superheated water are the most studied and promising under a certain point of view.

5.4.1 Steam Explosion

The Steam Explosion (SE) is a hydrothermal treatment that consists in the use of high pressure saturated steam to rapidly heat the biomass in a continuous or discontinuous reactor. The material is kept at the desired temperature (180–230 °C) for a short time (1–10 min), during which steam enters the material, dissolving it. At the end of treatment, the pressure is rapidly brought back to atmospheric value, obtaining an explosive decompression that further weakens the biomass. This process was first introduced by Mason in 1928, for explosion fibration of ligno-cellulose material (Mason 1928) and it was then developed for the conversion of biomasses, such as the pulping of lignocellulosic materials, but also the extraction of sugar from corn starch and other agricultural products for biofuels production (McKendry 2002).

The apparatus consists of a chamber with a valve connected to a steam generator and to a storage tank. The reaction occurs in the chamber where the material is inserted and the steam is injected through a pneumatic load valve for some seconds and until the desired temperature and pressure inside the chamber are reached. The material is soaked with saturated steam for a desired time, then the explosion is induced via the opening of the blow valve which induces a dramatic pressure decrease in a short time. A developed form of SE is the Steam Flash Explosion (SFE), for which, the explosion occurs in 0.0875 s (Zhao et al. 2012).

In a pioneering work by Tonin et al. (2006) the steam explosion process was applied to wool fibres in order to explore chemical-free methods to extract keratin derivatives (polypeptides and/or aminoacids) exploitable for the development of high-tech applications in several fields (cosmetic, packaging, textile finishing, filtration, etc.). In this work, wool fibres were treated with saturated steam at 220 °C for 10 min affording a dark-yellow slurry that was then filtered with a 120-mesh cut-off filter, to separate the solid fraction from the liquid one. The liquid phase was centrifuged, and sediment and the supernatant were recovered. Proteins dissolved in the supernatant were purified by precipitation with acetone followed by washing with water. Following analyses demonstrated that the morphological fraction (62.4%) was composed of both wool fragments and shapeless aggregates, while the supernatant (1.1%) was constituted of microparticles having diameters in the range of 0.5–3 μm. The 18.7% of the water-soluble proteins were recovered from the supernatant. Because of the high temperature reached during the treatment, most of the cystine was destroyed, with the lowest cystine content found in the protein recovered from the supernatant (1.12 mol%). Moreover, disulphide bonds breakage is in agreement with peptide bonds breaking; in fact, the protein recovered from the supernatant showed the lower molecular weight distributions (14–3 kDa).

The flash steam explosion was proposed as sustainable and practical pretreatment to increase the solubility of feathers in deionized water and other buffers or solvents, such as potassium phosphate (PBS) 0.01 M, pH 7.5, also with 2% urea and 0.2%NaOH (Zhao et al. 2012; Zhang et al. 2015). During the pretreatment step, feathers were put into the chamber of the steam explosion plant, treated at different steam pressure (1.4–2 MPa) for a desired time (0.5–5 min) and then exploded within 0.1 s. The protein extraction yield in different media was found to be correlated to process

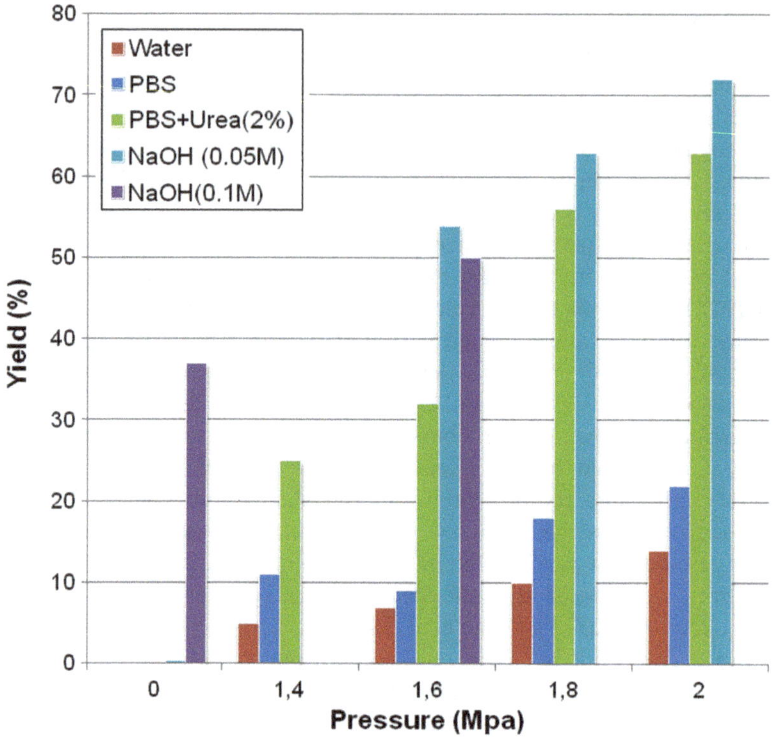

Fig. 9 Protein extraction yields in different media, from feathers treated with SFE at different pressures. The data are related to a SFE retention time of 3 min. Image reproduced according to the references (Zhao et al. 2012; Zhang et al. 2015)

parameters of the steam explosion. As shown in Fig. 9, the extraction yield increased with increasing the pressure of SFE treatment in all considered media. However, the effect of pressure increase was enhanced for the PBS + Urea (2%) buffer and NaOH (0.05 M).

Also in this case, the cystine amount of the recovered protein decreased with increasing the pressure of the SFE pretreatment (Zhang et al. 2015). Moreover, it was shown that the improvement of the protein extraction relates to an enhanced conversion of β-sheet structures into unfolded or disordered structures.

5.4.2 Microwave Treatment

In the last 20 years, microwave technology has been extensively developed as an alternative heating method for chemical reactions. The advantage of microwave as respect to conventional systems is related to a more homogenous heating of the solution in the fraction of time (Choi et al. 2006). Indeed, while in conventional processes the heating transfer starts from the wall of the reactor and diffuses into the inner part, in the microwave-assisted heating all molecules present in the reactor adsorb energy, thus providing a rapid and homogeneous rise of temperature; as a consequence, the reacting times are highly reduced with a consistent energy saving. As for SE and SFE, also microwave has been applied to the treatment of several biomasses, such as the induction of structural destruction of orange peels to increase pectin extraction yield (Wang et al. 2007).

Microwave irradiation has been investigated as a chemical-free method for extracting keratin from wool by Zoccola et al. (2012). In this work, microwave irradiation was applied for up to 60 min to a wool sample immersed in water, using different temperatures (150, 170 and 180 °C) and different liquor ratio (1:5, 1:30, 1:100). At the end of treatment, the yellowish-brown slurry was sifted at 120 mesh to recover the liquid fraction that was further filtered at 0.65 μm and then freeze-dried in order to obtain a powder made of protein, polypeptides and free amino acids. As expected, the extraction yield of the powder increased with increasing the temperature and the maximum extraction yield (62%) was obtained at 180 °C and with a liquor ratio of 1:30. Also in this case, higher temperatures induced a greater cystine loss: as respect to starting wool, 86% of cystine was lost at a working temperature of 150 °C, which increased to 99% when the treatment temperature was raised to 180 °C. The protein content of the recovered powder was analysed by Bradford assay for detecting polypeptides with molecular weights higher than 3 kDa (protein fraction). The authors describe that the powder containing the highest protein fraction (74%) was isolated by using the lowest liquor ratio (1:5) at 150 °C for 60 min. However, this protein fraction was characterized by a very low and broad molecular weight distribution in the 8–3 kDa range; moreover, this strong degradation results in a regenerated powder predominantly characterized by the random/coil structures, with no traces of α-helices structures.

The microwave-alkali process has been studied also for extraction of proteins and/or polypeptides from feathers (Lee et al. 2016). The highest extraction yield (26,74 mg/mL) was obtained using a power of 800 W, a NaOH concentration of 0.5 M, a liquor ratio of 1:50 and a treatment time of 10 min. As for wool, also in this case, the random coil and disordered structures prevail on the more structured β-sheet ones.

5.4.3 Superheated Water

Keratin hydrolysis with superheated water consists in the treatment of biomasses with water under specific conditions of pressure and temperature at which the scission of the protein into oligopeptides takes place. In a work by Bhavasar and co-workers, wool fibres were treated with superheated water together with water at 170 °C at a pressure of 7 bar for 1 h and the obtained solution was filtered three times to remove solid matter. Under those conditions, wool fibres were almost totally dissolved and the protein fraction in the liquid phase was characterized by a molecular weight

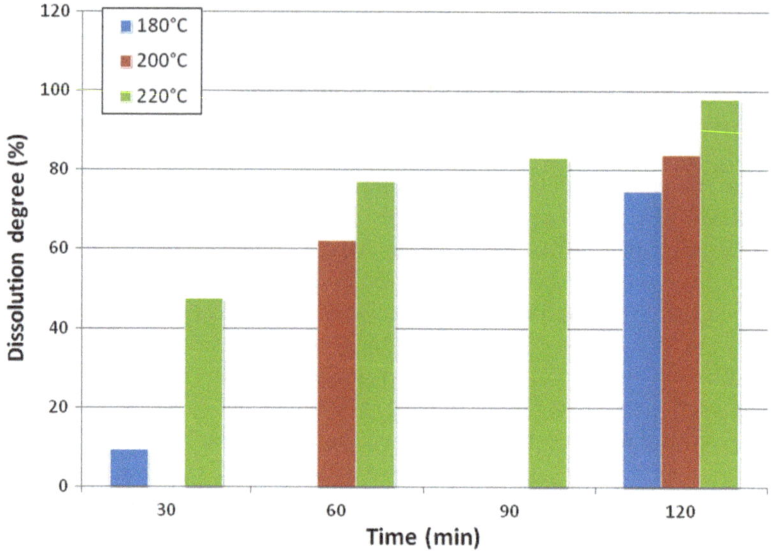

Fig. 10 Dissolution degree of feathers in superheated water, as function of time and temperature

distribution between 14 and 3 kDa (where 3 kDa is the lowest limit of detection) (Bhavsar et al. 2017).

Superheated water treatment was applied also to keratin extraction from feathers, which were sealed in a pressure cell with water at a concentration of 20 mg/mL and placed in a pre-heated oven. The pressure in the cell increased due to the partial pressure of water (22 bar at 220 °C); as expected, it was found that the dissolution degree depended on the temperature and time of treatment (Fig. 10). An extraction yield of about 98%, was found for 1 h treatment under these conditions and the majority of oligopeptides dissolved in the liquid fraction had a molar mass between 1 and 1.8 kDa (Yin et al. 2007).

6 Conclusion

Keratins extracted from different biomasses have high potential in several fields. Indeed, keratin proteins can be used for applications in tissue engineering, drug delivery, water depuration, air cleaning, textile finishing, packaging, as well as they can be used as fertilizers and animal feed. For the aforementioned applications, keratin-based materials are generally used in form of powders, micro-/nanoparticles, films, hydrogels, sponges and nanofibrous membranes. However, the potential applications of keratins extracted from different biomasses, depend on the nature and chemico-physical characteristics of the proteins, which depend on the kind of starting biomass, as well as from the method used for extraction.

This chapter describes the methods to valorize the keratinous wastes. Denaturing methods, such as reduction-, oxidation- and sulphitolysis-based methods, are the benchmark for obtaining good yields and undamaged keratins. These methods are based only on protein denaturation (intended as an alteration of supermolecular structures) which can be achieved by selectively breaking the inter-/intra-chain disulphide covalent bonds and hydrogen bonds, without breaking the peptide bonds. These methods allow obtaining keratins with the highest molecular weight distribution, that can be processed alone into several kinds of structural materials (e.g. hydrogels, films, sponges, nanofibres, etc.). However, the extraction processes based on these methods are time-consuming and require large amount of chemicals, sometimes also toxic and environmental polluting. Currently, the processes based on the reduction by l-cystine and by sulphitolysis seems to be the most industrial up-scaling ones.

On the other hand, the extraction methods based on the protein degradation such as the treatments with alkali, ionic liquids, enzymes, steam explosion, microwave and superheated water, are more eco-friendly. However, the extraction yield, as well as the protein chemical properties (e.g. molecular weight distributions and amino acid composition), strictly depend on the process parameters (e.g. temperature, pressure and time of treatment).

In general, keratins with molecular weights lower than 10 kDa are obtained with these methods. These kinds of proteins cannot be processed alone into structural materials, due to their low molecular weight. Therefore, they are often proposed as fertilizers or animal feed, as well as fillers for the development of composite materials. The ionic liquids are environmentally friendly solvents that can be totally recovered and the end of the extraction process and reused. However, their high price limits their industrial applications. The imidazole-based ILs are very common and relatively inexpensive, and this is a great advantage to decrease the obstacle of the high price for their industrial application.

Enzymatic processes are very costly, and it is not clear about its conversion and yield. Finally, the steam explosion, microwave and superheated water treatments are thermal methods which mainly employ water, high temperatures and high pressure to destroy the keratinous biomasses. However, by using these methods, lower protein extraction yield is obtained without the use of chemicals such as sodium hydroxide or sulphides. In order to make keratin a mainstream biomaterial, there is a need for the development of sustainable processes. Therefore, further investigations are required to set up green, cost-effective and less energy-intensive extraction processes which allow obtaining a keratin protein with reproducible chemico-physical properties in terms of molecular weight distribution and amino acid composition.

References

Aboushwareb T, Eberli D, Ward C et al (2008) A keratin biomaterial gel hemostat derived from human hair: evaluation in a rabbit model of lethal liver injury. J Biomed Mater Res Part B: Appl Biomater 90B:45–54. https://doi.org/10.1002/jbm.b.31251
Alexander P, Earland C (1950) Structure of wool fibres: isolation of an alpha and beta-protein in wool. Nature 166:396–397
Altman GH, Diaz F, Jakuba C et al (2003) Silk-based biomaterials. Biomaterials 24:401–416
Aluigi A, Vineis C, Tonin C et al (2009) Wool keratin-based nanofibres for active filtration of air and water. J Biobased Mater Bioenergy 3:311–319. https://doi.org/10.1166/jbmb.2009.1039
Aluigi A, Varesano A, Vineis C, Del Rio A (2017) Electrospinning of immiscible systems: the wool keratin/polyamide-6 case study. Mater Des 127:144–153. https://doi.org/10.1016/j.matdes.2017.04.045
Aluigi A, Tonetti C, Rombaldoni F et al (2014) Keratins extracted from Merino wool and Brown Alpaca fibres as potential fillers for PLLA-based biocomposites. J Mater Sci 49:6257–6269. https://doi.org/10.1007/s10853-014-8350-9
Arai KM, Takahashi R, Yokote Y, Akahane K (1983) Amino-acid sequence of feather keratin from fowl. Eur J Biochem 132:501–507. https://doi.org/10.1111/j.1432-1033.1983.tb07389.x
Bach E, Lopes FC, Brandelli A (2015) Biodegradation of α and β-keratins by Gram-negative bacteria. Int Biodeterior Biodegradation 104:136–141. https://doi.org/10.1016/j.ibiod.2015.06.001
Bach E, Daroit DJ, Corrêa APF, Brandelli A (2011) Production and properties of keratinolytic proteases from three novel Gram-negative feather-degrading bacteria isolated from Brazilian soils. Biodegradation 22:1191–1201. https://doi.org/10.1007/s10532-011-9474-0
Bhavsar PS, Zoccola M, Patrucco A et al (2017) Superheated water Hydrolyzed Keratin: a new application as a foaming agent in foam dyeing of cotton and wool fabrics. ACS Sustain Chem Eng 5:9150–9159. https://doi.org/10.1021/acssuschemeng.7b02064
Blackburn S, Lee GR (1956) The reaction of wool keratin with alkali. Biochim Biophys Acta 19:505–512. https://doi.org/10.1016/0006-3002(56)90474-7
Bomou M, Xue Q, Hou X, Yang Y (2016) Pure keratin membrane and fibers from chicken feather. Int J Biol Macromol 89:614–621. https://doi.org/10.1016/j.ijbiomac.2016.04.039
Borrelli M, Joepen N, Reichl S et al (2015) Keratin films for ocular surface reconstruction: evaluation of biocompatibility in an in-vivo model. Biomaterials 42:112–120. https://doi.org/10.1016/j.biomaterials.2014.11.038
Bragulla HH, Homberger DG (2009) Structure and functions of keratin proteins in simple, stratified, keratinized and cornified epithelia. J Anat 214:516–559. https://doi.org/10.1111/j.1469-7580.2009.01066.x
Brandelli A (2008) Bacterial Keratinases: useful enzymes for bioprocessing agroindustrial wastes and beyond. Food Bioprocess Technol 1:105–116. https://doi.org/10.1007/s11947-007-0025-y
Brennecke JF, Maginn EJ (2001) Ionic liquids: innovative fluids for chemical processing. AIChE J 47:2384–2389. https://doi.org/10.1002/aic.690471102
Burns JA, Butler JC, Moran J, Whitesides GM (1991) Selective reduction of disulfides by tris(2-carboxyethyl)phosphine. J Org Chem 56:2648–2650. https://doi.org/10.1021/jo00008a014
Cardamone JM (2010) Investigating the microstructure of keratin extracted from wool: peptide sequence (MALDI-TOF/TOF) and protein conformation (FTIR). J Mol Struct 969:97–105. https://doi.org/10.1016/j.molstruc.2010.01.048
Cavello IA (2013) Study of the production of Alkaline Keratinases in submerged cultures as an alternative for solid waste treatment generated in leather technology. J Microbiol Biotechnol 23:1004–1014. https://doi.org/10.4014/jmb.1211.11016
Cecil R, Mcphee JR (1959) The sulfur chemistry of proteins. Adv Protein Chem. 14:255–389
Chen K-N, Huang J-C, Chung C-I et al (2011) Identification and characterization of H10 enzymes isolated from Bacillus cereus H10 with keratinolytic and proteolytic activities. World J Microbiol Biotechnol 27:349–358. https://doi.org/10.1007/s11274-010-0465-9

Choi I, Choi SJ, Chun JK, Moon TW (2006) Extraction yield of soluble protein and microstructure of soybean affected by microwave heating. J Food Process Preserv 30:407–419. https://doi.org/10.1111/j.1745-4549.2006.00075.x

Clifford JBTG (1933) Sulfur in proteins. J Biol Chem 62:383–403

Corfield MC (1963) A new fraction from oxidized wool. Biochem J 86:125–129

de Guzman RC, Merrill MR, Richter JR et al (2011) Mechanical and biological properties of keratose biomaterials. Biomaterials 32:8205–8217. https://doi.org/10.1016/j.biomaterials.2011.07.054

DeFrates K, Markiewicz T, Callaway K et al (2017) Structure–property relationships of Thai silk–microcrystalline cellulose biocomposite materials fabricated from ionic liquid. Int J Biol Macromol 104:919–928. https://doi.org/10.1016/j.ijbiomac.2017.06.103

Dou Y, Huang X, Zhang B et al (2015) Preparation and characterization of a dialdehyde starch crosslinked feather keratin film for food packaging application. RSC Adv 5:27168–27174. https://doi.org/10.1039/C4RA15469J

Earland C, Knight CS (1955) Studies on the structure of keratin. Biochim Biophys Acta 17:457–461. https://doi.org/10.1016/0006-3002(55)90406-6

Eslahi N, Dadashian F, Nejad NH (2013a) An investigation on keratin extraction from wool and feather waste by enzimatic hydrolysis. Prep Biochem Biotechnol 43:624–648. https://doi.org/10.1080/10826068.2013.763826

Eslahi N, Dadashian F, Nejad NH (2013b) Optimization of enzymatic hydrolysis of wool fibers for nanoparticles production using response surface methodology. Adv Powder Technol 24:416–426. https://doi.org/10.1016/j.apt.2012.09.004

Eslahi N, Moshggoo S, Azar SK et al (2015) Application of extracted feather protein to improve the shrink resistance of wool fabric. J Ind Text 44:835–848. https://doi.org/10.1177/1528083713516666

Esparza Y, Ullah A, Boluk Y, Wu J (2017) Preparation and characterization of thermally crosslinked poly(vinyl alcohol)/feather keratin nanofiber scaffolds. Mater Des 133:1–9. https://doi.org/10.1016/j.matdes.2017.07.052

Feng L, Chen Z (2008) Research progress on dissolution and functional modification of cellulose in ionic liquids. J Mol Liq 142:1–5. https://doi.org/10.1016/j.molliq.2008.06.007

Fernández JF, Waterkamp D, Thöming J (2008) Recovery of ionic liquids from wastewater: aggregation control for intensified membrane filtration. Desalination 224:52–56. https://doi.org/10.1016/j.desal.2007.04.079

Ferrareze PAG, Correa APF, Brandelli A (2016) Purification and characterization of a keratinolytic protease produced by probiotic Bacillus subtilis. Biocatal Agric Biotechnol 7:102–109. https://doi.org/10.1016/j.bcab.2016.05.009

Forgács G, Lundin M, Taherzadeh MJ, Sárvári Horváth I (2013) Pretreatment of chicken feather waste for improved biogas production. Appl Biochem Biotechnol 169:2016–2028. https://doi.org/10.1007/s12010-013-0116-3

Fraser RDB, Parry DAD (2008) Molecular packing in the feather keratin filament. J Struct Biol 162:1–13. https://doi.org/10.1016/j.jsb.2008.01.011

Ghosh A, Clerens S, Deb-Choudhury S, Dyer JM (2014) Thermal effects of ionic liquid dissolution on the structures and properties of regenerated wool keratin. Polym Degrad Stab 108:108–115. https://doi.org/10.1016/j.polymdegradstab.2014.06.007

Gillespie JM (2007) The high-sulfur proteins of α-keratins: their relation to fiber structure and properties. J Polym Sci Part C: Polym Symp 20:201–214. https://doi.org/10.1002/polc.5070200117

Goddard DRM (1934) Studies on keratin. J Biol Chem 106:605–614

Gonzalo M, Jespersen CM, Jensen K et al (2016) Pig Bristles—an underestimated biomass resource, 1–3, 62nd International Congress of Meat Science and Technology, 14–19th August 2016, Bangkok, Thailand

Goo HC, Hwang YS, Choi YR et al (2003) Development of collagenase-resistant collagen and its interaction with adult human dermal fibroblasts. Biomaterials 24:5099–5113

Gousterova A, Braikova D, Goshev I et al (2005) Degradation of keratin and collagen containing wastes by newly isolated thermoactinomycetes or by alkaline hydrolysis. Lett Appl Microbiol 40:335–340. https://doi.org/10.1111/j.1472-765X.2005.01692.x

Grewther WG, Fraser RDB, Lennox FG, York N, Neurath H, Press A, York N, Jones CB, Mecham DK (1970) Alkali solubilization of chicken feather keratin. Agric Biol Chem 34:16–22

Gupta A (2014) Human hair "Waste" and its utilization: gaps and possibilities. J Waste Manag 2014:1–17. https://doi.org/10.1155/2014/498018

Gupta A, Perumal R, Yunus RBM, Kamarudin NB (2011) Extraction of keratin protein from chicken feather. In: Chemeca 2011: engineering a better world, pp 1–10

Gurav RG, Jadhav JP (2013) Biodegradation of keratinous waste by *Chryseobacterium* sp. RBT isolated from soil contaminated with poultry waste. J Basic Microbiol 53:128–135. https://doi.org/10.1002/jobm.201100371

Habbeche A, Saoudi B, Jaouadi B et al (2014) Purification and biochemical characterization of a detergent-stable keratinase from a newly thermophilic actinomycete Actinomadura keratinilytica strain Cpt29 isolated from poultry compost. J Biosci Bioeng 117:413–421. https://doi.org/10.1016/j.jbiosc.2013.09.006

Harris M (1935) Effect of alkalies on wool. J Res Natl Inst Stand Technol 15:63–71

Heid HW, Werner E, Franke WW (1986) The complement of native alpha-keratin polypeptides of hair-forming cells: a subset of eight polypeptides that differ from epithelial cytokeratins. Differentiation 32:101–119

Holkar CR, Jadhav AJ, Bhavsar PS et al (2016) Acoustic cavitation assisted alkaline hydrolysis of wool based keratins to produce organic amendment fertilizers. ACS Sustain Chem Eng 4:2789–2796. https://doi.org/10.1021/acssuschemeng.6b00298

Huang JCCM (2007) Selection and characterization of new keratin-degrading bacteria isolated from porcine hair. J Chin Soc Anim Sci 36:43–52

Idris A, Vijayaraghavan R, Rana UA et al (2013) Dissolution of feather keratin in ionic liquids. Green Chem 15:525. https://doi.org/10.1039/c2gc36556a

Isarankura Na Ayutthaya S, Tanpichai S, Wootthikanokkhan J (2015) Keratin extracted from chicken feather waste: extraction, preparation, and structural characterization of the keratin and keratin/biopolymer films and electrospuns. J Polym Environ 23:506–516. https://doi.org/10.1007/s10924-015-0725-8

Jayathilakan K, Sultana K, Radhakrishna K, Bawa AS (2012) Utilization of byproducts and waste materials from meat, poultry and fish processing industries: a review. J Food Sci Technol 49:278–293. https://doi.org/10.1007/s13197-011-0290-7

Ji Y, Chen J, Lv J et al (2014) Extraction of keratin with ionic liquids from poultry feather. Sep Purif Technol 132:577–583. https://doi.org/10.1016/j.seppur.2014.05.049

Kabir MM, Forgács G, Sárvári Horváth I (2013) Enhanced methane production from wool textile residues by thermal and enzymatic pretreatment. Process Biochem 48:575–580. https://doi.org/10.1016/j.procbio.2013.02.029

Kakkar P, Madhan B, Shanmugam G (2014) Extraction and characterization of keratin from bovine hoof: a potential material for biomedical applications. Springerplus 3:596. https://doi.org/10.1186/2193-1801-3-596

Katoh K, Tanabe T, Yamauchi K (2004) Novel approach to fabricate keratin sponge scaffolds with controlled pore size and porosity. Biomaterials 25:4255–4262. https://doi.org/10.1016/j.biomaterials.2003.11.018

Khardenavis AA, Kapley A, Purohit HJ (2009) Processing of poultry feathers by alkaline keratin hydrolyzing enzyme from *Serratia* sp. HPC 1383. Waste Manag 29:1409–1415. https://doi.org/10.1016/j.wasman.2008.10.009

Łaba W, Chorążyk D, Pudło A et al (2017) Enzymatic degradation of pretreated pig bristles with crude keratinase of *Bacillus cereus* PCM 2849. Waste and Biomass Valorization 8:527–537. https://doi.org/10.1007/s12649-016-9603-4

Lasekan A, Abu Bakar F, Hashim D (2013) Potential of chicken by-products as sources of useful biological resources. Waste Manag 33:552–565. https://doi.org/10.1016/j.wasman.2012.08.001

Lee YS, Phang L-Y, Ahmad SA, Ooi PT (2016) Microwave-Alkali treatment of chicken feathers for protein hydrolysate production. Waste Biomass Valorization 7:1147–1157. https://doi.org/10.1007/s12649-016-9483-7
Ma B, Qiao X, Hou X, Yang Y (2016) Pure keratin membrane and fibers from chicken feather. Int J Biol Macromol 89:614–621. https://doi.org/10.1016/j.ijbiomac.2016.04.039
Mai NL, Ha SH, Koo Y-M (2014) Efficient pretreatment of lignocellulose in ionic liquids/co-solvent for enzymatic hydrolysis enhancement into fermentable sugars. Process Biochem 49:1144–1151. https://doi.org/10.1016/j.procbio.2014.03.024
Mason WH (1928) Apparatus for and process of explosion fibration of lignocellulose material, Patent US1655618A
Mazotto AM, de Melo ACN, Macrae A et al (2011) Biodegradation of feather waste by extracellular keratinases and gelatinases from *Bacillus* spp. World J Microbiol Biotechnol 27:1355–1365. https://doi.org/10.1007/s11274-010-0586-1
Mazotto AM, Ascheri JLR, de Oliveira Godoy RL et al (2017) Production of feather protein hydrolyzed by B. subtilis AMR and its application in a blend with cornmeal by extrusion. LWT Food Sci Technol 84:701–709. https://doi.org/10.1016/j.lwt.2017.05.077
McKendry P (2002) Energy production from biomass (part 2): conversion technologies. Bioresour Technol 83:47–54. https://doi.org/10.1016/S0960-8524(01)00119-5
McKittrick J, Chen P-Y, Bodde SG et al (2012) The structure, functions, and mechanical properties of keratin. JOM 64:449–468. https://doi.org/10.1007/s11837-012-0302-8
Migliorati V, Ballirano P, Gontrani L et al (2011) Thermal and structural properties of ethylammonium chloride and its mixture with water. J Phys Chem B 115:4887–4899. https://doi.org/10.1021/jp2010138
Mokrejs P, Svoboda P, Hrncirik J et al (2011) Processing poultry feathers into keratin hydrolysate through alkaline-enzymatic hydrolysis. Waste Manag Res 29:260–267. https://doi.org/10.1177/0734242X10370378
Nakamura A, Arimoto M, Takeuchi K, Fujii T (2002) A rapid extraction procedure of human hair proteins and identification of phosphorylated species. Biol Pharm Bull 25:569–572
Oladele IO, Olajide JL, Daramola OO, Siaw KB (2017) Re-evaluation of bovine fiber biomass as exploitable keratinous bio-resource for biomedical and industrial applications. J Miner Mater Charact Eng 05:1–17. https://doi.org/10.4236/jmmce.2017.51001
Patterson WI, Geiger WB, Mizell LR (1951) The rôle of cystine in the structure of the fibrous protein, wool. J Res Natl Bur Stand 27:89–103
Pollitt CC (2004) Equine laminitis. Clin Tech Equine Pract 3:34–44. https://doi.org/10.1053/j.ctep.2004.07.003
Posati T, Sotgiu G, Varchi G et al (2016) Developing keratin sponges with tunable morphologies and controlled antioxidant properties induced by doping with polydopamine (PDA) nanoparticles. Mater Des 110:475–484. https://doi.org/10.1016/j.matdes.2016.08.017
Rajput R, Tiwary E, Sharma R, Gupta R (2012) Swapping of pro-sequences between keratinases of Bacillus licheniformis and Bacillus pumilus: altered substrate specificity and thermostability. Enzyme Microb Technol 51:131–138. https://doi.org/10.1016/j.enzmictec.2012.04.010
Ramakrishnan N, Sharma S, Gupta A, Alashwal BY (2018) Keratin based bioplastic film from chicken feathers and its characterization. Int J Biol Macromol 111:352–358. https://doi.org/10.1016/j.ijbiomac.2018.01.037
Ramirez DOS, Carletto RA, Tonetti C et al (2017) Wool keratin film plasticized by citric acid for food packaging. Food Packag Shelf Life 12:100–106. https://doi.org/10.1016/j.fpsl.2017.04.004
Rizvi TZ, Khan MA (2008) Temperature-dependent dielectric properties of slightly hydrated horn keratin. Int J Biol Macromol 42:292–297. https://doi.org/10.1016/j.ijbiomac.2008.01.001
Sanchez-Olivares G, Sanchez-Solis A, Calderas F, Alongi J (2017) Keratin fibres derived from tannery industry wastes for flame retarded PLA composites. Polym Degrad Stab 140:42–54. https://doi.org/10.1016/j.polymdegradstab.2017.04.011

Sarmani SB, Hassan RB, Abdullah MP, Hamzah A (1997) Determination of mercury and methylmercury in hair samples by neutron activation. J Radioanal Nucl Chem 216:25–27. https://doi.org/10.1007/BF02034489

Savige WE (1960) The dispersion of wool protein by thiols in acid solution. Text Res J 30:1–10. https://doi.org/10.1177/004051756003000101

Schrooyen PMM, Dijkstra PJ, Oberthür RC et al (2000) Partially carboxymethylated feather keratins. 1. Properties in aqueous systems. J Agric Food Chem 48:4326–4334. https://doi.org/10.1021/jf9913155

Schrooyen PMM, Dijkstra PJ, Oberthür RC et al (2001) Stabilization of solutions of feather keratins by sodium dodecyl sulfate. J Colloid Interface Sci 240:30–39. https://doi.org/10.1006/jcis.2001.7673

Sharma S, Gupta A (2016) Sustainable management of keratin waste biomass: applications and future perspectives. Braz Arch Biol Technol 59:1–14. https://doi.org/10.1590/1678-4324-2016150684

Sharma S, Gupta A, Kumar A et al (2018) An efficient conversion of waste feather keratin into ecofriendly bioplastic film. Clean Technol Environ Policy. https://doi.org/10.1007/s10098-018-1498-2

Shavandi A, Bekhit AE-DA, Carne A, Bekhit A (2017a) Evaluation of keratin extraction from wool by chemical methods for bio-polymer application. J Bioact Compat Polym 32:163–177. https://doi.org/10.1177/0883911516662069

Shavandi A, Silva TH, Bekhit AA, Bekhit AE-DA (2017b) Keratin: dissolution, extraction and biomedical application. Biomater Sci 5:1699–1735. https://doi.org/10.1039/C7BM00411G

Shui-qing J, Lin Z, Haixia W, Gang C (2017) Study on effective extraction of keratin from human hair wastes. Integr Ferroelectr 180:102–107. https://doi.org/10.1080/10584587.2017.1338506

Silva SS, Mano JF, Reis RL (2017) Ionic liquids in the processing and chemical modification of chitin and chitosan for biomedical applications. Green Chem 19:1208–1220. https://doi.org/10.1039/C6GC02827F

Sinkiewicz I, Śliwińska A, Staroszczyk H, Kołodziejska I (2017) Alternative methods of preparation of soluble keratin from chicken feathers. Waste Biomass Valorization 8:1043–1048. https://doi.org/10.1007/s12649-016-9678-y

Song N-B, Jo W-S, Song H-Y et al (2013) Effects of plasticizers and nano-clay content on the physical properties of chicken feather protein composite films. Food Hydrocoll 31:340–345. https://doi.org/10.1016/j.foodhyd.2012.11.024

Sporkert F, Pragst F (2000) Use of headspace solid-phase microextraction (HS-SPME) in hair analysis for organic compounds. Forensic Sci Int 107:129–148

Staroń P, Banach M, Kowalski Z et al (2014) Hydrolysis of keratin materials derived from poultry industry. Proc ECOpole 8:443–448. https://doi.org/10.2429/proc.2014.8(2)050

Swatloski RP, Spear SK, Holbrey JD, Rogers RD (2002) Dissolution of cellose with ionic liquids. J Am Chem Soc 124:4974–4975. https://doi.org/10.1021/ja025790m

Taskin M, Kurbanoglu EB (2011) Evaluation of waste chicken feathers as peptone source for bacterial growth. J Appl Microbiol 111:826–834. https://doi.org/10.1111/j.1365-2672.2011.05103.x

Tesfaye T, Sithole B, Ramjugernath D (2017) Valorisation of chicken feathers: a review on recycling and recovery route—current status and future prospects. Clean Technol Environ Policy 19:2363–2378. https://doi.org/10.1007/s10098-017-1443-9

Thilagar S, Jothi NA, Omar ARS et al (2009) Effect of keratin-gelatin and bFGF-gelatin composite film as a sandwich layer for full-thickness skin mesh graft in experimental dogs. J Biomed Mater Res Part B Appl Biomater 88B:12–16. https://doi.org/10.1002/jbm.b.31024

Tonin C, Zoccola M, Aluigi A et al (2006) Study on the conversion of wool keratin by steam explosion. Biomacromol 7:3499–3504. https://doi.org/10.1021/bm060597w

Tonin C, Aluigi A, Vineis C et al (2007) Thermal and structural characterization of poly(ethylene-oxide)/keratin blend films. J Therm Anal Calorim 89:601–608. https://doi.org/10.1007/s10973-006-7557-7

Tsuda Y, Nomura Y (2014) Properties of alkaline-hydrolyzed waterfowl feather keratin. Anim Sci J 85:180–185. https://doi.org/10.1111/asj.12093

Villa ALV, Aragão MRS, dos Santos EP et al (2013) Feather keratin hydrolysates obtained from microbial keratinases: effect on hair fiber. BMC Biotechnol 13:15. https://doi.org/10.1186/1472-6750-13-15

Wang QY, Liao MD (2014) Biodegradation of feather wastes and the purification and characterization of a concomitant keratinase from paecilomyces lilacinus. Прикладная биохимия и микробиология 50:311–317. https://doi.org/10.7868/S0555109914030313

Wang JB, Liu JY, Wang J, Bin DS (2012) The efficient wool keratin dissolution based on new reductant. Adv Mater Res 441:656–660. https://doi.org/10.4028/www.scientific.net/AMR.441.656

Wang B, Yang W, McKittrick J, Meyers MA (2016a) Keratin: structure, mechanical properties, occurrence in biological organisms, and efforts at bioinspiration. Prog Mater Sci 76:229–318. https://doi.org/10.1016/j.pmatsci.2015.06.001

Wang K, Li R, Ma JH et al (2016b) Extracting keratin from wool by using *l*-cysteine. Green Chem 18:476–481. https://doi.org/10.1039/C5GC01254F

Wang S, Chen F, Wu J et al (2007) Optimization of pectin extraction assisted by microwave from apple pomace using response surface methodology. J Food Eng 78:693–700. https://doi.org/10.1016/j.jfoodeng.2005.11.008

Welton T (1999) Room-temperature ionic liquids. solvents for synthesis and catalysis. Chem Rev 99:2071–2084. https://doi.org/10.1021/cr980032t

Widra A (1966) Ascosporogenesis by Nannizzia grubyia on a soluble fraction of keratin. Mycopathol Mycol Appl 30:141–144

Xie H, Li S, Zhang S (2005) Ionic liquids as novel solvents for the dissolution and blending of wool keratin fibers. Green Chem 7:606. https://doi.org/10.1039/b502547h

Xu J-T, Zhu P, Zhang L, Sui S-Y, Dong Z-H, Zuo Y (2015) Study of keratin extraction from human hair. 43:38–42

Yamauchi K, Yamauchi A, Kusunoki T et al (1996) Preparation of stable aqueous solution of keratins, and physiochemical and biodegradational properties of films. J Biomed Mater Res 31:439–444. https://doi.org/10.1002/(SICI)1097-4636(199608)31:4%3c439:AID-JBM1%3e3.0.CO;2-M

Yasutoyo B (1970) Solubilization of chicken feather keratin by ammonium copper hydroxide (Schweitzer's Reagent). Agric Biol Chem 34:575–584

Yin J, Rastogi S, Terry AE, Popescu C (2007) Self-organization of oligopeptides obtained on dissolution of feather keratins in superheated water. Biomacromol 8:800–806. https://doi.org/10.1021/bm060811g

Zeng CH, Lu Q (2014) Study on the recovery of waste wool by combining reduction and metallic salt methods. Adv Mater Res 881–883:551–555. https://doi.org/10.4028/www.scientific.net/AMR.881-883.551

Zhang Y, Yang R, Zhao W (2014) Improving digestibility of feather meal by steam flash explosion. J Agric Food Chem 62:2745–2751. https://doi.org/10.1021/jf405498k

Zhang Y, Zhao W, Yang R (2015) Steam flash explosion assisted dissolution of keratin from feathers. ACS Sustain Chem Eng 3:2036–2042. https://doi.org/10.1021/acssuschemeng.5b00310

Zhang Z, Nie Y, Zhang Q et al (2017a) Quantitative change in disulfide bonds and microstructure variation of regenerated wool keratin from various ionic liquids. ACS Sustain Chem Eng 5:2614–2622. https://doi.org/10.1021/acssuschemeng.6b02963

Zhang Z, Zhang X, Nie Y et al (2017b) Effects of water content on the dissolution behavior of wool keratin using 1-ethyl-3-methylimidazolium dimethylphosphate. Sci China Chem 60:934–941. https://doi.org/10.1007/s11426-016-9019-8

Zhao W, Yang R, Zhang Y, Wu L (2012) Sustainable and practical utilization of feather keratin by an innovative physicochemical pretreatment: high density steam flash-explosion. Green Chem 14:3352. https://doi.org/10.1039/c2gc36243k

Zoccola M, Aluigi A, Tonin C (2009) Characterisation of keratin biomass from butchery and wool industry wastes. J Mol Struct 938:35–40. https://doi.org/10.1016/j.molstruc.2009.08.036
Zoccola M, Aluigi A, Patrucco A et al (2012) Microwave-assisted chemical-free hydrolysis of wool keratin. Text Res J 82:2006–2018. https://doi.org/10.1177/0040517512452948
Zoccola M, Montarsolo A, Mossotti R et al (2015) Green hydrolysis as an emerging technology to turn wool waste into organic nitrogen fertilizer. Waste and Biomass Valorization 6:891–897. https://doi.org/10.1007/s12649-015-9393-0

Chapter 4
Keratin Processing

Diego Omar Sanchez Ramirez, Riccardo Andrea Carletto and Francesca Truffa Giachet

Abstract This chapter deals with the various ways in which keratin (extracted from different sources) can be processed to obtain different types of products. In the first section, solvents and polymers that must be employed to make this natural biopolymer usable are discussed. Sections 2–5 are mainly oriented in the transformations of keratin in processes such as spinning, electrospinning, casting, foaming, and freeze-drying. In addition, some products (fibers, nanofibers, films, coating, and sponge) and applications (filtration, adsorption, and scaffolds) corresponding to the procedures mentioned above are reported. The last section is related to the chemical treatments (e.g., crosslinking) applied to keratin to modify its properties.

Keywords Keratin · Films · Sponges · Fibers · Nanofibers · Crosslinking

1 Keratin Solutions

There are two main forms of employing keratin: it can be processed when it is still in extraction solution or it can be purified to be employed as a pure protein. In this section only the second case will be discussed, which means that solely the pure keratin (kerateina or keratose) extracted from different sources, without considering keratin obtained from processes that destroy or considerably modify the macromolecular structure, as acid or alkali hydrolysis, will be taken into account. The most important sources of keratin with their main aspects are reported in Tables 1, 2, 3, and 4.

As Tables 1, 2, 3, and 4 show, the most effective solvent for keratin obtained from different sources is water. This result does not surprise because, as is well known, the macromolecular structure and conformation of proteins is strongly bonded with biological processes which are realized in an aqueous medium in nature. Therefore, keratin stable forms in solution can be obtained by using water. Obviously, it is

D. O. Sanchez Ramirez (✉) · R. A. Carletto · F. Truffa Giachet
Institute for Macromolecular Studies, National Research Council of Italy (CNR-ISMAC), Corso Giuseppe Pella 16, 13900 Biella, Italy
e-mail: do.sanchezramirez@bi.ismac.cnr.it

S. Sharma and A. Kumar (eds.), *Keratin as a Protein Biopolymer*, Springer Series on Polymer and Composite Materials, https://doi.org/10.1007/978-3-030-02901-2_4

Table 1 Chicken feather keratin

Solvent	Co-solvent(s)	Polymer(s)	Other components	Product(s)	Reference(s)
Water	–	–	–	Films, NPs, cosmetics	Singh et al. (2017); Wang et al. (2016) and Sionkowska et al. (2011)
Water[a]	–	–	Pd NPs, glycerol, sorbitol	Powder/Pd NPs, sponges, films	Hengchang et al. (2013); Gao et al. (2014a, b); Placido Moore et al. (2006); Zhuang et al. (2013); Schrooyen et al. (2001) and Martelli et al. (2006a, b)
Water	–	PEO, alginate, gellan, agar	Glycerol, Ag NPs, papain	Films, fibers, Ag NPs or papain/sponges	Martelli et al. (2006a); Martelli and Borges Laurindo (2012); Wrześniewska-Tosik et al. (2007b); Gupta and Nayak (2015) and Nayak and Gupta (2017)
Water	–	Polyurethane (TDI + PPG or PEO)	H_2SO_4, NaOH	Sponges	Saucedo-Rivalcoba et al. (2011a, b) and Manrique-Juárez et al. (2013)
Water[a]	Buffer solution	–	SDS, glycerol, Irgacure/BSA/stem cell	Fibers, films, BSA or stem cell/hydrogels	Xu et al. (2014a); Xu and Yang (2014); Ma et al. (2016) and Barati et al. (2017)
Water	H_2O_2	–	–	Hydrogels, sponges	Wang et al. (2017)
Water	Acetic ac.	Chitosan	Glycerol	Fibers	Wawro et al. (2009)
TFA	–	–	–	Nanopowders	Pedram Rad et al. (2012)
Formic ac.	Chloroform/acetone	PLA	Clay (Na-montmorillonite)	NFs	Na Ayutthaya and Wootthikanokkhan (2013) and Na Ayutthaya et al. (2015, 2016)

[a]Keratin chemically modified (see Sect. 6)

Table 2 Wool keratin

Solvent	Co-solvent(s)	Polymer(s)	Other components	Product(s)	Reference(s)
Water[a]	–	–	–	Films, microparticles, hydrogels, sponges, colloidal solution	Tachibana et al. (2002, 2006); Ozaki et al. (2014); Cilurzoa et al. (2013); Sekimoto et al. (2013); Yang et al. (2009); Alemdar et al. (2005); Aluigi et al. (2007b) and Nakata et al. (2015)
Water[a]	–	–	PDA NPs, Ag NPs, glycerol, NaOH, $CaCl_2$, HA, citric acid	Sponges, PDA NPs/sponges, Ag NPs films, hydrogels	Posati et al. (2016); Zhang et al. (2011); Yamauchi et al. (1996); Tu et al. (2016); Dias et al. (2010); Nakata et al. (2014) and Sanchez Ramirez et al. (2017)
Water	–	PEO, fibroin, alginate, PVA or PVA/PEI	SDS/Ag NPs or TiO_2 NPs, BABP/Irgasan, urea/$K_2S_2O_8$/DOX · HCl/dimethyl sulfoxide, $Na_2S_2O_4$	NFs/Ag NPs or TiO_2 NPs, NFs/drug delivery, film/drug delivery, sponges, NPs/drug delivery, hydrogels, fibers	Aluigi et al. (2008b); Varesano et al. (2015); Tomaselli et al. (2016); Hamasaki et al. (2008); Li et al. (2012); Park et al. (2013) and Katoh et al. (2004)
Water[a]	Buffer solution	Alginate, PluronicF127–chitosan	$CaCl_2$/$MgCl_2$, GNP/laponite	Sponges, hydrogels	Tachibana et al. (2005); Singh et al. (2016) and Eslahi et al. (2016)
Water	Acetic ac.	Chitosan	EDGE, GDE	Films	Tanabe et al. (2002, 2004)
Formic ac.	–	–	–	NFs, NPs, films	Aluigi et al. (2013a, b); Ebrahimgol et al. (2014) and Varesano et al. (2014)
Formic ac.	–	Fibroin, PA6, cellulose acetate, PCL/PVP	–	NFs, films	Aluigi et al. (2009, 2011, 2017); Zoccola et al. (2008); Dickerson et al. (2013) and Zhu et al. (2017)
Formic ac.	Water	PVA	–	NFs	Li and Yang (2014)
HFIP	–	Polyurethane	–	NFs	Yanga et al. (2018)
1,4-Dioxane	–	PLLA	Paraffin (microspheres)	Sponges	Li et al. (2009)
Chloroform	DMF	PLGA	–	NFs	Zhang and Liu (2013)

[a]Keratin chemically modified (see Sect. 6)

Table 3 Human hair keratin

Solvent	Co-solvent(s)	Polymer(s)	Other components	Product(s)	Reference(s)
Water	–	–	–	Sponges, films/coating, hydrogels	Xu et al. (2013); Reichl (2009); Verma et al. (2008); Hill et al. (2010); Richter et al. (2012) and Burnett et al. (2013)
Water[a]	–	–	NaOH, glycerol, rhIGF-1, rhBMP-2, ciprofloxacin	Films/coating, hydrogels/drug delivery, sponges	de Guzman et al. (2015); Reichl et al. (2011); Ham et al. (2016) and Han et al. (2015)
Water	–	PEO, PVA	EGDE, GLX/$Na_2S_2O_5$/H_3PO_4	Sponges, NFs, films	Li et al. (2013); Liu et al. (2015); Fan et al. (2016); Park et al. (2015) and Choi et al. (2015)
Water	Buffer solution	–	rhBMP-2, riboflavin/hydroquinone, halofuginone	Hydrogels	Shen et al. (2011); Pace et al. (2013); de Guzman et al. (2013); Placone et al. (2017) and Peyton et al. (2012)
Water	Acetic ac.	Chitosan/gellan	–	Fibers	Hirao et al. (2005)
Water	TFE	PCL	–	NFs	Edwards et al. (2015)
Formic ac.	Water	Gelatin/PVA	GAL	NFs	Yao et al. (2017)

[a]Keratin chemically modified (see Sect. 6)

Table 4 Other sources of keratin

Keratin source	Solvent	Co-solvent(s)	Polymer(s)	Product(s)	Reference(s)
Pigeon feathers	Water	–	–	Sponges	Zhou et al. (2014)
Cattle and buffaloes horns	Water	–	–	Sponges	Srinivasan et al. (2010)
Hagfish	HFIP	–	PLA	NFs	Kim et al. (2013)
Hooves and horns	HFIP[a]	–	PHBV, PCL, PLA	NFs	Yuan et al. (2008, 2009) and Li et al. (2016)
Cattle and buffaloes horns	Water	Acetic ac.	Chitosan, gelatin	Sponges	Balaji et al. (2012)

[a]Keratin chemically modified (see Sect. 6)

necessary to consider the pH, the concentration of salts (ionic strength), the pI of protein and the presence of other compounds (e.g. surfactants—SDS) that can contribute to stabilize the protein structure. However, in case of using exclusively water as a solvent, only products with simple processing as films/coating, microparticles, NPs, colloidal solutions, hydrogels, and sponges can be obtained.

The materials made of only keratin have poor mechanical properties that limit their application. This characteristic is mainly a consequence of the absence of disulfide bonds between proteins, which are broken during the extraction of keratin. That absence hinders the stabilization and the formation of an ordered polymeric matrix in the protein structure (α helix, β sheet, β turn, and Ω loop). To improve the mechanical property after extraction process, it is necessary either to make chemicals modifications (see Sect. 6) or to add polymers soluble in water as PEO (Martelli et al. 2006a; Martelli and Borges Laurindo 2012; Aluigi et al. 2008b; Varesano et al. 2015; Tomaselli et al. 2016; Li et al. 2012, 2013; Liu et al. 2015; Fan et al. 2016), alginate (Wrześniewska-Tosik et al. 2007b; Gupta and Nayak 2015; Nayak and Gupta 2017; Hamasaki et al. 2008; Singh et al. 2016), gellan (Nayak and Gupta 2017; Hirao et al. 2005), agar (Nayak and Gupta 2017), fibroin (Vasconcelos et al. 2008) (after dissolution and purification process) or PVA (Park et al. 2013; Katoh et al. 2004; Park et al. 2015; Choi et al. 2015). When PPG or PEO with TDI are employed, it is possible to produce sponges of polyurethane/keratin with good mechanical properties (Saucedo-Rivalcoba et al. 2011a, b; Manrique-Juárez et al. 2013).

In addition, keratin is sometimes mixed in water with plasticizer compounds as glycerol (Gao et al. 2014a, b; Placido Moore et al. 2006; Zhuang et al. 2013; Schrooyen et al. 2001; Martelli et al. 2006a, b; Yamauchi et al. 1996; Tu et al. 2016; Dias et al. 2010; Reichl et al. 2011), sorbitol (Martelli et al. 2006a, b) or citric acid (Sanchez Ramirez et al. 2017) to reduce the fragility of films and sponges that occurs as an effect of the drying process. In this way, hydrogen bonds can plasticize the polypeptide matrix without applying any chemical reactions. In hydrogels, keratin, as any other protein, can form gel either when solution is heated to increase the protein concentration by the evaporation of water (this heating can also induce the denaturation of protein) or when there is an addition of compounds as laponite (Eslahi et al. 2016), PEI (Park et al. 2013), alginate (Singh et al. 2016), and chitosan (Eslahi et al. 2016). It is important to remember that polysaccharides (alginate and chitosan) as proteins can easily form gels in aqueous solutions. To make gels of keratin, it is not necessary to induce denaturation because after extraction the thiols groups of cysteine remain active and they can be reoxidized in air at low temperature (37 °C) (Wang et al. 2017; Eslahi et al. 2016; Hill et al. 2010; Richter et al. 2012; Burnett et al. 2013; Ham et al. 2016; Han et al. 2015; de Guzman et al. 2013). To accelerate the process of reoxidation, H_2O_2 (Wang et al. 2017) can be used as well. On the other hand, in some cases, keratin was chemically modified to avoid the reoxidation of thiols groups and the formation of aggregates during the dialysis phase (Schrooyen et al. 2001; Barati et al. 2017; Pedram Rad et al. 2012; Yang et al. 2009; Nakata et al. 2014, 2015; Tachibana et al. 2002, 2005, 2006; Zhang et al. 2011; Han et al. 2015; Yuan et al. 2008, 2009; Li et al. 2016). Then, the dialysis can be done against water with a low concentration of reducing agent ($Na_2S_2O_4$, 2-mercaptoethanol, or dithiothreitol) and after that, the thiol groups of dialysate proteins can be oxidized with compounds of small dimension (see Sect. 6).

Occasionally, the pH of water has to be modified at specific values and in these cases, an acid (H_2SO_4, H_3PO_4) (Manrique-Juárez et al. 2013; Park et al. 2015; Choi et al. 2015), an alkali (NaOH) (Manrique-Juárez et al. 2013; Dias et al. 2010; de Guzman et al. 2015; Reichl et al. 2011) or a buffer solution has to be employed (Xu et al. 2014a, Xu and Yang 2014; Ma et al. 2016; Barati et al. 2017; Tachibana et al. 2005; Singh et al. 2016; Eslahi et al. 2016; Shen et al. 2011; Pace et al. 2013; de Guzman et al. 2013; Placone et al. 2017; Peyton et al. 2012). In particular, buffer solutions are used to prevent the denaturation of keratin by variation of pH and to keep the pH value in specific ranges (e.g. physiological ranges); in this way, keratin-based products (hydrogels, sponges or films) can have biomedical applications (Xu et al. 2014a; Xu and Yang 2014; Ma et al. 2016; Barati et al. 2017; Tachibana et al. 2005; Singh et al. 2016; Eslahi et al. 2016; Shen et al. 2011; Pace et al. 2013; de Guzman et al. 2013; Placone et al. 2017; Peyton et al. 2012). Moreover, the variation of pH to acid conditions with acetic acid makes chitosan soluble in water; thereby chitosan can be mixed with an aqueous solution of keratin or keratin/gelatin to produce materials as films, sponges, and fibers (Wawro et al. 2009; Tanabe et al. 2002, 2004; Hirao et al. 2005).

When more polymers (PEO, alginate, gellan, agar, fibroin, and PVA) are mixed with keratin in water, more complex products as fibers and NFs can be produced (Wrześniewska-Tosik et al. 2007b; Aluigi et al. 2008b; Varesano et al. 2015; Tomaselli et al. 2016; Katoh et al. 2004; Liu et al. 2015; Fan et al. 2016; Park et al. 2015; Choi et al. 2015). In addition, these materials (films, sponges, hydrogels, fibers, and NFs) can be embedded with NPs [Pd (Hengchang et al. 2013), Ag (Nayak and Gupta 2017; Zhang et al. 2011; Varesano et al. 2015), TiO_2 (Varesano et al. 2015) and PDA (Posati et al. 2016)], chemical compounds [citric acid (Sanchez Ramirez et al. 2017), irgasan (Tomaselli et al. 2016), ciprofloxacin (Han et al. 2015), riboflavin (Placone et al. 2017), halofuginone (Peyton et al. 2012), hydroquinone (Placone et al. 2017), HA (Tu et al. 2016; Dias et al. 2010; Nakata et al. 2014) and DOX · HCl (Li et al. 2012)], microcapsules [pluronicF127 (Eslahi et al. 2016) and BABP (Tomaselli et al. 2016)], other proteins [papain (Nayak and Gupta 2017), rhIGF-1 (Ham et al. 2016), and rhBMP-2 (Han et al. 2015; de Guzman et al. 2013)]. All these allow the functionalization of keratin-based products with multiple properties such as catalytic (TiO_2, Pd and papain), biocidal (Ag, TiO_2, citric acid, irgasan, and ciprofloxacin), antioxidant (PDA), growth factor (rhIGF-1 and rhBMP-2), antibiotic (DOX · HCl), and other (riboflavin, halofuginone, hydroquinone, HA, etc.). It is important to highlight that the functional materials or compounds mentioned above are stable in aqueous solutions and do not modify considerably the structure of keratin.

Nevertheless, none of these avoids the redissolution of keratin or keeps the structure of keratin-based products when they are immersed in water. Consequently, it is necessary to employ compounds that can create new chemical bonds between protein chains (see Sect. 6). On the other hand, organic solvents as formic acid (Na Ayutthaya and Wootthikanokkhan 2013; Na Ayutthaya et al. 2015, 2016; Aluigi et al. 2009, 2011, 2013a, b, 2017; Ebrahimgol et al. 2014; Varesano et al. 2014; Zoccola et al. 2008; Dickerson et al. 2013; Zhu et al. 2017; Li and Yang 2014; Yao et al. 2017), HFIP (Yanga et al. 2018; Kim et al. 2013; Yuan et al. 2008, 2009; Li et al. 2016), TFA (Pedram Rad et al. 2012), TFE (Edwards et al. 2015), DMF (Alemdar et al. 2005; Zhang and Liu 2013), 1,4-dioxane (Li et al. 2009), acetone and chloroform (Na Ayutthaya and Wootthikanokkhan 2013; Na Ayutthaya et al. 2015, 2016) can be used to prepare keratin solutions or dispersions with other polymers that are insoluble in water. These solvents allow the application of polymers that increases considerably the mechanical properties of keratin and its stability in many more solvents than water does; these are: PLA or PLLA (Na Ayutthaya and Wootthikanokkhan 2013; Na Ayutthaya et al. 2015, 2016; Li et al. 2009; Kim et al. 2013; Yuan et al. 2008), fibroin (Vasconcelos et al. 2008; Zoccola et al. 2008), PA6 (Aluigi et al. 2009, 2011, 2017), cellulose acetate (Dickerson et al. 2013), PCL (Zhu et al. 2017; Edwards et al. 2015; Li et al. 2016), PVP (Zhu et al. 2017), polyurethane (Yanga et al. 2018), PLGA (Zhang and Liu 2013) and PHBV (Yuan et al. 2009).

From the organic solvents mentioned above only formic acid, HFIP, and TFA can dissolve keratin. When keratin is dissolved in formic acid, resulting solution can be processed alone or mixed in the same solvent with fibroin (Vasconcelos et al. 2008; Zoccola et al. 2008), PA6 (Aluigi et al. 2009, 2011, 2017), cellulose acetate (Dickerson et al. 2013) and PCL/PVP (Zhu et al. 2017). These polymers improve the processing of keratin, which allows to obtain NFs membranes with better mechanical properties; in addition, reduce the redissolution of keratin in aqueous mediums. Keratin in formic acid can be mixed with solutions of PVA (or PVA/gelatin) in water (Li and Yang 2014; Yao et al. 2017) or with solutions of PLA in chloroform/acetone (Na Ayutthaya and Wootthikanokkhan 2013; Na Ayutthaya et al. 2015, 2016). This complex mix of polymers and solvents is also used to obtain NFs membranes. Keratin dissolved in HFIP can be mixed with solutions of polyurethane (Yanga et al. 2018), PLA (Kim et al. 2013; Yuan et al. 2008), PHBV (Yuan et al. 2009) and PCL (Li et al. 2016) in the same solvent to obtain NFs as well. Moreover, keratin can be solubilized in TFA, but in this case, its processing is limited to the production of nanopowders (Pedram Rad et al. 2012). In addition, solvents as chloroform/DMF are used to solubilize PLGA; in this case, keratin can be added to the PLGA solution to generate a dispersion of keratin (Zhang and Liu 2013). Successively this solution can also be used to produce NFs. On the other hand, 1,4-dioxane can solubilize PLLA and be mixed with paraffin and keratin; in this case, the solution can produce sponges instead of NFs from this dispersion (Li et al. 2009). Finally, TFE can be employed to solubilize PCL and be mixed with a solution of keratin in water to generate other complex solutions that also allow the production of NFs (Edwards et al. 2015).

It is evident when organic solvents are necessary, in the processing of keratin and in the mixing of keratin with other polymers, most keratin-based products are NFs membranes. As a result, keratin solutions with organic solvents permit to obtain a higher concentration of protein, solutions with more viscosity, more complex and stable polymeric matrix. Nevertheless, it is important to say that the mix of keratin with an organic solvent hinders the processing of keratin due to the fact that most of these solvents are unhealthy, harmful and hazardous: from a technical and environmental point of view, more cautions must be made to keep the productions of keratin-based products harmless. It is worth noting that the functionalization of keratin could be limited by using these types of organic solvents because they can destroy, oxidize, reduce or change the nature of functional compounds, NPs and other materials with the functional property. These problems are a consequence of the strong acid behavior of solvents. Therefore, in organic solvent cases, the functional properties of keratin-based products can be obtained only by keratin itself or by the property of any other polymer used.

2 Casting

2.1 Solvent Cast Process

Solvent cast process is the oldest technique in plastic film manufacturing: its development corresponded to the birth of the photographic industry as an answer to its unprecedented needs. The invaluable set of outstanding advantages related to the products of this technique makes it nowadays useful for niche products with high-quality features.

Some of the undeniable qualities of solvent cast films are uniform thickness distribution, extremely high optical purity, absence of gels or specks, remarkable transparency, isotropic orientation, low optical retardation, and uncommon flatness. The performance and the high-quality of these films cannot be achieved through other processes. Moreover, advantages do not regard only the products, but also the process, which is the only one that dries a liquid on a surface without employing thermal or mechanical force. On the other hand, in general, solvent cast films are expensive for solvent recovery and for the investments in equipment for handling solvents and solutions.

Solvent cast technology is not employable with all raw materials, but only with ones that possess some peculiar prerequisites. First, the polymer must be soluble in a volatile solvent or water; the resulting solution should be stable, with a reasonably low amount of solid content and viscosity and, in the end, the formation of a uniform film and removal from the casting support must be feasible. Satisfying all these requisites is not a simple issue. To solve this question, some collateral processes have been developed during the years: dissolution at overpressure, co-solvent systems, use of molecular weight distributions of polymers or copolymers, additives such as release agents, plasticizers, etc. Typical additives employed to supply specific film qualities are chelating agents, antiblocking, and antistatic substances, colors, electrical conductive materials, pigments, etc. Viscosity is an important parameter; for this reason, rheological conditions must be carefully controlled, in particular when adding solid compounds that can be processed as highly loaded dispersions. When putting soluble solid or liquid substances such as plasticizers, the attention must be focused to solubility and stability in solution and in film formation procedure. Interestingly, cast films can in many times be redissolved and reemployed as raw materials. This is because solution and polymer film are subjected to reasonably low thermal or mechanical stress and, consequently, degradation and adverse side reactions of film forming process are irrelevant.

The first step of solvent cast technology is dope preparation, which consists in dissolution in solvents of the solid polymer in granules, flakes, chips, or powder. This step requires monitoring of solution viscosity, degassing of the solution in order to avoid air bubbles formation and solution clarification and filtration. The whole dope line including tanks, pipes, filters, etc., needs precise temperature monitoring. The dope is pumped by a specific pump through heat exchangers into the spreader, which distributes the liquid homogeneously on casting support. Casting supports can be of different types: a supporting belt, a supporting drum, or a moving plastic film. It should be noted that films formed by solvent cast process have surfaces that replicate the casting support surface, so the choice of casting support is essential.

The last step of the process is drying to remove the solvents. This step consists of two different phases and can include different processes such as indirect heating, air-stream drying, and heating by radiation. Finally, it is to note that the film removal from the support needs precision and regulation of process conditions (Siemann 2005).

2.2 Keratin Films

In the last years, an increased interest has been addressed to the employment of proteins as a renewable source to produce compostable films. Several fields of application have been investigated, including biodegradable packaging, agricultural film, or edible film (Schrooyen et al. 2001). Several proteins have been employed for these purposes such as collagen, albumin, gelatin, fibroin, and keratin. Of these, keratin has shown promising properties for the biomedical field, due to its characteristic biocompatibility, biodegradability, natural abundance, and mechanical durability. For years researchers have employed keratin films preparation to investigate structural and biological features of self-assembled keratins (Rouse and Van Dyke 2010). In wide terms, operating principles and method fundamentals of solvent casting are described in the previous paragraph. Depending on raw materials, applications, situation needs, and necessities, variations and adjustments are applied to this general procedure. Figure 1 reports an example of wool keratin/citric acid film formed by casting-solution (Sanchez Ramirez et al. 2017).

For example, three main casting methods were developed for preparing films from keratin extracted especially from human hair by using Shindai solution: Pre-cast, Post-cast, and soft Post-cast method. All these procedures are based on the characteristic ability of keratin solutions: these solutions, with different denaturant conditions or with reduced pH, would make aggregates and mold them to produce a film (Fujii 2012). The obtained protein films turn out to be water insoluble and mainly consisted of α-keratins (Fujii et al. 2008).

In the Pre-cast method, this first solution is mixed with the following denaturants such as trichloroacetic acid (TCA), perchloric acid (PCA), or guanine-HCl (GHA), and then quickly poured onto tissue-cultured dishes containing distilled water (Fujii and Ide 2004). The resulting protein aggregates remain for 1–2 h at room temperature, then they are washed over 12 h and replaced with deionized water for 3 h (Fujii

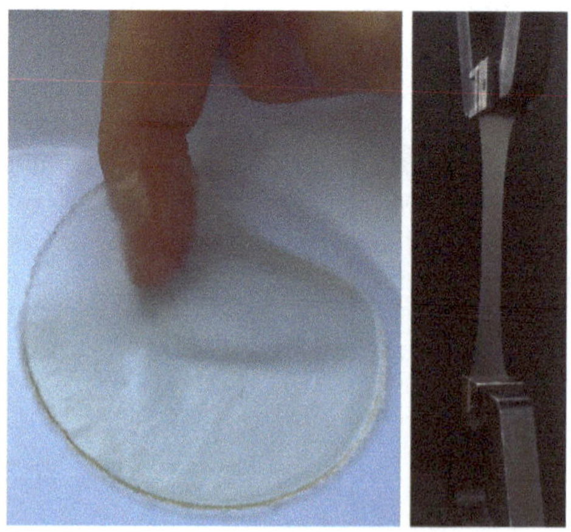

Fig. 1 Picture of wool keratin film: before (left) and during (right) tensile strength test

et al. 2004). Water-insoluble protein films are formed. Post-cast method involves the direct casting of protein in the second denaturant solution containing TCA, PCA, GHA, HCl, H_2SO_4, or acetate buffer. The formed protein aggregates are washed as mentioned above. Light brown films are immediately formed. Both in Pre- and in Post-cast method, the maximum protein recovery in the films, calculated by weighing with an electronic balance, is 60–85% and the main component is α-keratin.

As regards soft Post-cast method, the protein solution is directly exposed in dedicated dishes containing $MgCl_2$, $CaCl_2$, NaCl, and KCl. After standing for 1–2 h at room temperature, a membrane-like protein aggregate is formed, and it is washed by rinsing with water and then by distilled water. The film is lastly dried in a silica gel box. The maximum protein recovery in the films depends on the used developing solvent: by using $MgCl_2$ or $CaCl_2$, the maximum yield is approximately 70–80%, while when NaCl or KCl are used, the maximum yield is greater than 50–60%. Divalent cations such as Mg^{2+} or Ca^{2+} are essential for protein film formation and this is confirmed because similar results are obtained if $Mg(CH_3COO)_2$ or $Mg(NO_3)_2$ are used instead of $MgCl_2$.

In comparison with the films obtained by both the Pre- and Post-cast method, the ones obtained by the soft Post-cast method show more flexibility. The durability of these films was tested by folding and unfolding them more than 20 times without any noteworthy damage. These films by soft Post-cast methods are clear enough to show the bottom of the dishes. The flexibility and the flat surface of these films may depend on the matrix protein content, which is higher than that of the films prepared by Post-cast method. It is possible to state this because films by soft Post-cast method can be re-extracted using the Shindai solution and so the protein composition can be analyzed by SDS–PAGE. The result is that the extracted solution of all the films mainly consisted of α-keratins with molecular masses of 40–60 kDa and matrix proteins of 10–30 kDa, which are known as keratin-associated proteins. Interestingly, as already mentioned, the amount of matrix proteins in films by soft Post-cast method is higher than the one of films by Post-cast method (Fujii and Ide 2004).

A variation of solvent casting technique is solvent cast particulate leaching. In this method, porogen particles with controlled size and shape (e.g. inorganic salts) are used during the phase of dissolution of the polymer in an organic solvent. In this way, by monitoring the size of the porogen and the polymer/porogen ratio, it is possible to control both the final size of pores and the quantity of porosity. It is an easily possible methodology that does not need a specialized apparatus.

The main disadvantages of the solvent casting technique are the few possible shapes, the potential presence of a retained solvent in the polymer and denaturation of the proteins in the solution (Puglia et al. 2015). However, solvent casting is not the only technique to produce protein films: compression molding is another common method that overcomes main disadvantages of solvent casting. Even if it is less commonly used than solvent casting because most biopolymers are non-thermoplastic, compression molding was used to produce keratin films from chicken feathers chemically modified, plasticized, or grafted with acrylates. Using this method, thermoplastic films are produced, and their tensile strength and stability are higher than those of films made by solvent casting (Reddy et al. 2013).

2.2.1 Keratin Films Characterization

Keratin films produced through the solvent casting method appear yellowish and flexible (He et al. 2017). The surfaces of the films look very flat, without any visible grainy and porous structure. It is to note that, as regards pure keratin films, the two sides of the films look identical. This is not true when keratin is blended with other polymers, such as PVA or sodium alginate, probably this is due to the hindered arrangement of polymer chains during solvent evaporation.

SEM observations were carried out by He et al. (2017) to evaluate the homogeneity and the microstructure of feathers keratin films. The surface of the films is very smooth and dense without cracks or pores. This result agrees with the study of Ma et al. (2016) on pure chicken feather keratin films. The surfaces and cross sections of these films are homogeneous with a roughness on the micro level (Ma et al. 2016). He et al.'s (2017) investigation presents transmittance in visible range and transparency values of pure feather keratin films. Transmittance is greater than 80% at 700–800 nm and transparency is less than 1: this value indicates that pure keratin film is highly transparent.

Being a very important characterization for any materials, which directly affects their applications, mechanical properties were also evaluated within this study: single pure keratin film exhibits average tensile strength value of 2.83 MPa, which is lower than values that other authors reported for other protein films. This result may suggest that feather keratin cannot build strong networks (He et al. 2017). This agrees with Ma et al. results, which report 3.5 MPa as tensile strength value of pure keratin films and 127% as break elongation value (Ma et al. 2016).

Feather keratin promotes water affinity thanks to its hydroxyl groups. Water sorption ability and mass transfer rate of water molecules depend on the density and distribution of hydrophilic groups in the chains, while film morphology, chemical structure, and environment conditions establish permeability of the film. It is reported that blending is a valid method to decrease water vapor permeability and to increase the moisture content of the films (He et al. 2017).

Also, moisture regain is a relevant factor, which has a huge effect on the comfortability of textile materials. Previously mentioned work (Ma et al. 2016) illustrates that moisture regains curves of raw chicken feathers and of keratin film in specified conditions show first a rapid increase and they tend to remain stable. Moisture regain of the chicken feathers is about 7.7%, while the value of keratin film can reach 14.7%. The high moisture regain of keratin film is strongly related to its low crystallinity. In fact, it is well known that the two forms of incorporated water in materials, namely "free water" and "bound water", are easier to penetrate the amorphous region than crystalline area. In addition, the specific surface of keratin film is restricted, and this causes few exposures of the hydrophilic groups in keratin macromolecular chains, that influences moisture adsorption rate. Therefore, moisture adsorption rate of keratin film is lower than the one of raw chicken feathers. So, the moisture regain is not only related to crystallinity, but also to the specific surface of materials.

The same work reports TGA analysis results conducted on the raw feather and on feather keratin films (Ma et al. 2016). TGA profiles present a similar trend with two phases of decomposition, indicating that the thermal stability of the original chicken feather has been preserved. There is 5–8% weight loss before 100 °C and this is attributed to the evaporation of incorporated water including "free occur water" and "bounded water". Samples are stable until 200 °C, and a sharp weight loss occurs from 200 °C to 400 °C with a reduction of up to 70%. This is related to the denaturation of helix structure and to the disruption of chain connections, peptide bridges, and the skeletal degradation. Several chemical reactions take place in this phase where the keratin is decomposed to lighter products and volatile compounds such as CO_2, H_2S, H_2O, and HCN. From the DTG profiles, raw chicken feathers reach a maximum decomposition rate at 340 °C, while keratin film at 329 °C. This minimal difference indicates that the extraction procedure and shaping process slightly affect the thermal stability of keratin.

In previously mentioned work (He et al. 2017), the thermal properties of feather keratin films were studied by DSC technique, revealing an endothermic peak at 130–150 °C, which is due to the evaporation of adsorbed moisture that could not be completely removed by drying. In agreement with TGA and DSC results, Fujii (2012) states that keratin films can resist to thermal treatment at a temperature of 200 °C, and treatments at temperatures higher than 180 °C for 10 min cause color changes in keratin films: white to light beige ones turn to light yellow and then to light brown. In this work, Fujii introduces an original method to precisely evaluate hair damages using two-dimensional and uniform hair keratin film samples as a novel substitute tool for hair in the field of hair science and, in particular, in hair damage analysis. The observed color changes due to heat treatment are similar to those of Japanese gray hair samples exposed to the same treatment.

In order to understand the chemical structure of raw chicken feather and of regenerated keratin film, FT-IR measurement was employed (Ma et al. 2016). Comparing the FT-IR spectra of samples, similar characteristic peaks can be found. This means that keratin content in raw feather is so high that its characteristic peaks are in compliance with the extracted and processed keratin. On the other hand, methods employed to obtain keratin film slightly influence the chemical structure of protein: keratin film shows characteristic transmission bands attributed predominantly to the peptide bonds.

2.3 *Improving Keratin Films Properties: Crosslinking and Blending*

Feather keratin has the prospective to be used as a commercial and industrial material because it is inexpensive, produced in a large amount in centralized locations, it is eco-friendly and it has suitable chemical and physical features (Poole and Church 2015; Sharma et al. 2018). Yet, difficulties exist in processing the native feather structure and in cleaving and reforming the disulfide crosslinks that occur between the protein chains and in a loss of the semicrystalline structure in the regenerated product. These complexities restrict feather keratin possible applications and explain the reason why widespread industrial use has not been realized, in fact, the result is materials with inadequate mechanical strength for the industrial purpose.

Some studies have been focused on improving the keratin film properties by crosslinking agents. Crosslinking involves the formation of intra- and/or intermolecular covalent bonds between protein chains (Poole and Church 2015). Aldehyde crosslinking agents are often added to improve tensile properties, water barrier property, and thermal stability and to reduce the solubility of protein-based films. However, aldehydes have low molecular weight and, consequently, their use for enhancing the properties of protein-based films applied in the biofield and food packaging is limited (Dou et al. 2014). Traditional aldehyde crosslinking methods form strong covalent bonds by the reaction with the side chain groups of lysine, arginine, tyrosine, tryptophan, histidine, cysteine, asparagine, and glutamine (Poole and Church 2015). FAL is also known to perform condensation reactions with amide groups. GAL is bifunctional, so it forms a crosslink of five carbon atoms in length. The result is a statistically significant enhancement in conditioned Young's modulus and tensile strength. All aldehyde treatments significantly enhance the wet tensile strength. Metal-catalyzed photo-assisted crosslinking of keratin generally gives poorer mechanical properties to the control. For more information about crosslinking processes see Sect. 6.

Crosslinking is not the only validated method to enhance keratin film properties: the large potential application can be envisaged also for blending. Blends of keratin with appropriate polymers can give the expected performances and have the advantage of overcoming the main drawbacks of keratin films, which are fragility and poor mechanical properties (Ramakrishnan et al. 2018). However, also plasticized keratin films have quite poor mechanical properties and water resistance, if compared to synthetic films (Dou et al. 2014). Therefore, a huge number of works have been carried out to enhance the properties of keratin films through the blending technique by the addition of natural and synthetic polymers. PVA is synthetic, nonhazardous, biodegradable, water-soluble with great chemical stability and good film forming, which has been employed frequently to blend with biopolymers for food packaging, tissue engineering, and controlled drug delivery systems. In the work of Dou et al. (2014), PVA is used to enhance mechanical properties, stability, and water resistance of keratin films, but its high free hydroxyl content could increase also water vapor permeability.

Another example of a polymer–keratin blend is reported by Ramakrishnan et al. (2018), who conducted a study on the evaluation of biodegradability and mechanical properties of bioplastic film produced by dissolving chicken feathers and then by mixing different concentrations of glycerol. It results that glycerol at rising concentrations decreases the maximum tensile strength of chicken feather keratin films. In this way, monitoring of glycerol concentration can be useful to control tensile strength and elongation at break of the films. The surface of keratin films modified by the addition of glycerol is more uniform than one film without plasticizer. Moreover, it has good morphologies without edge and holes and bioplastic produced is biodegradable. Therefore, these films can be employed as a suitable substitute of fossil oil–based materials which have demostrated to be dangerous for the environment (Ramakrishnan et al. 2018).

Plasticizers are often added to coating formulations to improve the elasticity of film formers (Martelli et al. 2006a; Martelli and Borges Laurindo 2012). This reduces coating fragility and breakage, which is shown as flaking of coatings from the coated surface. Plasticizers are generally small molecules such as polyols like sorbitol, glycerol, and PEG that intercalate among polymer chains, destroying hydrogen bonding, and spreading the chains apart. This behavior improves flexibility, water vapor, and gas permeability. Martelli et al. (2006a, b) and Martelli and Borges Laurindo (2012) conducted a study to compare different kinds of plasticizers and their concentrations applied to keratin films, in order to determine how the plasticizer affects film properties. First of all, the authors noted the great influence of plasticizers on the microstructure of feather keratin films, for example, the use of sorbitol makes the film surfaces more homogeneous. Films made with glycerol and sorbitol are more flexible than a film of keratin and PEG, probably because of glycerol and sorbitol lower molecular weights. Moreover, films with PEG are less stiff and they result more difficult to handle compared with the ones made with glycerol and sorbitol. The plasticizers increase the water solubility of keratin films because they increase the equilibrium moisture content. This is particularly true for glycerol, because its high hydrophilicity causes the strongest effect on film moisture content and solubility. An increase in the glycerol concentration improves water vapor permeability, mainly due to the water solubility increase, which improves with glycerol concentration.

Even if keratin film containing glycerol shows adequate strength and flexibility, glycerol dissolves in aqueous solution, resulting in the fragile film again. Tanabe et al. (2002) solved this problem reinforcing the mechanical strength of keratin film by mixing with chitosan. The blend was cast and dried to obtain a transparent film. The film prepared from keratin mixed with 10 wt% chitosan was reasonably flexible and strong. Keratin–chitosan composite film shows remarkably improved waterproof properties; it shows suppressed swelling at the acidic condition and mechanical properties turn out to be almost the same before and after swelling. At alkaline condition, keratin–chitosan film shows less swelling than keratin film. Keratin–chitosan composite film also possesses antibacterial activity. Thus, chitosan could considerably affect the properties of keratin film when the biological or mechanical properties are considered.

Finally, the addition of NPs as nanofillers to improve mechanical properties of protein film is well known (Poole and Church 2015). Chemical interactions between the host material and the filler enhance specific properties of the product.

2.4 *Keratin Films Applications: Focus on Biomedical Field*

In the last years, many studies have been conducted on extraction and purification of protein fractions in order to produce novel biomaterials such as films appropriate for new large-scale uses in conventional fields (i.e., textiles, packaging, sanitation, and filtration) or high-quality applications in niche sectors, such as biotechnology and medical supplies and devices, for which biocompatibility is critical (Aluigi et al. 2007a). Regenerated keratin films degrade in vitro and in vivo and they well support cells growth and proliferation, so they are expected to be desirable for biomedical purposes. In addition, keratinous materials can absorb toxic compounds such as heavy-metal ions, FAL, and other dangerous VOC, so possible applications can be seen also in water purification and air filtration.

Thanks to their biodegradability, biocompatibility, and capacity to support cell growth, one of the most innovative and attractive applications of keratin films is in the field of tissue engineering (Puglia et al. 2015). Tissue engineering is a field based on the employment of degradable materials as scaffolds, which need to have peculiar features, e.g., degradation rate, mechanical strength, porosity, monitored microstructure, and well-defined shape and size, such that they will support cell adhesion and growth, and organ regeneration. Moreover, this particular field of application requires additional specific scaffold's features: for example, in order to promote tissue growth, scaffolds degradation rate has to be similar to the tissue regeneration rate, and their surface area must be sufficient to allow cell attachment, ingrowth, and diffusion.

As previously reported, keratin has great practical applications in the sector of bioengineering, and recently research is in progress in all sorts of keratin hybrid materials with potentiality in drug delivery, wound healing, tissue engineering, and medical devices, in the forms of keratin sponge scaffolds and keratin films (Tu et al. 2016). Tu et al. (2016) conducted a study on biomedical employments of keratin/HA films formed via a simple solution casting method. HA is known as inducing factor for bone and teeth with great biocompatibility and bioactivity. The authors obtained a series of keratin/HA blending films by controlling the weight ratios of keratin and HA, in order to identify the most suitable conditions for biomedical applications. They found that the agglomeration of excessive HA particles causes strength decrease and breaking strain of composite membranes simultaneously, enhancing wettability, biocompatibility, and bioactivity of the films, compared to pure keratin film. The composite film with 5% wt of HA granules appeared to be the optimum composition, considering the enhancement of mechanical properties and resulting in better cell viability. In this composition, keratin/HA film can be considered as a promising membrane material for biomedical purposes.

In another study, Tanabe et al. (2002) prepared a keratin/chitosan composite film by casting the mixed solutions of both biopolymers in acetic acid and considered its physicochemical and antibacterial properties on Gram-negative bacteria *Escherichia coli*. The result is that keratin/chitosan composite films significantly reduce bacteria number (54–62%) and, moreover, these films were shown to be a good substrate for mammalian cells.

Zhang et al. (2016) considered the employment of PLGA/keratin composite membranes as guided tissue regeneration (GTR) membranes by solvent casting. Regenerative processes using GTR membranes are currently well established in periodontal therapy. In this technology, GTR membrane is the main factor that directly influences the final regeneration result. Therefore, finding suitable barrier membrane material is one of the critical and not easy issues in this research field. PLGA and wool keratin are two excellent medical biomaterials with potential applications in the field of GTR because on the one hand, wool keratin promotes cell affinity and bioactivity of PLGA and decreases the amount of PLGA-induced aseptic inflammation; on the other hand, PLGA enhances mechanical properties of wool keratin.

Wool keratin films are also highly used in wound care products, as reported by Ganesan (2017). Textile materials and, in general, products engineered for medical aims have to be resistant, flexible, biocompatible, biodegradable, nontoxic, and oxygen permeable. Wool keratin is a common textile material employed in wound healing application. In this study, it has been used in combination with honey and chitosan to form different films that have been considered as potential medical textiles for wound healing. All the films showed good antimicrobial activity against Gram-positive bacteria (*Staphylococcus aureus*) and Gram-negative bacteria (*Escherichia coli*). Moreover, from cytotoxicity test, it results that the developed composite films exhibited no cytotoxic reactivity to fibroblast cells after 24 h contact time. This indicates that these films have a good potential to be employed for wound dressing.

Standard cultivation of 12 different cell lines has been involved to evaluate hair keratin films as coatings of cell culture plates (Reichl 2009). The results show that keratin coating supports the attachment and proliferation of most cell types and seems to offer advantages over the traditional approaches such as low cost, high seeding efficiency, mechanical stability, transparency, and growth substrate properties.

Those are only a few examples of the copious studies reported in the literature that highlight the potentialities and the advantages of using keratin films as biomaterials or, in general, for biomedical applications. As conclusive consideration, it is possible to state that keratin films are suitable materials for this kind of applications because of their biodegradability, biocompatibility, and capacity to support cell growth. The main drawbacks of these films are fragility and, in general, poor mechanical properties. Crosslinks and blends are two of the main methods for improving these properties; through an accurate selection of the materials to use in combination with keratin, the production of films with excellent biomedical properties is possible and, in this way, the biomedical applications that could be covered are very numerous and various.

3 Foaming and Freeze-Drying

3.1 Foam Technology and Lyophilization

Foam technology is an efficient low add-on technology, due to expanded volume and great internal surface area of the foam bubbles (Bhardwaj et al. 2015). Due to its exceptional performance, foam technology finds its application in different textile wet processing such as sizing, dyeing, mercerization, finishing, and printing. In general, a wide range of synthetic surfactants can be used as foaming agents.

One of the main foam technology techniques is lyophilization, also called freeze-drying. It is a drying process for the conservation of thermolabile materials (Salazar et al. 2018). It is based on the process of sublimation, which minimizes structural changes and conserves bioactive substances of the dried product. Lyophilization involves three main phases: product freezing, primary drying (direct sublimation under reduced pressure is employed to remove ice), and secondary (desorption and diffusion are used to release unfrozen water). Particular critical is freezing, which is the first step of lyophilization and it involves ice formation. Freezing procedure may be divided into three steps: nucleation or initial crystal seed creation, propagation or growth of ice-crystals, recrystallization or maturation. Freezing rate strongly affects the final properties of the dried product, as it is critical for pore dimensions after ice crystal sublimation (Salazar et al. 2018; Geidobler and Winter 2013).

3.2 Keratin Sponges Characterization

Lyophilization is a suitable method to obtain keratin sponges, thanks to the capacity of keratin to self-assemble into 3D porous structures (Vasconcelos and Cavaco-Paulo 2013). SEM analyses demonstrate that different regions of keratin sponges exhibit different structures: while the top and the sides of the sponges show an ordered fully exposed micropore architecture, the bottom is composed by dense keratin with an internal porous network (Wang et al. 2017). Puglia et al. (2015) report FESEM images of freeze-dried keratin sponges obtained from Merino wool and from Brown Alpaca fibers. In both the cases, flake-like structure is showed. As an example, Fig. 2 shows a wool keratin sponge obtained by means of freeze-drying process.

To investigate chemical modifications of keratin sponges, FT-IR analysis can be performed on feather keratins and on keratin sponge (Wang et al. 2017). The chemical organization of the sponge is similar to that of feather keratin, with a difference: the reduction of –SH due to oxidation. Both the spectra show the characteristic

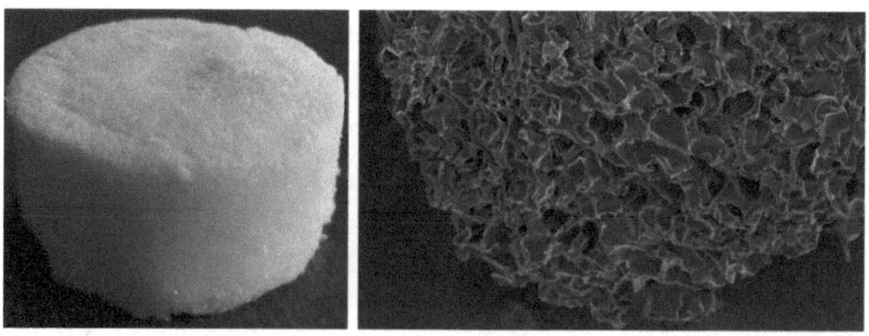

Fig. 2 Picture of wool keratin sponge (left) was kindly provided by Dr. Claudia Vineis (CNR-ISMAC Biella). Typical flake-like structure of keratin sponges (right), this picture was kindly provided by Annalisa Aluigi (CNR-ISOF Bologna)

absorbance peaks of proteins, while the peak related to –SH (2400 cm^{-1}) disappeared in the sponge spectrum.

DSC analysis was performed by Patrucco et al. (2016), who compared DSC line of keratin sponge with that of the original wool fibers. In both cases, water evaporation was registered below 100 °C. The denaturation of wool fibers turned out to be at a lower temperature than the denaturation of sponge keratin. Keratin denaturation peaks are associated with the presence of different contents of lanthionine and isopeptides crosslinked keratins. Degradation of all the fiber components occurs at temperatures higher than 300 °C. These results are confirmed by Posati et al. (2016), who performed DSC analyses on keratin sponges obtained from keratin solutions at different keratin concentrations. They discovered that all DSC curves showed three major thermal steps: a peak below 100 °C related to water evaporation, a transition between 195 °C and 200 °C related to degradation of small molecular weight keratin fractions with low thermal stability, and an endothermic peak between 210 °C and 250 °C, related to the denaturation of crystalline areas.

3.3 Effects of Freeze-Drying Conditions on Keratin Sponges

There are many parameters governing the freeze-drying process, such as the choice of substrate, the way of deposition of particles on the substrate, the freezing rate, the sublimation temperature, etc., and many possible artifacts can take place (Fowler and Aebi 1983). Several studies in literature compare different freeze-drying processes in order to investigate the effect of these processes on the properties of resulting scaffolds (Li et al. 2013; Bhardwaj et al. 2015).

Bhardwaj et al. (2015) fabricated through freeze-drying a bioactive 3D silk fibroin/keratin-blended scaffold as a dermal substitute for skin repair and regeneration and they evaluated effects of changing parameters in pre-freezing process. They controlled pore formation during the pre-freezing process by adjusting the temperature gradients. Bigger the temperature gradient, longer the phase of ice crystal formation, and therefore larger the pore dimension. For skin tissue engineering, matrices high porosity and 3D porous structure is a requirement as it significantly affects cellular activity. By artificially adjusting pre-freezing parameters, in this study, they produced matrices with pore sizes of 100–120 μm with interconnected porous microstructure first by pre-freezing at −20 °C then by sublimation of crystals.

Vasconcelos and Cavaco-Paulo (2013) discovered that by adjusting the freezing rate and pH it is possible to control pore dimensions: slow freezing rate produces bigger pores. Other effects can be achieved by mixing the keratin solution with other compounds. For example, after lyophilization and leaching of a keratin solution mixed with calcium alginate beads, a keratin sponge with high porosity and interconnectivity can be obtained and this result is suitable for example in the biomedical field. In another study, Fowler and Aebi (1983), evaluated the effects of the presence of glycerol in freeze-dried protein scaffolds. The presence of glycerol seems to partially preserve protein structure from distortion and collapse.

Moreover, monomer–polymer equilibrium and other protein–protein interactions may be affected by the presence of glycerol.

Tachibana et al. (2002) studied the response of different keratin freeze-dried scaffolds to treatments and manipulations. They discovered that keratin sponges prepared from keratin solution by 3-day freezing at −20 °C and lyophilization was resistant against the heat treatment (60 °C, 1 h). The sponges prepared by lyophilization after overnight freezing at −20 °C or rapid freezing with dry ice-methanol were not stable and they were destroyed and partially dissolved after heat treatment (60 °C, 1 h).

According to Posati et al. (2016), scaffolds with dissimilar morphologies and adjustable chemical–physical properties were obtained by a precise control of freezing temperature gradient and of protein concentration. In particular, in their study, they revealed that the increase of keratin concentration makes the pore size distribution narrower and shifted to a lower value. In particular, increasing keratin concentration from 1 to 10% wt causes a significant decrease of both the mean pore size and porosity, while the mean values observed for higher keratin concentrations did not show statistically significant differences. This phenomenon can be explained because the greater intermolecular packing of the protein at higher concentrations could support the formation of smaller ice seeds during the freezing process, causing the production of sponges with lower pore size and porosity. As regards the rheological behavior, it has been shown that protein concentration directly affects the viscosity of keratin-based solutions, because of an increase of keratin chains entanglements, which could cause also the higher stability of the sponges. All keratin sponges have a viscoelastic behavior and the material response is mainly elastic. DSC curves reveal that the systems were similar in terms of secondary structure organization, independently from keratin content. TG curves demonstrate a limited shift toward higher temperatures and little degradation rate in sponges with higher keratin amount. Moreover, the increased amount of keratin in the sponges causes an improvement of residual material at the end of the tests.

As regards freezing conditions, they highly affect pores morphology: while pores obtained at −18 °C appear almost round, pores produced at −196 °C turn out to be randomly distributed, elongated, and radially oriented. Moreover, sponges obtained by freezing at a lower temperature present significantly lower pore dimensions and a pore distribution shifted to a lower value. These results depend on two different elements: the higher cooling rates that cause the production of smaller ice-crystals and the nucleation that starts in the cold surface and continues toward the center of the sample forming leaf-like structures. In contrast, the secondary structure of keratin does not depend on freezing rate.

Investigating the effect of freeze-drying process on the properties of keratin scaffolds was also the aim of Li et al. (2013), who compared three processes. In the first process, concentrated keratin solution was poured into the cold trap and, after the freeze, the vacuum pump was opened for dry processing. In the second process, concentrated keratin solution remained at room temperature for 4 h and then it was put into a refrigerator at −20 °C for 4 h; then it was put into the cold trap for 24 h at −56 °C; the vacuum pump was opened during the last 4 h and the frozen temperature increased gradually. In the third process, concentrated keratin solution remained at room temperature for 4 h, then it was put into refrigerator to −20 °C for 4 h and in the end, the vacuum pump was opened for dry processing for 24 h. Keratin scaffold obtained with the first freezing process was keratin powder with an irregular surface and heterogeneous internal structure. Keratin scaffold was not produced with the last freezing process. In general, keratin scaffold produced with the second freezing process had satisfying mechanical properties and uniform internal structure, but it also had some drawbacks such as low flexibility.

3.4 *Keratin Sponges: Not Only Lyophilization*

Lyophilization is not the only method to obtain keratin sponges. For example, Thermally Induced Phase Separation (TIPS) technique is a suitable method to fabricate sponges with high porosity values, anisotropic tubular morphology and large pore interconnectivity (Puglia et al. 2015). This technique is based on a phase-separation mechanism that leads to the porous morphology typical of sponges and that can be of two different types: solid–liquid or liquid–liquid phase separation. Different results in terms of pore morphology and size, mechanical properties, and other features can be obtained by controlling processing parameters.

Keratin can be organized in a structure similar to foams for the three-dimensional network of polymer chains and for the presence of interconnected pores: this structure is named hydrogel. Keratin hydrogels can be obtained through different techniques, such as tube inversion test (Kakkar and Madhan 2016), reduction or oxidation hydrolysis (Cardamone et al. 2013) or microfabrication techniques, like photopatterning and wet spinning (Yue et al. 2018). Keratin hydrogels are different depending on the fabricating technique. For example, while keratin hydrogels formed by reduction hydrolysis are homogeneous, smooth, soft and with no well-defined features, those formed by oxidation hydrolysis are well defined and they show a mat-like structure (Cardamone et al. 2013). It is possible to obtain another morphology by photopatterning, that gives rise to strong, transparent keratin hydrogels (Yue et al. 2018), or by tube inversion test, that gives rise to homogeneous hydrogels with porous microarchitecture (Kakkar and Madhan 2016).

3.5 *Keratin Sponges Applications: Focus on Biomedical Field*

Freeze-drying is one of the most suitable methods to obtain scaffolds. Scaffolds are structures capable of supporting three-dimensional tissue formation, thanks to their porous architecture that allows the diffusion of nutrients and metabolites for cells. One of the requisites of scaffolds is the ability to function as a synthetic ECM that facilitates cell–cell and cell–matrix interactions. Scaffolds fabrication requires carefully chosen materials (Verma et al. 2008), so nowadays various natural and synthetic biomacromolecules have been investigated in order to develop various ECM.

Biologically derived protein biopolymers have a central role in the production of biomimetic ECM scaffolds (Xu et al. 2013). Among these, keratin-based materials have emerged as promising candidates because of their biocompatibility, biodegradability, and mechanical strength (Kakkar and Madhan 2016). The ability of extracted keratin proteins to self-assemble and polymerize into complex 3D structures has led to their development as scaffolds for tissue engineering and regenerative medicine (Rouse and Van Dyke 2010; Xu et al. 2013). Wang et al. (2017) studied keratin sponges applications in one of the main fields of regenerative medicine: wound healing. A suitable biomaterial for wound healing should preserve a moist environment around the wound, absorb the excess wound exudates, allow for gaseous exchange, prevent microorganisms from infecting the wound, be nontoxic, nonallergenic, biocompatible, and be easy to remove or be biodegradable. In this study, authors demonstrated that keratin sponges obtained by lyophilization possess all these features: while their highly porous structure allows the absorption of tissue fluid and wound exudates in vivo, other characteristics like mimicking of the function of ECM, gaseous exchange, rapid hemostasis allow keratin scaffolds to accelerate wound healing.

In another study, Tachibana et al. (2002) fabricated, by means of lyophilization, wool keratin sponge scaffolds for long-term and high-density cell cultivation. They obtained encouraging results, finding that these scaffolds are suitable for cell adhesion, spreading and growth. Moreover, in contrast with other protein-based biomaterials, keratin sponges without crosslinking reagent are resistant against heat treatment. This is a remarkable property because it makes more easy chemical modification and the following washing to remove cytotoxic reagents and it makes possible the introduction of bioactive substances into keratin sponges. Because of all these reasons, keratin sponges may be suitable for long-term aims in the field of biomedical, including tissue engineering.

Another investigation regarding the employ of keratin freeze-drying to fabricate scaffolds for stimulating cells growth was conducted by Xu et al. (2013). They developed well-interconnected, hydrophilic and cytocompatible human hair keratin scaffolds for subcutaneous implantation and for treating skin defects. In agreement with the previous reported study of Wang et al. (2017), the biocompatibility, the biodegradation, and wound healing properties of keratin scaffolds are here confirmed.

Those are only some examples of the numerous investigations regarding keratin scaffolds applications in regenerative medicine and in tissue engineering. However, the biomedical field is not the only sector of employment of keratin foams and sponges. Zhou et al. (2014) utilized highly porous keratin sponges as a possible adsorbent for oil spill clean-up, in order to face current issues concerning environment, increasing oil prices, and the restricted oil reserves. All these questions are driving rising demands for new products with recyclable and sustainable bio-based materials, which are compostable, non-petroleum based and non-hazardous for the environment. In the present study, keratin sponges were realized by freeze-drying treatment of a dissolved pigeon feathers keratin solution. Good results were obtained: it was found that the highly interconnected porous structures of keratin sponges adsorbed instantly oil droplets placed on them, while water droplets showed a form with an acute contact angle. Moreover, after adsorption, the adsorbed oil could be maintained steadily by the sponge. Despite regenerated keratin sponge exhibited a high oil adsorption capacity in the pure oil medium, it demonstrated a relatively low value in oil/water medium, because of its poor hydrophobicity. However, chemical modifications might improve this property for a high selectivity adsorption ability of the sponge in oil/water medium, making it suitable for potential use in clean-up.

Another possible application of keratin foams is evaluated by Bhavsar et al. (2017), who employed keratin foams to apply chemicals and dyes on the textiles substrate where water is replaced with air. This method, named foam dyeing, allows a lower wet pickup than conventional pad dyeing and it results in economic advantages, such as low drying duration, less water, and energy employment, enhancement of the productivity and low load on effluents. After the generation of foams, foam dyeing of cotton and wool was performed on laboratory scale horizontal padding mangle and the results were promising: keratin showed satisfying foaming ability and foam stability both for reactive and acid dye solution, respectively, for cotton and wool dyeing. The percentage of dye fixation on the cotton fabric was found similar to conventional padding samples. As regards wool dyeing, foam dyed wool samples exhibited better results than the ones obtained with the conventional pad-steam process, with higher color strength values. Moreover, rubbing and washing fastness properties of foam dyed cotton and wool samples resulted in the same of conventional dyed samples. The advantages of keratin foam dyeing are related to its sustainability and to the biodegradability of its materials.

In conclusion, it is possible to state that the combination of keratin properties with the several possible applications of foams and sponges, characterized by an interesting porous structure, gives rise to promising products for different uses. Moreover, the possibility of employing different techniques to obtained keratin foams and sponges, such as lyophilization, hydrolysis, microfabrication, and the opportunity of changing parameters and operating conditions lead to the prospect of obtaining very different results, depending on the aim. These characteristics make the field of keratin foams and sponges attractive and fascinating and, despite the yet obtained results, a lot of possibilities are so far to be discovered and investigated.

4 Spinning

Wet spinning is a traditional fiber-forming technique used by the textile industry for manufacturing synthetic fibers, however, it has recently been employed also to produce single-fiber biomaterials. The method consists of an extrusion of a solution from a spinneret into a coagulation bath and after that drawing and stretching to promote polymer chain alignment and consequently the fiber formation.

First reports concerning the development of pure keratin fibers from chicken feathers via wet spinning had been published in the 1940s (Harris and Brown 1947; Evans and Shore 1948; Wormell and Happey 1949), however, they did not report fibers mechanical properties. After several years, in 1996, the use of keratin for the production of fibrous materials was reported by McCurry and other researchers for products of personal hygiene such as towels and children's diapers. It was used to produce fibers with a high content of keratin from chicken feathers (Siller-Jackson et al. 2003; McCurry 1996; Schmidt et al. 1998). It is important to highlight that the keratin content of feathers is near 90% of their weight, so they are a valuable and cheap source of protein.

Another source for cheap raw material is waste wool: it contains about more than 80 wt% of keratin with molecular weight higher than 10 kDa, which is the premise for fibers spinning (Poole et al. 2008). Waste wool can be mainly obtained from textile processing: scouring, carding, combing, yarn spinning, generate large quantities of waste wool that are too short for direct textile processing (Chen and Burns 2006). Moreover, other short and coarse fibers not suitable for high value-added products in textiles (FAO-United Nations 2018) come from meat goats and sheep breeding. The used wool textiles discarded as municipal solid wastes that end up in incineration or landfill represents the last source of waste wool, potentially the available quantity corresponds to half the total amount (EPA-United States 2018). The production of pure keratin fibers has been hindered by the physical limitations of keratin materials. For this reason, research has overcome these challenges by blending synthetic and natural polymers to improve material properties.

The fiber-forming capabilities of aqueous keratin solutions have been improved by Katoh et al. (2004) using PVA. The aim is to increase the viscosity of the spinning solution, allowing the spinning of fibers with a keratin content ranging from 13% to 46%. The maximum keratin content for sufficient fiber formation is determined to be 30%, due to the fragility determined by the high amounts of keratin. Moreover, the blending of keratin with PVA promoted improvements in mechanical strength, waterproof characteristics, and adsorption of toxic substances, making this material interesting for industrial applications as an absorbent of heavy metals ions and FAL gas.

Novel fibrous keratin-based materials that have potential application as hygienic fabrics produced by wet spinning techniques have been developed by Wrześniewska-Tosik et al. (2007a). Fibers with better sorption properties, higher hygroscopicity, and a smaller wetting angle than cellulose-only fibers were obtained by blending keratin extracted from chicken feathers with bio-modified cellulose. Composite fibrous materials were produced despite the presence of keratin that decreased the mechanical properties of the fibers; however, the presence of keratin promoted the biodegradation of the cellulose-keratin fibers compared to the cellulose fibers only.

Wet spinning of pure regenerated keratin fibers via ionic liquid dissolution was reported by Idris et al. (2013). Moreover, keratin has been blended with PEO (Tonin et al. 2007), cellulose acetate (Aluigi et al. 2008a), cellulose via ionic liquid dissolution (Hameed and Guo 2010), silk (Iridag and Kazanci 2006; Vasconcelos et al. 2008), PHBV (Yuan et al. 2009), etc., as well as extruded with plasticizers (Barone et al. 2005, 2006). To conclude, the spinning of pure keratin remains challenging, but indicates the possibility of successful development of macromolecules in films, sponges, and many other forms (Xu et al. 2014b).

5 Electrospinning

Since 1990s (Formhals 1934; Doshi and Reneker 1993), the process of fiber formation from polymer within an electric field was named electrostatic spinning or electrospinning. This technique was discovered in the 1930s (Doshi and Reneker 1995) and recently has been rediscovered because it is a simple, versatile and low-cost method to produce ultrafine (<1 μm) polymer fibers with an intermediate production rate (Li and Xia 2004; Subbiah et al. 2005; Ramakrishna et al. 2006).

There are several other methods to produce small diameter fibers such as fibrillation (Homonoff 2008; Perez et al. 2000) island in the sea (Pourdeyhimi et al. 2006) and melt-blowing systems (Bryner et al. 2006; Ellison et al. 2007) for high-volume production, nanolithography (Tseng et al. 2005; Xie et al. 2005) and self-assembly (Zhang 2002) for high accuracy. The physical phenomenon of electrospinning is a branch of electro-fluid-dynamics, it is based on the electrostatic repulsion of charges according to Coulomb law: an electrospinning apparatus, by means of an electrostatic field, generates an electrically driven polymer jet that solidifies into sub-micron-scale filament (called nanofiber).

The process is usefully employed for producing porous membranes and nanostructured nonwovens consisting of submicron-scale fibers suitable for a wide range of applications such as membrane separation (Shin et al. 2005; Gopal et al. 2006; Yoon et al. 2006), filtration (Tsai et al. 2002; Qin and Wang 2006; Li et al. 2006; Barhate and Ramakrishna 2007; Heikkilä et al. 2007; Yun et al. 2007), breathable garments and protective clothing (Gibson et al. 2001), thermal insulation (Gibson et al. 2007), drug delivery (Sill and Von Recum 2008), wound dressings, and scaffolds for tissue engineering (Sill and Von Recum 2008; Pham et al. 2006; Lee et al. 2008). Nowadays, the last application is widely studied because the electrospun mats, characterized by a small pore size, high porosity, 3D features, and high surface area-to-volume ratio, promote cell attachment, growth, and proliferation.

Electrospinning works with both polymer solutions (simply referred to as "electrospinning") and molten polymers (so-called "melt electrospinning"). Melt electrospinning requires several additional equipment (e.g., heating devices, vacuum chamber, higher voltage power supplies), although it could be more appealing for industrial production because of the absence of solvents. A basic electrospinning setup consists of three elements: an electrical generator (high voltage supply), a capillary (jet source), and a metal collector (target). Usually, the generator electrically charges the solution while the collector is grounded, but it is also possible to invert the system charging the collector and grounding the feeding solution (Kilic et al. 2008).

The process operates according to the following operations: a pump pushes the polymer solution through the capillary (metal tip) with a constant flow rate by means of a syringe. The generator, connected to the capillary by a conductive wire, electrically charge the solution. A stream of solution is ejected from the tip (jet) toward the grounded metal collector (flat surface or rotary drum) where electrospun NFs are collected.

When the voltage generator is turned on, the electrospinning process starts. At the capillary tip, the electrical potential of the droplet surface increases to a sufficiently high value that allows the droplet to assume a steady conical shape (the Taylor cone), instead of a spherical shape. The formation of the cone is promoted by Coulomb repulsion between the charges; contemporarily the electrostatic forces act in opposition to the surface tension of the fluid, favoring sphere-like shapes. As the electrostatic forces overcome the surface tension, a charged jet of fluid is ejected from the top of the cone (Yarin et al. 2001; Shin et al. 2001), with an initial rectilinear trajectory. The diameter of the jet is one or two orders of magnitude smaller than the inner diameter of the capillary from which the jet is generated and decreases progressively moving towards the collector.

Two mechanisms are responsible for NFs formation: first, the expansion in the stream of solution that is realized by electrical charges; secondly, the polymer solidification in the stream of solution induced by a sharp increase in the solvent evaporation (which is a result of convective forces acting during the stream motion and the tiny diameter of the stream). Furthermore, it is important to highlight that the concentration and the viscoelastic properties of the liquid jet change as the solvent evaporates. These phenomena are a consequence of a stream motion called whipping motion (oscillations that form a stream of solution into the shape of spiral) which appears after the initial rectilinear trajectory of the jet. As a result, dried NFs obtained from a continuous fiber are deposited on the collector in a disordered structure. The onset of whipping motion is an intricate issue that has given rise to multiple and in-depth analyses to set out this phenomenon (Shin et al. 2001; Deitzel et al. 2001; Reneker and Yarin 2008).

The formation of fiber hinges on different variables that have to be considered. These are voltage, tip-collector distance, the volumetric flow of pump, diameter needle gauge and its shape, the temperature and the relative humidity of the electrospinning process (Theron et al. 2004). Furthermore, it is necessary to consider the type of solvent(s) and polymer(s), the molecular weight(s) and concentration(s) of polymer(s) in solution and the properties of resulting solution (dynamic viscosity, electrical conductivity, and surface tension). On the other hand, in electrospinning process could sometimes appear instabilities that bring about either the formation of NFs with flat ribbons and buckling structures or the production of NFs with several defects (beads and branches) (Reneker and Yarin 2008).

The production rate of a conventional electrospinning system is less than 10 g h^{-1} of NFs (Tsai et al. 2002), depending on process conditions (mainly flow rate of solution) and on polymer concentration. For large-scale nanofiber production, can be designed multiple jet electrospinning systems to increase both productivity and covering area (Li et al. 2006; Ding et al. 2004; Tomaszewski and Szadkowski 2005; Theron et al. 2005; Kim et al. 2006; Varabhas et al. 2008; Varesano et al. 2009). Figure 3 shows examples of multiple jet electrospinning systems with two different configurations using PEO solutions (Varesano et al. 2009).

Fig. 3 Pictures of multiple jet electrospinning system. These pictures were kindly provided by Dr. Alessio Varesano (CNR-ISMAC Biella)

5.1 Keratin Nanofibers for Biomedical Applications

Thanks to tiny diameter of NFs, it is possible to generate structures with high porosity and high surface area-to-volume ratio that can be employed as scaffolds in tissue engineering. However, it is important to highlight that the selection of a polymer with excellent biocompatibility is fundamental, given that this property is influenced by the chemical properties of polymer (Li et al. 2002). A common biopolymer that can be used to made scaffolds is keratin, which has been deeply investigated to demonstrate its application in tissue engineering. These studies have shown that keratin can mimic ECMs due to its content of intrinsic sites recognized by cells (Furth et al. 2007). In addition, several cell types have demonstrated mutagenic and chemotactic responses to some keratin materials by modifications in gene expressions that promote wound healing processes. For these reasons, keratin is an excellent option to be taken into account in tissue engineering, since it can be used as a scaffold during cell cultivation.

After extraction processes, pure keratin has intrinsically poor mechanical properties which is a consequence of lack of disulfide bonds. It limits its application so it is necessary to employ other polymers (synthetic or natural) to improve the mechanical structure of keratin materials. For this reason, many researchers have resorted to the addition of synthetic or natural polymers in order to increase the processability of keratin for fiber formation. A great amount of work has been done for the characterization of the intermolecular interactions between keratin and the additivated macromolecules in order to correlate the properties of the blend solution to the properties of the NFs. PEO is an amphiphilic water-soluble and nondegradable polymer, with and low toxicity (Bergsma et al. 1995) and good biocompatibility (Desai and Hubbel 1991), often it is used as an ideal model in electrospinning (Theron et al. 2004; Son et al. 2004) because it can be electrospun from aqueous solutions without defects.

Aqueous solutions of keratin and PEO has been combined to create keratin/PEO materials by Aluigi et al. (2007a, 2008b). Electrospinning parameters (voltage and working distance) has been investigated to create defect-free fibrous materials. Blended solutions with a keratin/PEO weight ratio of 50:50 and 5, 7% and 10% total polymer concentrations were tested but only 7% and 10% solution concentrations demonstrated suitable viscosities to obtain NFs with few defects. Spectroscopic and thermal analyses showed that the electrospinning process destabilized the natural self-assembly of keratin and promoted a less complex protein conformation (Aluigi et al. 2007a).

Rheological, chemical and physical properties of keratin/PEO blend solutions in different proportions has been studied in order to be correlated with the morphological, structural, thermal and mechanical properties of the electrospun mats. Keratin/PEO solutions show increased viscosities in comparison to pure PEO and keratin, moreover, the blends exhibit a non-Newtonian flow behavior with strong shear-thinning properties depending on PEO concentration. The low viscosity of blends with high keratin content blends show low viscosity that greatly hinder their ability to form fibers, whilst, solutions with a lower concentration of keratin were successfully electrospun without defects (Varesano et al. 2008). However, the employ of the keratin/PEO NFs is limited by their water instability and poor mechanical properties (Aluigi et al. 2008b).

Blends of keratin and fibroin regenerated from silk has been used to improve electrospinning processability (Zoccola et al. 2008). Macromolecule interactions promoted the formation of network structures with maximum synergy at a 50/50 (w/w) blend ratio as shown by characterizations of the rheological behavior of keratin/fibroin solutions: the synergistic effects result in the formation of finer NFs as compared to fibers formed using other ratios. The prevalence of β-sheet secondary structure in keratin/fibroin films, except at the 50/50 blend in which the proteins showed a propensity to assemble in the α-helix-coiled structure, has been confirmed by conformational analyses.

The electrospinning process, on the contrary, induce changes in secondary structure at all keratin/fibroin blend ratios by preventing β-sheet formation and promoting a random coil or α-form structure. Moreover, the α-crystallites formed by electrospinning are less thermally stable, probably because the high rate of fiber formation limits the molecular rearrangement and crystallization of the keratin chains (Zoccola et al. 2008).

Modified keratin with IAC, to enhance its stability was electrospun using PHBV in HFIP solution (Yuan et al. 2009). Keratin is a polar polymer containing both carboxyl and amine groups, it has higher conductivity than non polar ones. Blending PHBV with keratin, increase the conductivity leading to smaller diameter of fibers. Then the fibers are crosslinked with GAL vapor and saturated with 25% GAL aqueous solution to reduce their solubility. The ratio of (7/3) PHBV/keratin must be employed in order to preserve the mechanical properties.

When compared to a PHBV control, modified keratin increased the wettability and accelerated cell spreading and proliferation of fibroblasts on nanofiber mats. Moreover, the structure and characteristics of keratin improved cell viability and accelerated wound healing as confirmed by in vitro and in vivo experiments (Yuan et al. 2012).

5.2 Keratin Nanofibers for Filtration and Pollutant Removal

The potential for using NFs as a filtering medium is highly promising: since mid-1990s, the interest in the application for high efficiency filtration increased (Montefusco 2005; Gopal et al. 2006; Dotti et al. 2007). However due to poor mechanical properties of thin NFs membranes, they need a porous depositon substrate suitable enough to be employed as a mechanical support in filtration process. The small fiber diameters cause slip flows at fiber surfaces, creating an increase in the interception and inertial impaction efficiencies of these composite filter media. The enhancement in filtration efficiency at the same pressure drop is possible with fibers having diameters less than 0.5 micron (Subbiah et al. 2005). Small pore dimensions of NFs and high surface area associated to great surface cohesion allow the capture of submicron particles improving both filtration efficiency and cleaning operations with an increase of filter life.

Keratin can be considered in active filtration for its properties of adsorption of toxic, sensitizing and suspected carcinogenic agents substances such as heavy-metal ions (copper, chromium, lead, mercury, etc.) (Aluigi et al. 2009, 2011, 2012a, b, 2013a; Kar and Misra 2004; Ki et al. 2007;), FAL (Wortmann et al. 1999) and other hazardous VOCs. These adsorption properties are a consequence of the high content of amino acids with polar and ionizable side chains that are able to bind charged species such as metal ions. Is worth to note that free carboxyl groups of aspartyl and glutamyl residues are the more suitable binding sites over a wide pH range, having a p*K*a about 4.5 (Maclaren and Milligan 1981; Taddei et al. 2003).

Between the solvents, formic acid is a good media for keratin dissolution having strong solvation properties for regenerated keratin from wool (Alemdar et al. 2005; Aluigi et al. 2007b). It is also a solvent for polyamides so that it can be used to prepare keratin/PA6 blends in different proportions. Literature reports that solutions of pure keratin and pure PA6 with a final concentration of 15% (w/w) have been tested by dissolving the polymers in concentrated formic acid. Moreover, several keratin/PA6 solutions blend (100/0, 90/10, 70/30, 50/50, 30/70, 10/90 and 0/100), have been obtained; NFs have been produced from these solutions by electrospinning, with different results according to their viscosity and conductivity (Aluigi et al. 2011).

Despite the similar chemical structure characterized by amide bonds, negligible interactions between the protein and the synthetic polymer were highlighted from viscosity and infrared spectroscopy analysis. NFs obtained from keratin/PA6 blends by electrospinning are thin and free from defects, on the contrary, pure keratin NFs show many bead defects due to the jet breaks up into droplets as a result of the lower viscosity of the solution. With the blend composition, the mean diameter of the NFs does not change significantly (from 70 nm to 300 nm, with a mean diameter of about 150 nm); however, NFs rich in PA6 appear more homogeneous than keratin-rich ones. Blend NFs containing more than 50% wt keratin are not stable in water, while a slight swelling can be observed in the 30% keratin NFs water immersion test for 24 h. The morphology results unchanged for 10% keratin and pure PA6 NFs (Aluigi et al. 2009).

It is worth to note that keratin NFs show a good capacity as a medium for water purification from chromium ions: a low amount of keratin (10% wt) is enough to improve the chromium uptake with respect to pure PA6. The adsorption of Cr^{3+} ions by keratin can be studied only at low pH values in order to avoid the precipitation of chromium hydroxide above pH 4 (Maclaren and Milligan 1981). Only the pure PA6 and PA6 rich NFs have been tested for the chromium adsorption because of their good stability in water. In any case, the chromium adsorption capacity of keratin-based NFs is at least one order of magnitude higher than that of the films with the same composition and this is due to the higher exposed surface of NFs (Aluigi et al. 2009).

The keratin-based nanofiber has been tested as adsorbent also for another heavy metal using Cu(II) ion solutions. NFs rich in keratin content demonstrated interesting adsorption performances showing that the adsorption capacity increases by increasing the keratin percentage. NFs containing 50%, 70%, and 90% of keratin show maximum adsorption capacities of 61.7 (mg/g), 90 (mg/g), and 103.5 (mg/g), respectively, which are better than that of commercially activated carbon (50 mg/g). Furthermore, the keratin-rich NFs adsorption of Cu^{2+} is pH dependent, test results highlighted that the optimum pH was found above the isoelectric point of keratin (Aluigi et al. 2011).

Furthermore, wool-derived keratin nanofiber membranes, produced almost defect free, with a mean diameter of NFs of 240 nm, have been characterized and tested for the adsorption of Cu(II) ions in coexistence with Ni(II) and Co(II) metal ions, through dynamic adsorption experiments (Aluigi et al. 2013a). Keratin nanofiber membranes have shown good adsorption performances for Cu(II) ions in the mixed solutions Cu(II)–Ni(II)–Co(II). However, the presence of Ni(II) and Co(II) disturbed the process kinetics: the Cu(II) initial adsorption rate from a mono-element solution was higher than that from a multi-element solution. Moreover, the selectivity order toward the considered metal ions shown by keratin nanofiber membranes has been Cu(II) > Ni(II) > Co(II) (Aluigi et al. 2013a).

Keratin-based NFs were tested also as FAL absorbers. The high reactivity of FAL towards proteins is known; it is able to react with different side chains of amino acids as arginine, glutamine, asparagine, and amino groups of lysine. As a consequence, experimental results showed good performance in FAL adsorption tests. Multi-component filters of PP and keratin/PA6 blend NFs have been produced by electrospinning the solutions directly onto the surface of the PP filters. The adsorption performances have been compared with that of PP filters and pure PA6 NFs/PP filters. Because of the selectivity of keratin towards FAL, only the NFs containing at least the 90% wt of keratin have been considered. It was found that PP filters and multi-component filters of PP sheets and PA6 NFs are able to reduce the FAL concentration of about 30% and 40%, respectively, but the reduction reaches 70% in the presence of keratin-based NFs (Aluigi et al. 2009).

The main reason is that PP and PA6 NFs/PP filters adsorb FAL only by physisorption. On the other hand, both physisorption and chemisorption occur simultaneously when keratin-based NFs are employed because of their stronger capability of adsorbing FAL. In particular, the physisorption occurs when the interaction forces between adsorbent and adsorbate are intermolecular attractions (Van der Waals forces), while the chemisorption is a type of adsorption in which the adsorbent adheres to the adsorbate through the formation of chemical bonds. As a consequence, chemisorption is more selective than physisorption.

6 Chemical Modifications

Chemical modifications of keratin can be made either to stabilize the macromolecular structure of a protein or to increase the number of inter or intramolecular covalent bonds (crosslinking) in the polymeric matrix of final products. In Figs. 4 and 5 and Table 5, the most popular chemical modifiers used in keratin-based products and keratin sources, solvents, polymers, and other components are reported. In the stabilization case, small molecules with a reactive group react with an active (residual) one in the protein chain. Especially extracted keratin has free thiol groups that remain active and which are able to restore disulfide bonds. The oxidation of thiols groups without a specific arrangement of any protein structures (e.g., α helix and β sheet) bears a casual formation of aggregate between protein chains during dialysis and storing of keratin after lyophilization.

In order to avoid casual linkages between proteins, which is a result of oxidation of thiols groups, it is possible to employ compounds such as IAC (Nakata et al. 2015; Tachibana et al. 2006; Schrooyen et al. 2000), IAA (Nakata et al. 2015; Schrooyen et al. 2000), CAA (Wrześniewska-Tosik and Adamiec 2007), BEA (Nakata et al. 2015; Tachibana et al. 2006) and BSCA (Schrooyen et al. 2000). These compounds contain halogen groups that are able to oxidize and cap thiol groups even under environmental conditions (e.g. 20 °C). These halogen compounds can specifically react by nucleophilic substitution with thiol groups at pH values between physiological range (pH 7.5–8.5) and form stable thioether bonds (Thermo Scientific 2018); also, halo derivatives compounds can show a relative reactivity as follows: I > Br > Cl > F (Hermanson 2013a). In addition, these reactions can be carried out with a small concentration of reducing agent in order to keep a stable form of protein and to prevent the oxidation of cysteine. In this way, keratin can be processed to make keratin-based products such as stable solutions, hydrogels, sponges, films, nanopowders, NFs by using water, buffers, organic solvents (TFA and HFIP) and other polymers.

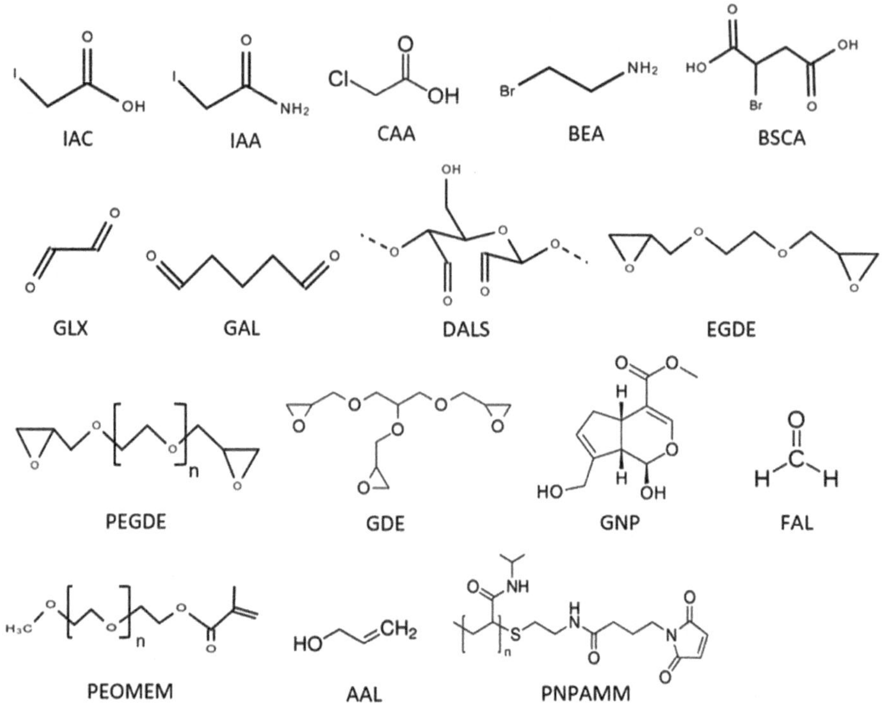

Fig. 4 Chemical modifiers (or crosslinking agents)

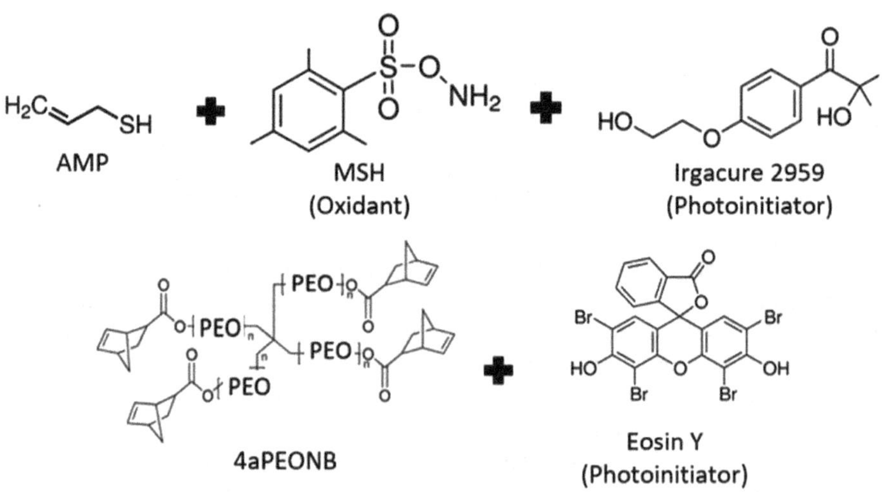

Fig. 5 Chemical modifier with photoinitiator

Table 5 Chemical modifiers applied in keratin-based products

Chemical modifier	Keratin source(s)	Solvent(s)	Polymer(s)	Other component(s)	Product(s)	Reference(s)
IAC, IAA, CAA, BEA, BSCA	Chicken feathers, wool, human hair, hooves and horns	Water, Buffer, TFA, HFIP	PHBV, PCL, PLA	Glycerol, Shindai solution, Trizma-base, NaOH/Na_2S, Ag NPs, HA, rhBMP-2, rhIGF-1, ciprofloxacin	Stable solutions, hydrogels, sponges, films, nanopowders, NFs	Nakata et al. (2015); Tachibana et al. (2006); Schrooyen et al. (2000) and Wrześniewska-Tosik and Adamiec (2007)[a]
GLX, GAL, DALS	Human hair, wool, chicken feathers, hooves and horns	Formic acid, water, HFIP	Gelatin/PVA, PVA, PCL, PHBV	$Na_2S_2O_5/H_3PO_4$, glycerol, HA, NaOH	NFs, films	Park et al. (2015); Choi et al. (2015); Yao et al. (2017); Yuan et al. (2009); El-Kheir et al. (2012); Zhao et al. (2015); Thonpho et al. (2016) and Dou et al. (2015)
FAL	Wool	Formic ac.	–	–	NFs, films	Aluigi et al. (2013b)
EGDE, PEGDE, GDE	Human hair, wool, chicken feathers	Water	PEO	Na_2SO_4/NaOH/HCl	NFs, films, powders	Tanabe et al. (2004); Liu et al. (2015); Fan et al. (2016) and Thonpho et al. (2016)
GNP	Wool, human hair	Water/buffer	PluronicF127–chitosan	Laponite	Films, hydrogel	Eslahi et al. (2016) and Thonpho et al. (2016)
PEOMEM, AAL, PNPAMM	Wool, chicken feathers	Water	PEO	Urea/$K_2S_2O_8$, DOX · HCl,	NPs/drug delivery, powders	Li et al. (2012) and Khosa and Ullah (2014)
AMP+MSH+irgacure	Chicken feathers	Water	–	BSA or stem cell	Hydrogels	Barati et al. (2017)
4aPEONB+Eosin Y	Human hair	Water/buffer	Alginate	–	Hydrogels, fibers	Yue et al. (2018)
Clorox®	Wool	Water/acetic ac.	Cellulose acetate	NaCl	NFs, films	Dickerson et al. (2013)

[a]See also Tables 1, 2, 3 and 4

In crosslinking case, reactions between a chemical modifier (crosslinking agent) and two residual groups in different protein chains are carried out. Chemical modifiers, in general, have two functional groups which react readily with functional groups in the polypeptide chain. On one hand, the most characteristic functional groups in modifiers are epoxides [EGDE (Tanabe et al. 2004; Liu et al. 2015; Fan et al. 2016), PEGDE (Thonpho et al. 2016; Khosa and Ullah 2014) and GDE (Na Ayutthaya and Wootthikanokkhan 2013)], aldehydes [GLX (Park et al. 2015; Choi et al. 2015), GAL (Yao et al. 2017; Yuan et al. 2009; El-Kheir et al. 2012; Zhao et al. 2015; Thonpho et al. 2016), DALS (Dou et al. 2015), FAL (Aluigi et al. 2013b)] or carbon–carbon double bonds [GNP (Eslahi et al. 2016; Thonpho et al. 2016), PEOMEM (Li et al. 2012), AAL (Khosa and Ullah 2014), PNPAMM (Khosa and Ullah 2014), AMP (Barati et al. 2017) and 4aPEONB (Yue et al. 2018)] which lie as terminal groups. On the other hand, thiols, amines, alcohol, and carboxyl groups are presented along keratin chain as terminal groups which are able to react with epoxide, aldehyde, or carbon–carbon double bond. As a result, when crosslinking reactions are carried out, it is possible to obtain keratin-based products (NFs, sponges, hydrogels, NPs, films, and powders) with better mechanical and water-insoluble properties in comparison with keratin materials without crosslinking agents.

Compounds with aldehyde groups in terminal position can be employed to crosslinkage: keratin and PVA in aqueous solution at pH 2–3 with GLX (Park et al. 2015; Choi et al. 2015); keratin/gelatin/PVA or keratin/PCL with a vapor or an aqueous solution of GAL (Yao et al. 2017; El-Kheir et al. 2012; Zhao et al. 2015); keratin and PVA in aqueous solution at pH 9.5 at 40 °C with DALS (Dou et al. 2015). Aldehydes compounds are able to react with amines groups to form Schiff base intermediates in protein chains and in polymers as PVA (at pH 4 and 80 °C), and chitosan (at pH 4 and 40 °C) are able to react with hydroxyl groups (Hermanson 2013a; Zhang et al. 2010; Yang et al. 2005). It is important to highlight that crosslinking agents with aldehyde groups are pH dependent. On the other hand, FAL crosslinking is a process that is carried out using its own vapor, which is possible due to the low boiling point of this compound. FAL is the simplest crosslinking agent and its reaction with amine groups in proteins occurs in two-step pattern: first, FAL behaves as a strong nucleophile to form Schiff base intermediates; second, this Schiff intermediate reacts with another nucleophile (amino group) (Hermanson 2013b; Hoffman et al. 2015) to form secondary amines. As a result, it is possible to employ aldehydes crosslinking process in keratin materials (NFs and films) obtained from different sources (human hair, wool, chicken feathers, hooves, and horns).

When compounds with epoxide as terminal group are employed, it is possible to make crosslinking process with the vapor of epoxide at 80 °C (Liu et al. 2015) or to prepare an epoxide aqueous solution at a temperature of 40–60 °C (Tanabe et al. 2004; Liu et al. 2015; Fan et al. 2016; Thonpho et al. 2016). Moreover, epoxide groups are known for their high reactivity, which is a consequence of the strained three-membered ring system whose nucleophilic attack (in a ring-opening) can release the ring strain (Obach and Kalgutkar 2018). However, the reactions that epoxide groups can realize with primary amines, thiols or hydroxyl groups are pH dependent. In hydroxyl case, the reaction requires pH conditions between 11 and 12 (Hermanson

2013a). In amine case, the reaction needs buffer environments with a pH value close to 9 (Hermanson 2013a). In thiols case, it is necessary to employ buffer solutions with pH value closer to physiological ranges (7.5–8.5) (Hermanson 2013a). In particular, the crosslinking of epoxide compounds with keratin obtained from different sources (human hair, wool, and chicken feathers) can be used to produce materials as NFs, films, and powders.

Compounds with carbon–carbon double bonds can be used to crosslinkage keratin in different conditions. In GNP case, macromolecular structures with amine groups as keratin and chitosan in buffer solution (at pH 7.4 and 37 °C) can form amide and tertiary amines by the nucleophilic attack to carboxyl and double bond in GNP (Eslahi et al. 2016), respectively. In PEOMEM case, the thiols groups of keratin in aqueous and oxidant ($K_2S_2O_8$) medium at 70 °C are able to produce thioether bonds (Li et al. 2012). In PNPAMM case, the thiols groups of keratin can also alkylate the double bond (at pH 8) in the methacrylate group (Khosa and Ullah 2014); a similar situation can be observed with AAL and keratin (at pH 8) which are used together (Khosa and Ullah 2014). The double bond of compounds mentioned above undergoes a nucleophilic attack of thiol groups which makes thioether bonds stable under alkaline conditions of pH (Hermanson 2013a). All these crosslinking processes allow the production of keratin materials as films, hydrogels, NPs/drug delivery, and powders by using keratin obtained from different sources (human hair, wool, and chicken feathers).

Sometimes the crosslinking of keratin is done by photocatalytic reactions of double bonds (Barati et al. 2017; Yue et al. 2018). When the reaction of keratin, MSH and AMP is considered, crosslinking involves the formation of two thioethers groups between protein chains by using AMP and double bonds of allyl groups in modified keratin (Barati et al. 2017). The allyl groups in keratin are obtained by oxidative elimination of thiols groups with MSH at 4 °C (Barati et al. 2017). After that, the thiols groups of AMP react with the allyl group of modified keratin (at pH 8) to produce an allyl thioether group (Barati et al. 2017). Finally, the photoinitiator (Irgacure) can be added to the solution that contains modified keratin; in this way, the double bonds of allyl thioether groups can react by illumination with UV irradiation to produce hydrogels of keratin (Barati et al. 2017). On the other hand, when the reaction of keratin with 4aPEONB is considered, the thiol groups of keratin can react with the double bonds of norbornene by using Eosin Y as photoinitiator, which illuminates with visible light permits to obtain hydrogels or fibers with thioether bonds (Yue et al. 2018).

Other modifications of keratin involve the formation of halamine compound as a consequence of the reaction between the nitrogen of amine groups in keratin and the hypochlorous acid content in a commercial cleaning product (Clorox®) (Dickerson et al. 2013). This modification can be applied to keratin-based product (NFs and films) with a buffer solution of sodium acetate at pH 4 (Dickerson et al. 2013).

Finally, the majority of chemical modifiers employed in keratin crosslinking processes are harmful and generate different technical problems for an operator. In cases in which it is not possible to use the compounds mentioned above, keratin proteins can be also crosslinked at 180 °C during at least 2 hours in air to induce the reaction between residual carboxylic and amine groups (Aluigi et al. 2013b; Varesano et al. 2014). In this way, it is possible to produce new amide bonds in keratin chain of NFs and films (Aluigi et al. 2013b; Varesano et al. 2014).

7 Conclusions

The most common solvents used in keratin-based products are water, formic acid, and HFIP. All of them can be employed to produce fibers and NFs membranes from complex solutions of keratin with synthetic or natural polymers. Considering the synthetic polymer cases, the processing of keratin is considerably improved, thanks to the fact that the organic solvents are commonly required. When fibers are produced an additional coagulation system is needed to stabilize the polymeric matrix before the drying process. Moreover, this step can be also applied in the production of films, sponges or hydrogels. The keratin-based products mentioned above are obtained from complex solutions by using mainly water or aqueous mediums instead of organic solvents. In particular, keratin films is the most simple and common applications to valorize keratin considering that this process can use either aqueous or organic solvents.

In general, when are needed in the final keratin-based product good mechanical performances and water insolubility any other polymer should be added to keratin matrix. Additional properties can be obtained by means of the addition of functional materials with properties such as catalytic, biocidal, antioxidant, growth factors, antibiotics, etc. However, in order to preserve the attributes of the functional materials that are embedded an appropriate solvent system has to be used.

Finally, it is important to highlight that the keratin-based products (films, NFs, and sponges) have great potential for biomedical applications. For this reason, the selection of the solvent system and the crosslinking process involved in keratin-based products are relevant. Consequently, it is fundamental to consider the concentration of organic solvents and crosslinking agents in the final products in order to assure their biocompatibility and cytotoxicity.

References

Alemdar A, Iridag Y, Kazanci M (2005) Flow behavior of regenerated wool-keratin proteins in different mediums. Int J Biol Macromol 35:151–153

Aluigi A, Corbellini A, Rombaldoni F et al (2013a) Wool-derived keratin nanofiber membranes for dynamic adsorption of heavy-metal ions from aqueous solutions. Text Res J 83:1574–1586

Aluigi A, Corbellini A, Rombaldonia F et al (2013b) Morphological and structural investigation of wool-derived keratin nanofibres crosslinked by thermal treatment. Int J Biol Macromol 57:30–37

Aluigi A, Tonetti C, Vineis C et al (2011) Adsorption of copper(II) ions by keratin/PA6 blend nanofibres. Eur Polym J 47:1756–1764

Aluigi A, Tonetti C, Vineis C et al (2012a) Study on the adsorption of chromium(VI) by hydrolyzed keratin/polyamide 6 blend nanofibers. J Nanosci Nanotechnol 12:7250–7259
Aluigi A, Tonetti C, Vineis C et al (2012b) Wool keratin nanofibres for copper(II) adsorption. J Biobased Mater Bioenergy 6:1–7
Aluigi A, Varesano A, Montarsolo A et al (2007a) Electrospinning of keratin/poly(ethylene oxide) blend nanofibers. J Appl Polym Sci 104:863–870
Aluigi A, Varesano A, Vineis C et al (2017) Electrospinning of immiscible systems: the wool keratin/polyamide-6 case study. Mater Des 127:144–153
Aluigi A, Vineis C, Ceria A et al (2008a) Composite biomaterials from fibre wastes: characterization of wool–cellulose acetate blends. Compos A 39:126–132
Aluigi A, Vineis C, Tonin C et al (2009) Wool keratin-based nanofibres for active filtration of air and water. J Biobased Mater Bioenergy 3:311–319
Aluigi A, Vineis C, Varesano A et al (2008b) Structure and properties of keratin/PEO blend nanofibres. Eur Polym J 44:2465–2475
Aluigi A, Zoccola M, Vineis C et al (2007b) Study on the structure and properties of wool keratin regenerated from formic acid. Int J Biol Macromol 41:266–273
Balaji S, Kumar R, Sripriya R et al (2012) Preparation and comparative characterization of keratin–chitosan and keratin–gelatin composite scaffolds for tissue engineering applications. J Mater Sci Eng C 32:975–982
Barati D, Kader S, Pajoum Shariati SR et al (2017) Synthesis and characterization of photo-crosslinkable keratin hydrogels for stem cell encapsulation. Biomacromolecules 18:398–412
Barhate RS, Ramakrishna S (2007) Nanofibrous filtering media: filtration problems and solutions from tiny materials. J Membr Sci 296:1–8
Barone JR, Schmidt WF, Gregoire N (2006) Extrusion of feather keratin. J Appl Polym Sci 100:1432–1442
Barone JR, Schmidt WF, Liebner CF (2005) Thermally processed keratin films. J Appl Polym Sci 97:1644–1651
Bergsma JE, Rozema FR, Bos RRM et al (1995) In vivo degradation and biocompatibility study of in vitro pre-degraded as-polymerized polylactide particles. Biomaterials 16:267–274
Bhardwaj N, Sow WT, Devi D et al (2015) Silk fibroin-keratin based 3D scaffolds as a dermal substitute for skin tissue engineering. Integr Biol 7:53–63
Bhavsar PS, Zoccola M, Patrucco A et al (2017) Superheated water hydrolyzed keratin: a new application as a foaming agent in foam dyeing of cotton and wool fabrics. ACS Sustainable Chem Eng 5:9150–9159
Bryner MA, Armantrout JE, Johnson BS (2006) Electroblowing web formation process. US Patent 2006/0138710, 29 Jun 2006
Burnett LR, Rahmany MB, Richter JR et al (2013) Hemostatic properties and the role of cell receptor recognition in human hair keratin protein hydrogels. Biomaterials 34:2632–2640
Cardamone JM, Tunick MH, Onwulata C (2013) Keratin sponge/hydrogel: I. Fabrication and characterization. Tex Res J 83:661–670
Chen HL, Burns LD (2006) Environmental analysis of textile products. Cloth Text Res J 24:248–261
Choi J, Panthi G, Liu Y et al (2015) Keratin/poly(vinyl alcohol) blended nanofibers with high optical transmittance. Polymer 58:146–152
Cilurzoa F, Selmina F, Aluigi A et al (2013) Regenerated keratin proteins as potential biomaterial for drug delivery. Polym Adv Technol 24:1025–1028
de Guzman RC, Saul JM, Ellenburg MD et al (2013) Bone regeneration with BMP-2 delivered from keratose scaffolds. Biomaterials 34:1644–1656
de Guzman RC, Tsuda SM, Ton MTN et al (2015) Binding interactions of keratin-based hair fiber extract to gold, keratin, and BMP-2. PLoS ONE 10:1–12
Deitzel JM, Kleinmeyer J, Hirvonen JK et al (2001) Controlled deposition of electrospun poly(ethylene oxide) fibers. Polymer 42:8163–8170
Desai NP, Hubbel JA (1991) Solution technique to incorporate polyethylene oxide and other water-soluble polymers into surfaces of polymeric biomaterials. Biomaterials 12:144–153

Dias GJ, Mahoney P, Swain M et al (2010) Keratin–hydroxyapatite composites: biocompatibility, osseointegration, and physical properties in an ovine model. J Biomed Mater Res, Part A 95A:1084–1095
Dickerson MB, Sierra AA, Bedford NM et al (2013) Keratin-based antimicrobial textiles, films and nanofibers. J Mater Chem B 1:5505–5514
Ding B, Kimura E, Sato T et al (2004) Fabrication of blend biodegradable nanofibrous nonwoven mats via multi-jet electrospinning. Polymer 45:1895–1902
Doshi J, Reneker DH (1993) Electrospinning process and applications of electrospun fibers. In: Conference record of the 1993 IEEE industry applications conference twenty-eighth IAS annual meeting, Toronto, October 1993. Lecture notes in industrial application society annual meeting, vol 3. IEEE, pp 1698–1703
Doshi J, Reneker DH (1995) Electrospinning process and applications of electrospun fibers. J Electrost 35:151–160
Dotti F, Varesano A, Montarsolo A et al (2007) Electrospun porous mats for high efficiency filtration. J Ind Text 37:151–162
Dou Y, Zhang B, He M et al (2014) Preparation and physicochemical properties of dialdehyde starch crosslinked feather keratin/PVA composite films. J Macromol Sci Part A Pure Appl Chem 51:1009–1015
Dou Y, Zhang B, He M et al (2015) Keratin/polyvinyl alcohol blend films cross-linked by dialdehyde starch and their potential application for drug release. Polymers 7:580–591
Ebrahimgol F, Tavanaia H, Alihosseinia F et al (2014) Electrosprayed recovered wool keratin nanoparticles. Polym Adv Technol 25:1001–1007
Edwards A, Jarvis D, Hopkins T et al (2015) Poly(ε-caprolactone)/keratin-based composite nanofibers for biomedical applications. J Biomed Mater Res, Part B 103B:21–30
El-Kheir AA, Mowafi S, Abou Taleb M et al (2012) Preparation and characterization of keratin-polyvinyl alcohol composite film. Egypt J Chem 55:491–507
Ellison CJ, Phatak A, Giles DW et al (2007) Melt blown nanofibers: fiber diameter distributions and onset of fiber break-up. Polymer 48:3306–3316
EPA-United States (Environmental Protection Agency U.S.). http://www.epa.gov/. Accessed 17 May 2018
Eslahi N, Simchi A, Mehrjoo M et al (2016) Hybrid cross-linked hydrogels based on fibrous protein/block copolymers and layered silicate nanoparticles: tunable thermosensitivity, biodegradability and mechanical durability. RSC Adv 6:62944–62957
Evans RL, Shore B (1948) Regenerated keratin. US Patent 2,434,688, 20 Jan 1948
Fan J, Lei TD, Li J et al (2016) High protein content keratin/poly(ethylene oxide) nanofibers crosslinked in oxygen atmosphere and its cell culture. Mater Des 104:60–67
FAO-United Nations (Food and Agricultural Organizations of the United Nations). http://faostat.fao.org/. Accessed 17 May 2018
Formhals A (1934) Process and apparatus for preparing artificial threads. US Patent 1,975,504, 2 Oct 1934
Fowler WE, Aebi U (1983) Preparation of single molecules and supramolecular complexes for high resolution metal shadowing. J Ultrastruct Res 83:319–334
Furth ME, Atala A, Van Dyke ME (2007) Smart biomaterials design for tissue engineering and regenerative medicine. Biomaterials 28:5068–5073
Fujii T (2012) Hair keratin film as a substitute device for human hair. J Biol Macromol 12:3–5
Fujii T, Murai S, Ohkawa K et al (2008) Effects of human hair and nail proteins and their films on rat mast cells. J Mater Sci Mater Med 19:2335–2342
Fujii T, Ide Y (2004) Preparation of translucent and flexible human hair protein films and their properties. Biol Pharm Bull 27:1433–1436
Fujii T, Ogiwara D, Arimoto M (2004) Convenient procedures for human hair protein films and properties of alkaline phosphatase incormporated in the film. Biol Pharm Bull 27:89–93
Ganesan P (2017) Natural and bio polymer curative films for wound dressing medical applications. Wound Med 18:33–40

Gao P, Liu Z, Wu X et al (2014a) Biosorption of chromium(VI) ions by deposits produced from chicken feathers after soluble keratin extraction. Clean: Soil, Air, Water 42:1558–1566
Gao P, Li K, Liu Z et al (2014b) Feather keratin deposits as biosorbent for the removal of methylene blue from aqueous solution: equilibrium, kinetics, and thermodynamics studies. Water Air Soil Pollut 225:1946
Geidobler R, Winter G (2013) Controlled ice nucleation in the field of freeze-drying: fundamentals and technology review. Eur J Pharm Biopharm 5:214–222
Gibson P, Schreuder-Gibson H, Rivin D (2001) Transport properties of porous membranes based on electrospun nanofibers. Colloid Surf A 187–188:469–481
Gibson PW, Lee C, Ko F et al (2007) Application of nanofiber technology to nonwoven thermal insulation. J Eng Fibers Fabr 2:32–40
Gopal R, Kaur S, Ma Z et al (2006) Electrospun nanofibrous filtration membrane. J Membr Sci 281:581–586
Gupta P, Nayak KK (2015) Compatibility study of alginate/keratin blend for biopolymer development. J Appl Biomater Funct Mater 13:e332–e339
Hamasaki S, Tachibana A, Tada D et al (2008) Fabrication of highly porous keratin sponges by freeze-drying in the presence of calcium alginate beads. J Mater Sci Eng C 28:1250–1254
Hameed N, Guo Q (2010) Blend films of natural wool and cellulose prepared from an ionic liquid. Cellulose 17:803–813
Ham TR, Lee RT, Han S et al (2016) Tunable keratin hydrogels for controlled erosion and growth factor delivery. Biomacromolecules 17:225–236
Han S, Hama TR, Haque S et al (2015) Alkylation of human hair keratin for tunable hydrogel erosion and drug delivery in tissue engineering applications. Acta Biomater 23:201–213
Harris M, Brown AE (1947) Natural and synthetic protein fibers. Tex Res J 17:323–330
Heikkilä P, Sipilä A, Peltola M et al (2007) Electrospun PA-66 coating on textile surfaces. Tex Res J 77:864–870
He M, Zhang B, Dou Y et al (2017) Blend modification of feather keratin-based films using sodium alginate. J Appl Polym Sci 2017:44680–44688
Hengchang M, Zhikang B, Guobin H et al (2013) Nanoparticulate palladium catalyst stabilized by supported on feather keratin for Suzuki coupling reaction. Chin J Catal 34:578–584
Hermanson GT (2013a) Chapter 3—the reactions of bioconjugation. In: Audet J (ed) Bioconjugate techniques, 3rd edn. Elsevier, Inc., pp 229–258
Hermanson GT (2013b) Chapter 24—bioconjugation in the study of protein interactions. In: Audet J (ed) Bioconjugate techniques, 3rd edn. Elsevier, Inc., pp 989–1016
Hill P, Brantley H, Van Dyke M (2010) Some properties of keratin biomaterials: kerateines. Biomaterials 31:585–593
Hirao Y, Ohkawa K, Yamamoto H et al (2005) A novel human hair protein fiber prepared by watery hybridization spinning. Macromol Mater Eng 290:165–171
Hoffman EA, Frey BL, Smith LM et al (2015) Formaldehyde crosslinking: a tool for the study of chromatin complexes. J Biol Chem 290:26404–26411
Homonoff E (2008) Nanofibrillated fibers: opening new markets to nano-fibre usage. Int Fibers J 23:22–24
Idris A, Vijayaraghavan R, Rana UA et al (2013) Dissolution of feather keratin in ionic liquids. Green Chem 15:525–534
Iridag Y, Kazanci M (2006) Preparation and characterization of Bombyx mori silk fibroin and wool keratin. J Appl Polym Sci 100:4260–4264
Kakkar P, Madhan B (2016) Fabrication of keratin-silica hydrogel for biomedical applications. J Mater Sci Eng C 66:178–184
Kar P, Misra M (2004) Use of keratin fiber for separation of heavy metals from water. J Chem Technol Biotechnol 79:1313–1319
Katoh K, Shibayama M, Tanabe T et al (2004) Preparation and properties of keratin–poly(vinyl alcohol) blend fiber. J Appl Polym Sci 91:756–762

Khosa MA, Ullah A (2014) In-situ modification, regeneration, and application of keratin biopolymer for arsenic removal. J Hazard Mater 278:360–371
Ki CS, Gang EH, Um IC et al (2007) Nanofibous membrane of wool keratose/silk fibroin blend for heavy metal ion adsorption. J Membr Sci 302:20–26
Kilic A, Oruc F, Demir A (2008) Effects of polarity on electrospinning process. Tex Res J 78:532–539
Kim BS, Park KE, Park WH et al (2013) Fabrication of nanofibrous scaffold using a PLA and hagfish thread keratin composite; its effect on cell adherence, growth, and osteoblast differentiation. Biomed Mater 8:1–9
Kim G, Cho YS, Kim WD (2006) Stability analysis for multi-jets electrospin-ning process modified with a cylindrical electrode. Eur Polym J 42:2031–2038
Lee J, Cuddihy MJ, Kotov NA (2008) Three-dimensional cell culture matrices: state of the art. Tissue Eng Part B 14:61–86
Li D, Xia Y (2004) Electrospinning of nanofibers: reinventing the wheel. Adv Mater 16:1151–1170
Li J, Li Y, Li L et al (2009) Fabrication and degradation of poly(L-lactic acid) scaffolds with wool keratin. Compos. B 40:664–667
Li J, Yu LH, Fan J et al (2013) Fabrication of three-dimensional porous keratin/PEO biological scaffolds. Adv Mater Res 821:1035–1038
Li L, Frey MW, Green TB (2006) Modification of air filter media with nylon-6 nanofibers. J Eng Fibers Fabr 1:1–24
Li Q, Zhu L, Liu R et al (2012) Biological stimuli responsive drug carriers based on keratin for triggerable drug delivery. J Mater Chem 22:19964–19973
Li S, Yang XH (2014) Fabrication and characterization of electrospun wool keratin/poly(vinyl alcohol) blend nanofibers. Adv Mater Sci Eng 2014:1–7
Li W, Laurencin CT, Caterson EJ et al (2002) Electrospun nanofibrous structure: a novel scaffold for tissue engineering. J Biomed Mater Res, Part B 60:613–621
Li Y, Wang Y, Ye J et al (2016) Fabrication of poly(ε-caprolactone)/keratin nanofibrous mats as a potential scaffold for vascular tissue engineering. J Mater Sci Eng C 68:177–183
Liu Y, Yu X, Li J et al (2015) Fabrication and properties of high-content keratin/poly(ethylene oxide) blend nanofibers using two-step cross-linking process. J Nanomater 2015:1–7
Ma B, Qiao X, Hou X et al (2016) Pure keratin membrane and fibers from chicken feather. Int J Biol Macromol 89:614–621
Maclaren JA, Milligan B (1981) Wool science: the chemical reactivity of the wool fibre. Science Press, Marrickville, Australia, pp 109–127
Manrique-Juárez MD, Martínez-Hernández AL, Olea-Mejía OF et al (2013) Polyurethane-keratin membranes: structural changes by isocyanate and pH, and the repercussion on Cr(VI) removal. Int J Polym Sci 2013:1–12
Martelli SM, Borges Laurindo J (2012) Chicken feather keratin films plasticized with polyethylene glycol. Int J Polym Mater 61:17–29
Martelli SM, Moore G, Paes SS et al (2006a) Influence of plasticizers on the water sorption isotherms and water vapor permeability of chicken feather keratin films. LWT 39:292–301
Martelli SM, Plácido Moore GR, Borges Laurindo J (2006b) Mechanical properties, water vapor permeability and water affinity of feather keratin films plasticized with sorbitol. J Polym Environ 14:215–222
McCurry J (1996) Fibres, yarns and fabrics. Tex World 28
Montefusco F (2005) The use of nonwovens in air filtration. Filtr Sep 42:30–31
Na Ayutthaya SI, Tanpichai S, Sangkhun W et al (2016) Effect of clay content on morphology and processability of electrospun keratin/poly(lactic acid) nanofiber. Int J Biol Macromol 85:585–595
Na Ayutthaya SI, Tanpichai S, Wootthikanokkhan J (2015) Keratin extracted from chicken feather waste: extraction, preparation, and structural characterization of the keratin and keratin/biopolymer films and electrospuns. J Polym Environ 23:506–516
Na Ayutthaya SI, Wootthikanokkhan J (2013) Extraction of keratin from chicken feather and electrospinning of the keratin/PLA blends. Adv Mater Res 747:711–714

Nakata R, Osumi Y, Miyagawa S et al (2015) Preparation of keratin and chemically modified keratin hydrogels and their evaluation as cell substrate with drug releasing ability. J Biosci Bioeng 120:111–116
Nakata R, Tachibana A, Tanabe T (2014) Preparation of keratin hydrogel/hydroxyapatite composite and its evaluation as a controlled drug release carrier. J Mater Sci Eng C 41:59–64
Nayak KK, Gupta P (2017) Study of the keratin-based therapeutic dermal patches for the delivery of bioactive molecules for wound treatment. J Mater Sci Eng C 77:1088–1097
Obach RS, Kalgutkar AS (2018) 1.15–Reactive electrophiles and metabolic activation. In: McQuieen CA (ed) Reference module in biomedical sciences, from Comprehensive Toxicology, 3rd edn. Elsevier, Inc., pp 295–331
Ozaki Y, Takagi Y, Mori H et al (2014) Porous hydrogel of wool keratin prepared by a novel method: an extraction with guanidine/2-mercaptoethanol solution followed by a dialysis. J Mater Sci Eng C 42:146–154
Pace LA, Plate JF, Smith TL et al (2013) The effect of human hair keratin hydrogel on early cellular response to sciatic nerve injury in a rat model. Biomaterials 34:5907–5914
Patrucco A, Cristofaro F, Simionati M et al (2016) Wool fibril sponges with perspective biomedical applications. J Mater Sci Eng C 61:42–50
Park M, Kima BS, Shin HK et al (2013) Preparation and characterization of keratin-based biocomposite hydrogels prepared by electron beam irradiation. J Mater Sci Eng C 33:5051–5057
Park M, Shin HK, Panthi G et al (2015) Novel preparation and characterization of human hair-based nanofibers using electrospinning process. Int J Biol Macromol 76:45–48
Pedram Rad Z, Tavanai H, Moradi AR (2012) Production of feather keratin nanopowder through electrospraying. J Aerosol Sci 51:49–56
Perez MA, Swan D, Louks JW (2000) Microfibers and method of making. US Patent 6,110,588, 29 Aug 2000
Peyton CC, Keys T, Tomblyn S et al (2012) Halofuginone infused keratin hydrogel attenuates adhesions in a rodent cecal abrasion model. J Surg Res 178:545–552
Pham QP, Sharma U, Mikos AG (2006) Electrospinning of polymeric nanofibers for tissue engineering applications: a review. Tissue Eng 12:1197–1211
Placido Moore GR, Martelli SM, Gandolfo C et al (2006) Influence of the glycerol concentration on some physical properties of feather keratin films. Food Hydrocolloids 20:975–982
Placone JK, Navarro J, Laslo GW et al (2017) Development and characterization of a 3D printed, keratin-based hydrogel. Ann Biomed Eng 45:237–248
Poole AJ, Church JS (2015) The effects of physical and chemical treatments on Na_2S produced feather keratin films. Int J Biol Macromol 73:99–108
Poole AJ, Church JS, Huson MG (2008) Environmentally sustainable fibers from regenerated protein. Biomacromolecules 10:1–8
Posati T, Sotgiu G, Varchi G et al (2016) Developing keratin sponges with tunable morphologies and controlled antioxidant properties induced by doping with polydopamine (PDA) nanoparticles. Mater Des 110:475–484
Pourdeyhimi B, Fedorova NV, Sharp SR (2006) High strength, durable micro & nano-fiber fabrics produced by fibrillating biocomponent islands in the sea fibers. US Patent 2006/0292355, 28 Dec 2006
Puglia D, Ceccolini R, Fortunati et al (2015) Effect of processing techniques on the 3D microstructure of poly (L-lactic acid) scaffolds reinforced with wool keratin from different sources. J Appl Polym Sci 132:42890
Qin XH, Wang SY (2006) Filtration properties of electrospinning nanofibers. J Appl Polym Sci 102:1285–1290
Ramakrishnan N, Sharma S, Gupta A et al (2018) Keratin based bioplastic film from chicken feathers and its characterization. Int J Biol Macromol 111:352–358
Ramakrishna S, Fujihara K, Teo WE et al (2006) Electrospun nanofibers: solving global issues. Mater Today 9:40–50

Reddy N, Jiang Q, Jin E et al (2013) Bio-thermoplastics from grafted chicken feathers for potential biomedical applications. Colloids Surf B 100:51–58
Reichl S (2009) Films based on human hair keratin as substrates for cell culture and tissue engineering. Biomaterials 30:6854–6866
Reichl S, Borrelli M, Geerling G (2011) Keratin films for ocular surface reconstruction. Biomaterials 32:3375–3386
Reneker DH, Yarin AL (2008) Electrospinning jets and polymer nanofibers. Polymer 49:2387–2425
Richter JR, de Guzman RC, Greengauz-Roberts OK et al (2012) Structure–property relationships of meta-kerateine biomaterials derived from human hair. Acta Biomater 8:274–281
Rouse JG, Van Dyke ME (2010) A review of keratin-based biomaterials for biomedical applications. Materials 3:999–1014
Salazar NA, Alvarez C, Orrego CE (2018) Optimization of freezing parameters for freeze-drying mango (*Mangifera indica* L.) slices. Drying Technol 36:192–204
Sanchez Ramirez DO, Carletto RA, Tonetti C et al (2017) Wool keratin film plasticized by citric acid for food packaging. Food Packag Shelf Life 12:100–106
Saucedo-Rivalcoba V, Martínez-Hernández AL, Martínez-Barrera G et al (2011a) Chicken feathers keratin)/polyurethane membranes. Appl Phys A Mater Sci Process 104:219–228
Saucedo-Rivalcoba V, Martínez-Hernández AL, Martínez-Barrera G et al (2011b) Removal of hexavalent chromium from water by polyurethane–keratin hybrid membranes. Water Air Soil Pollut 218:557–571
Schmidt FW, Waters RM, Gassner G (1998) Chemical engineering news, p 23
Schrooyen PMM, Dijkstra PJ, Oberthür RC et al (2000) Partially carboxymethylated feather keratins. 1. Properties in aqueous systems. J Agric Food Chem 48:4326–4334
Schrooyen PMM, Dijkstra PJ, Oberthür RC et al (2001) Partially carboxymethylated feather keratins. 2. Thermal and mechanical properties of films. J Agric Food Chem 49:221–230
Sekimoto Y, Okiharu T, Nakajima H et al (2013) Removal of Pb(II) from water using keratin colloidal solution obtained from wool. Environ Sci Pollut Res 20:6531–6538
Sharma S, Gupta A, Kumar A et al (2018) An efficient conversion of waste feather keratin into ecofriendly bioplastic film. Clean Techn Environ Policy 1–11
Shen D, Wang X, Zhang L et al (2011) The amelioration of cardiac dysfunction after myocardial infarction by the injection of keratin biomaterials derived from human hair. Biomaterials 32:9290–9299
Shin C, Chase GG, Reneker DH (2005) The effect of nanofibers on liquid–liquid coalescence filter performance. AIChE J 51:3109–3113
Shin YM, Hohman MM, Brenner MP et al (2001) Experimental characterization of electrospinning: the electrically forced jet and instabilities. Polymer 42:9955–9967
Siemann U (2005) Solvent cast technology—a versatile tool for thin film production. Prog Colloid Polym Sci 130:1–14
Siller-Jackson AJ, Mark Van Dyke ME, Timmons SF et al (2003) Keratin-based powder and hydrogel for pharmaceutical applications. US Patent 6,544,548, 8 Apr 2003
Sill TJ, Von Recum HA (2008) Electrospinning: Applications in drug delivery and tissue engineering. Biomaterials 29:1989–2006
Singh R, Lin YT, Chaung WL et al (2017) A new biodegradable gate dielectric material based on keratin protein for organic thin film transistors. Org Electron 44:198–209
Singh R, Sarker B, Silva R et al (2016) Evaluation of hydrogel matrices for vessel bioplotting: vascular cell growth and viability. J Biomed Mater Res, Part A 104A:577–585
Sionkowska A, Skopinska-Wiśniewska J, Kozłowska J et al (2011) Photochemical behaviour of hydrolysed keratin. Int J Cosmet Sci 33:503–508
Son WK, Youk JH, Lee TS et al (2004) The effects of solution proper-ties and polyelectrolyte on electrospinning of ultrafine poly(ethylene oxide) fibers. Polymer 45:2959–2966
Srinivasan B, Kumar R, Shanmugam K et al (2010) Porous keratin scaffold–promising biomaterial for tissue engineering and drug delivery. J Biomed Mater Res, Part B 92B:5–12

Subbiah T, Bhat GS, Tock RW et al (2005) Electrospinning of nanofibers. J Appl Polym Sci 96:557–569

Tachibana A, Furuta Y, Takeshima H et al (2002) Fabrication of wool keratin sponge scaffolds for long-term cell cultivation. J Biotechnol 93:165–170

Tachibana A, Kaneko S, Tanabe T et al (2005) Rapid fabrication of keratin–hydroxyapatite hybrid sponges toward osteoblast cultivation and differentiation. Biomaterials 26:297–302

Tachibana A, Nishikawa Y, Nishino M et al (2006) Modified keratin sponge: binding of bone morphogenetic protein-2 and osteoblast differentiation. J Biosci Bioeng 102:425–429

Taddei P, Monti P, Freddi G et al (2003) Binding of Co(II) and Cu(II) cations to chemically modified wool fibres: an IR investigation. J Mol Struct 650:105–113

Tanabe T, Okitsu N, Tachibana A et al (2002) Preparation and characterization of keratin–chitosan composite film. Biomaterials 23:817–825

Tanabe T, Okitsu N, Yamauchi K (2004) Fabrication and characterization of chemically crosslinked keratin films. J Mater Sci Eng C 24:441–446

Thermo Scientific (2018). Easy molecular bonding, crosslinking technology–Reactivity chemistries, applications and structure references. In: Thermo scientific crosslinking technical handbook. Thermo Scientific, pp 1–56. https://tools.thermofisher.com/content/sfs/brochures/1602163-Crosslinking-Reagents-Handbook.pdf. Accessed 15 Oct 2018

Theron SA, Yarin AL, Zussman E et al (2005) Multiple jets in electrospinning: experiment and modelling. Polymer 46:2889–2899

Theron SA, Zussman E, Yarin AL (2004) Experimental investigation of the governing parameters in the electrospinning of polymer solutions. Polymer 45:2017–2030

Thonpho A, Songeon W, Srihanam P (2016) Effect of cross-linked agents on keratin films property. Int J GEOMATE 11:2866–2869

Tomaselli S, Sanchez Ramirez DO, Carletto RA et al (2016) Electrospun lipid binding proteins composite nanofibers with antibacterial properties. Macromol Biosci 2016:1–6

Tomaszewski W, Szadkowski M (2005) Investigation of electrospinning with the use of a multi-jet electrospinning head. Fibres Text East Eur 52:22–26

Tonin C, Aluigi A, Vineis C et al (2007) Thermal and structural characterization of poly(ethylene-oxide)/keratin blend films. J Therm Anal Calorim 89:601–608

Tsai PP, Schreuder-Gibson H, Gibson P (2002) Different electrostatic methods for making electret filters. J Electrost 54:333–341

Tseng AA, Notargiacomo A, Chen TP (2005) Nanofabrication by scanning probe microscope lithography: a review. J Vac Sci Technol B 23:877–894

Tu H, Yu W, Duan L (2016) Structural studies and macro-performances of hydroxyapatite-reinforced keratin thin films for biological applications. J Mater Sci 51:9573–9588

Varabhas JS, Chase GG, Reneker DH (2008) Electrospun nanofibers froma porous hollow tube. Polymer 49:4226–4229

Vasconcelos A, Cavaco-Paulo A (2013) The use of keratin in biomedical applications. Curr Drug Targets 14:612–619

Vasconcelos A, Freddi G, Cavaco-Paulo A (2008) Biodegradable materials based on silk fibroin and keratin. Biomacromolecules 9:1299–1305

Varesano A, Aluigi A, Vineis C et al (2008) Study on the shear viscosity behavior of keratin/PEO blends for nanofibre electrospinning. J Polym Sci Part B Polym Phy 46:1193–1201

Varesano A, Carletto RA, Mazzuchetti G (2009) Experimental investigations on the multi-jet electrospinning process. J Mater Process Technol 209:5178–5185

Varesano A, Vineis C, Tonetti C et al (2014) Chemical and physical modifications of electrospun keratin nanofibers induced by heating treatments. J Appl Polym Sci 2014:1–7

Varesano A, Vineis C, Tonetti C et al (2015) Multifunctional hybrid nanocomposite nanofibers produced by colloid electrospinning from water solutions. Curr Nanosci 11:41–48

Verma V, Verma P, Ray P et al (2008) Preparation of scaffolds from human hair proteins for tissue-engineering applications. Biomed Mater 3:1–12

Wang J, Hao S, Luo T et al (2016) Development of feather keratin nanoparticles and investigation of their hemostatic efficacy. J Mater Sci Eng C 68:768–773
Wang J, Hao S, Luo T et al (2017) Feather keratin hydrogel for wound repair: preparation, healing effect and biocompatibility evaluation. Colloids Surf B 149:341–350
Wawro D, Stęplewski W, Wrześniewska-Tosik K (2009) Preparation of keratin-modified chitosan fibres. Fibres Text East Eur 17:37–42
Wormell RL, Happey F (1949) Regenerated keratin fibres. Nature 163:18
Wortmann G, Zwiener G, Sweredjiuk R et al (1999) Sorption of indoor air pollutants by sheep's wool: formaldehyde as an example. In: Proceedings of the international wool textile organization, Florence, 1999
Wrześniewska-Tosik K, Adamiec J (2007) Biocomposites with a content of keratin from chicken feathers. Fibres Tex East Eur 15:106–112
Wrześniewska-Tosik K, Wawro D, Ratajska M et al (2007a) Novel composites with feather keratin. Fibres Tex East Eur 15:157–162
Wrześniewska-Tosik K, Wawro D, Stęplewski W et al (2007b) Fibrous products with keratin content. Fibres Text East Eur 15:30–35
Xie H, Li S, Zhang S (2005) Ionic liquids as novel solvents for the dissolution and blending of wool keratin fibres. Green Chem 7:606–608
Xu H, Cai S, Xu L et al (2014a) Water-stable three-dimensional ultrafine fibrous scaffolds from keratin for cartilage tissue engineering. Langmuir 30:8461–8470
Xu H, Ma Z, Yang Y (2014b) Dissolution and regeneration of wool via controlled disintegration and disentanglement of highly crosslinked keratin. J Mater Sci 49:7513–7521
Xu H, Yang Y (2014) Controlled de-cross-linking and disentanglement of feather keratin for fiber preparation via a novel process. ACS Sustainable Chem Eng 2:1404–1410
Xu S, Sang L, Zhang Y et al (2013) Biological evaluation of human hair keratin scaffolds for skin wound repair and regeneration. J Mater Sci Eng C 33:648–655
Yanga G, Yao Y, Wang X (2018) Comparative study of kerateine and keratose based composite nanofibers for biomedical applications. J Mater Sci Eng C 83:1–8
Yang Q, Dou F, Liang B et al (2005) Studies of cross-linking reaction on chitosan fiber with glyoxal. Carbohydr Polym 59:205–210
Yang X, Zhang H, Yuan X et al (2009) Wool keratin: a novel building block for layer-by-layer self-assembly. J Colloid Interface Sci 336:756–760
Yamauchi K, Yamauchi A, Kusunoki T et al (1996) Preparation of stable aqueous solution of keratins, and physiochemical and biodegradational properties of films. J Biomed Mater Res, Part B 31:439–444
Yarin AL, Koombhongse S, Reneker DH (2001) Taylor cone and jetting from liquid droplets in electrospinning of nanofibers. J Appl Phys 90:4836–4846
Yao CH, Lee CY, Huang CH et al (2017) Novel bilayer wound dressing based on electrospun gelatin/keratin nanofibrous mats for skin wound repair. J Mater Sci Eng C 79:533–540
Yoon K, Kim K, Wang X et al (2006) High flux ultrafiltration membranes based on electrospun nanofibrous PAN scaffolds and chitosan coating. Polymer 47:2434–2441
Yuan J, Geng J, Xing Z et al (2012) Novel wound dressing based on nano fibrous PHBV–keratin mats. J Tissue Eng Regener Med 9:1027–1035
Yuan J, Shen J, Kang IK (2008) Fabrication of protein-doped PLA composite nanofibrous scaffolds for tissue engineering. Polym Int 57:1188–1193
Yuan J, Xing ZC, Park SW et al (2009) Fabrication of PHBV/keratin composite nanofibrous mats for biomedical applications. Macromol Res 17:850–855
Yue K, Liu Y, Byambaa B et al (2018) Visible light crosslinkable human hair keratin hydrogels. Bioeng Transl Med 3:37–48
Yun KM, Hogan CJ Jr, Matsubayashi Y et al (2007) Nanoparticle filtration by electrospun polymer fibers. Chem Eng Sci 62:4751–4759
Zhang H, Liu J (2013) Electrospun poly(lactic-coglycolic acid)/wool keratin fibrous composite scaffolds potential for bone tissue engineering applications. J Bioact Compat Polym 28:141–153

Zhang H, Wang J, Ma H et al (2016) Bilayered PLGA/wool keratin composite membranes support periodontal regeneration in beagle dogs. ACS Biomater Sci Eng 2:2162–2175
Zhang H, Yub Y, Cuia S (2011) Multilayer fluorescent thin films based on keratin-stabilized silver nanoparticles. Colloids Surf A 384:501–506
Zhang S (2002) Emerging biological materials through molecular self-assembly. Biotechnol Adv 20:321–339
Zhang Y, Zhu PC, Edgren D (2010) Crosslinking reaction of poly(vinyl alcohol) with glyoxal. J Polym Res 17:725–730
Zhao X, Lui YS, Chuen Choo CK et al (2015) Calcium phosphate coated keratin–PCL scaffolds for potential bone tissue regeneration. J Mater Sci Eng C 49:746–753
Zhuang Y, Wu X, Cao Z et al (2013) Preparation and characterization of sponge film made from feathers. J Mater Sci Eng C 33:4732–4738
Zhu H, Li R, Wu X et al (2017) Controllable fabrication and characterization of hydrophilic PCL/wool keratin nanonets by electronetting. Eur Polym J 86:154–161
Zhou LT, Yang G, Yang XX et al (2014) Preparation of regenerated keratin sponge from waste feathers by a simple method and its potential use for oil adsorption. Environ Sci Pollut Res 21:5730–5736
Zoccola M, Aluigi A, Vineis C et al (2008) Study on cast membranes and electrospun nanofibers made from keratin/fibroin blends. Biomacromolecules 9:2819–2825

Chapter 5
Degradation of Keratin Biomass by Different Microorganisms

I. A. Adelere and A. Lateef

Abstract Keratins are insoluble, fibrous, and structural proteins that are present in the epidermis and its appendages and these include feather, hair, wool, nail, hoof, and horns. Keratins adhere epidermal cells to one another and provide protection on the skin. They are structurally stabilized by their tightly packed peptide chains and the existence of several cross-linkages by disulphide bonds, hydrogen bonding, and hydrophobic interactions. Keratin-containing materials are generated abundantly as by-products of agro-industrial processing and constitute nuisance in the environment as a result of their recalcitrance to degradation by regular proteolytic enzymes like pepsin, trypsin, and papain. The traditional physical and chemical techniques for their treatment are expensive, energy consuming, can damage some essential amino acids, and non-environmentally benign. However, degradation by a variety of microorganisms had proven to be a viable alternative means of keratin treatment. A vast variety of bacteria, fungi, and actinomycetes have been recognized as keratin degraders. They degrade keratins mainly with their keratinases, which sometimes act synergistically with other enzymes like disulfide reductases and cysteine dioxygenase for effective degradation of keratins. The microbial keratinases hydrolyze keratins into soluble proteins, peptides, and amino acids. They are utility enzymes with very diverse biotechnological applications. Biodegradation of keratin-rich wastes by microorganisms is therefore an efficient, cheap, and eco-friendly method of waste management and production of products of high biotechnological value. The present review examines the trends in the role of microorganisms for the biotechnological treatment of keratin-rich wastes.

I. A. Adelere · A. Lateef (✉)
Laboratory of Industrial Microbiology and Nanobiotechnology, Microbiology Unit, Department of Pure and Applied Biology, Ladoke Akintola University of Technology, Ogbomoso PMB 4000, Nigeria
e-mail: alateef@lautech.edu.ng; agbaje72@yahoo.com

I. A. Adelere
e-mail: isiaka.ade@futminna.edu.ng

I. A. Adelere
Department of Microbiology, Federal University of Technology, Minna, Nigeria

S. Sharma and A. Kumar (eds.), *Keratin as a Protein Biopolymer*, Springer Series on Polymer and Composite Materials, https://doi.org/10.1007/978-3-030-02901-2_5

Keywords Keratin · Keratinase · Disulphide bond · Hydrogen bonding
Agro-industrial processing

1 Introduction

Keratins are largely released to environments as waste products of agro-industrial processing and are one of the major threats to the ecosystem owing to their resistance to common biodegradation processes. They are structurally characterized by cross linkage with numerous disulfide and hydrogen bonds, hydrophobic interactions, and tightly packed peptides chains that account for their recalcitrant property. Keratins are found as insoluble proteins in feathers, skin, hairs, nails, wool, hoof, horns, and other skin appendages. Feathers contain the largest quantity of keratin of all keratinous wastes and are generated substantially from poultry processing. Keratin biomass accumulates and constitutes nuisance in the environments. Apart from being a component of environmental pollution, keratins harbor essential amino acids that can be utilized for biotechnological processes. The conventional physical and chemical techniques for their treatment are expensive, energy consuming, non-eco-friendly, and also capable of destroying some essential amino acids in the proteins. Hence, there is the need to develop very efficient alternative means of recycling these wastes.

Keratins are insoluble and fibrous proteins that are the main constituent of the animal integumentary system. Their polypeptide chains are compactly packed into supercoiled α-helix (α-keratin) and β-sheet (β-keratin) structures (Parry and North 1998; Kreplak et al. 2004; Wang et al. 2016). Keratins are classified into hard and soft keratins on the basis of the amount of sulfur present. Hard keratins have high disulfide bonds which make them tougher than soft keratins with low disulfide bonds. Hard keratins are mostly found in appendages like feathers, hair, hooves, and nails, while soft keratins are present in the skin and callus. In spite of their resistance to hydrolysis by common proteases like pepsin, trypsin, and papain, quite a number of species of bacteria, fungi, and actinomycetes have been documented to efficiently degrade keratin-rich wastes (Sharma and Devi 2017).

Keratinolytic microorganisms use keratins as their sole source of carbon, nitrogen, and energy for growth after hydrolysis mainly by their keratinases. There are several species of keratinolytic bacteria but members of *Bacillus* are the most predominant (Lange et al. 2016). *Streptomyces* have also been found to be one of the prominent keratinolytic actinomycetes (Allure et al. 2015), while the common keratinolytic fungi are distributed in the genera *Aspergillus*, *Penicillium*, and *Fusarium* (Kannahi and Ancy 2012; Nwadiaro et al. 2015). It has been established that biodegradation of keratins through the activities of keratinolytic microorganisms is an efficient alternative means to manage keratin wastes and way of improving their biotechnological value (Jayalakshmi et al. 2012). These microorganisms have been isolated from diverse habitats such as keratin dumping sites, keratin wastes, hot spring, livestock processing plants, ponds among others (Lateef et al. 2010; Sekar et al. 2016; Kumar et al. 2016; Mortuza et al. 2017). Also, a few of these keratinolytic strains colonize skin where they cause superficial mycoses. Pathogenic keratinolytic fungi cause skin impairment by hydrolyzing keratin content in the skin (Sarkar et al. 2014).

Though the exact mechanism of biodegradation of keratin by microorganisms is not well understood, it had been proposed that keratinase plays a very prominent role in the process. Keratinases hydrolyze keratins to release soluble proteins, peptides, and amino acids that are trapped in them. These enzymes are predominantly extracellular and produced by microorganisms in the presence of keratin wastes (Adelere and Lateef 2016a). Keratinases are robust enzymes that display activities over a wide range of temperature and pH. They are most popular and important among proteases due to their versatility at degrading keratins and have found applications in numerous areas such as leather and detergent industries, development of cost-effective protein-rich products for animal feeds and fertilizers, production of personal care products, pharmaceuticals, and keratin waste management (Gupta and Ramnani 2006; Sharma and Gupta 2016).

Microbial degradation of keratinous wastes has been extensively researched in a decade past and is gaining unprecedented momentum in the recent times owing to huge advantages that are embedded in the process over the conventional means of treatment. Studies on the recycling of various keratinous wastes into biotechnologically important products using microorganisms isolated from diverse habitats have been documented. Parts of the studies put emphasis on the current and potential applications of keratinolytic enzymes in several areas of human activities. To complement these works, this chapter attempts to discuss sources and physiochemical properties of keratins, process of keratins degradation by different microorganisms, and some biotechnological importance of microbial degradation of keratinous wastes.

2 Structure and Chemical Composition of Keratin

Keratins are the main constituents of the horny layer of epidermis and its appendages (Wang et al. 2016). They are formed in epithelial cells to produce physical support and protection to the parts that are constantly exposed to considerable mechanical forces. They are tough, insoluble, and resistant to hydrolysis by common proteolytic enzymes which suggest the adaptation of vertebrates to live on land (Korniłłowicz-Kowalska and Bohacz 2011). Keratin is primarily made up of chains of amino acids which may vary in the number, sequence, charge, size, and polarity. It consists of at any rate a minimum of 17 amino acids (Holkar et al. 2017). The sequence and position of a particular amino acid within the keratin chain influence the properties, functions, and the entire three-dimensional structure of the keratin molecule (Wu et al. 2000). High cysteine content in keratin differentiates it from other structural proteins like elastin and collagen. The richness in cysteine contributes immensely to its resistance to degradation as it causes many cross-linking by disulfide bonds (Wang et al. 2016).

Keratins are filament-forming proteins that are classified as α- and β-keratin based on secondary structural confirmation (Voet and Voet 1995). The keratin fibrils in both groups are twisted in a parallel manner to form micro and macro fibrils that make great impact in their stability (Kreplak et al. 2004). Keratin filaments may consist of approximately 15 acidic classes and 15 neutral and alkaline keratin polypeptides (Korniłłowicz-Kowalska and Bohacz 2011). The α-keratin has 7 nm diameter intermediate filaments, while β-keratin has 3 nm diameter filaments. Both are rooted in an amorphous keratin matrix. The molecular unit of intermediate filaments is a coiled-coil heterodimer and that of β-keratin filament is a pleated sheet (Wang et al. 2016). β-keratin is easily degraded by keratinolytic enzymes than α-keratin because β-keratin has less disulfide bonds and exhibits the fibril and porosity structure (Gupta and Ramnani 2006). In addition, α-keratin is composed of specialized constituent proteins. Its intermediate filaments are composed of numerous types of low-sulfur proteins (Gillespie 1990), while the matrix consists of high-sulfur and high glycine—tyrosine proteins (Fraser et al. 1972). In the case of β-keratin, there are no different types of proteins (Fraser et al. 1972), as the filaments and matrix are incorporated into one single protein (Fraser and Parry 2011). The molecular mass of α-keratin ranged from 40 kDa to 68 kDa that is to a large extent bigger than that of β-keratin that is 10–22 kDa (Alibardi et al. 2006).

3 Keratin Degradation by Microorganisms

Keratinous wastes in the environment constitute a great threat to the ecosystem owing to their recalcitrant nature to biodegradation. The common methods for their treatment such as burning, landfilling, steam pressure cooking, and chemical hydrolysis consume enormous energy, non-eco-friendly, and also destroy some essential amino acids. In view of these problems, the development of biological/green method for the degradation of keratinous wastes has become imperative. The degradation of keratin by microorganisms has immense advantages as it reduces the environmental problem posed by these wastes in an eco-friendly manner; it is an economical and simple method that heralds the formation of biotechnologically important products like keratinolytic enzymes, peptides, and amino acids-rich keratin hydrolysates. Microbial degradation of keratin had been extensively studied in recent times as several species of bacteria, fungi, and actinomycetes that were isolated largely from soil and keratin-rich wastes have been identified as vast keratin degraders (Table 1).

3.1 Degradation by Bacteria

The keratinolytic bacteria are commonly Gram-positive, but a few strains of keratinolytic Gram-negative bacteria are also known. Keratinolytic bacteria that belonged to vast varieties of genera, viz., *Bacillus*, *Stenotrophomonas*, *Pseudomonas*,

Table 1 Diversity of some keratin degrading microorganisms

Keratinolytic microorganisms	Sources	Keratin wastes	Growth pH	Growth temperature (°C)	Agitation speed (rpm)	Fermentation period	References
Acinetobacter sp. R-1	Duck ranch soil	Wool	–	37	220	28 h	Zhang et al. (2016a)
B. licheniformis	Culture collection	Feather and animal hair	–	37	–	15 days	Poovendran et al. (2011)
Bacillus cytotoxicus and *B. licheniformis*	Soil	Feather	–	50 and 65	150	5 days	Cavello et al. (2018)
Aphanoascus fulvescens B21/4–5 and *Chrysosporium articulatum* B25/1	Soil	Feather	7.58	28.7	–	42 days	Bohacz (2017)
Lysinibacillus, Nocardiopsis, and *Micrococcus*	Coast	Feather	10	37	170	2 days	Osman et al. (2017)
Bacillus sp. 50-3	Animal feces	Feather	7.0	37	150	72 h	Yue et al. (2017)
Proteobacteria and Firmicutes	Bird nests	Feather	–	30	170	2 days	Saarela et al. (2017)
Chryseobacterium sp. kr6 and *Bacillus* sp. kr16	Decomposing feather	Feather	–	30 and 37	120	5 days	Maciel et al. (2017)
Fusarium sp. 1A	Soil	Horsehair	7.5	27	100	21 days	Călin et al. (2017)
Cochliobolus hawaiiensis AUMC 8606	Soil	Feather	9.0	30	–	21 days	Isaac and Abu-Tahon (2016)

(continued)

Table 1 (continued)

Keratinolytic microorganisms	Sources	Keratin wastes	Growth pH	Growth temperature (°C)	Agitation speed (rpm)	Fermentation period	References
Bacillus pumilus GRK	Soil	Feather	10	37	200	1 day	Reddy et al. (2017)
Chrysosporium tropicum and *Malbranchea* sp	Soil	Feather	–	28		20 days	Kumar et al. (2017)
Actinomycete strains	Soil	Feather	7.2	55	130	72 h	Vasileva-Tonkova et al. (2009)
Trichophyton mentagrophytes	Clinical sample	Horsehair strand	7	27	80	3 weeks	Călin et al. (2016)
Bacillus subtilis and *Bacillus licheniformis*	Soil and feather wastes	Feather	–	40	150	7 days	Veenayohini and Sangeetha (2016)
Bacillus cereus NYA101	Mushroom farm	Feather	8.5	22	150	96 h	Ahmadpour et al. (2017)
Bacillus subtilis	Soil	Feather	7.5	37	–	21 days	Sekar et al. (2016)
Bacillus sp. MBRL 575	Limestone habitat	Feather	–	30	150	2 days	Kshetri and Ningthoujam (2016)
Pseudomonas sp. LM19,	Soil	Feather	–	30–60	180	96 h	Mohamad et al. (2017)
Nocardiopsis sp. 28ROR	Soil	Feather	9	35	–	10–14 days	Omran (2017)
Aspergillus sp. DHE7	Soil	Feather and wool	6	28	120	4 days	El-Ghonemy and Ali (2017)
Bacillus safensis LAU13	Soil	Feather	6.5	30	100	6 days	Lateef et al. (2015a)
Bacillus sp. CL18	Soil	Feather	7	30	125	7 days	Rieger et al. (2017)

(continued)

Table 1 (continued)

Keratinolytic microorganisms	Sources	Keratin wastes	Growth pH	Growth temperature (°C)	Agitation speed (rpm)	Fermentation period	References
Bacillus sp. CL33A	Soil	Feather	7	30	125	11 days	de Oliveira et al. (2017)
Bacillus spp. and *Serratia marcescens*	Soil	Feather	–	37	–	7 days	Iruolaje et al. (2016)
Meiothermus taiwanensis WR-220	Hot spring	Feather	–	55 and 65	–	2 days	Wu et al. (2017)
Pseudomonas sp. P5	Poultry bedding	Feather	–	37	130	5 days	Stiborova et al. (2016)
Chryseobacterium sp. RBT	Soil	Feather		37	140	48 h	Gurav et al. (2016)
Aspergillus flavus S125	Soil	Feather	9	55	–	4 days	Mini et al. (2015)

Chryseobacterium, *Keratinibaculum*, *Paenibacillus*, *Meiothermus*, *Rhodococcus*, *Achromobacter*, *Exiguobacterium*, *Rummeliibacillus*, *Sporosarcina*, *Brevibacillus*, and *Aeromonas* (Lateef et al. 2010; Gareth et al. 2010; Kuo et al. 2012; Gurav and Jadhav 2013; Huang et al. 2013; Paul et al. 2014a; Saarela et al. 2017) have been isolated from soils, hot spring, birds' nests, and keratin wastes. Species of *Bacillus* are the predominant keratinolytic bacteria with very high efficiency to degrade keratins. Several authors have documented very remarkable keratinolytic activities of their strains, namely *B. licheniformis*, *B. pumilus*, *B. cereus*, and *B. subtilis* (Lin et al. 1999; Suh and Lee 2001; Kim et al. 2001; Werlang and Brandelli 2005; Lateef et al. 2010).

In our laboratory, we have recently isolated a novel strain of keratinolytic *Bacillus safensis* LAU 13 from feather dump site (Lateef et al. 2015a, b). The strain demonstrated very remarkable keratinolytic activity by completely degrading native chicken feather under ambient condition (Fig. 1), while Ahmadpour et al. (2017) isolated a strain of *B. cereus* from a mushroom farm for the degradation of feather wastes under ambient conditions. In a similar study, Veenayohini and Sangeetha (2016) isolated keratinolytic strains of *Bacillus subtilis* and *Bacillus licheniformis* from soil and feather wastes. The isolates effectively degraded feather keratin at a wide range of pH and temperature. The authors concluded that the characteristics demonstrated by these strains suggest their promising application to convert keratin wastes into high value products on a large scale.

A study was carried out to screen soil samples obtained from 15 different poultry feather dump sites in Tamil Nadu, India for the incidence of keratinolytic bacteria (Sekar et al. 2016). Forty-one keratinolytic bacterial strains were confirmed after primary and secondary screening using casein and feather meal agar, respectively. Five strains among the isolate efficiently degraded feather and were identified on the basis of biochemical and molecular characterization as *Bacillus licheniformis, B. megaterium, B. foramins, B. subtilis,* and *B. cereus.* The effectiveness of feather degradation was observed under microscope; *B. subtilis* showed highest degradation index by degrading feather keratin efficiently within 21 days. Similarly, Kumar et al. (2016) isolated 24 bacterial strains from a feather dump site at Kurukshetra, India using enrichment technique. The isolates were screened for proteolytic activity using skim milk agar plate method, out of which six isolates were positive for proteolytic activity by creating zone of clearance on the skim milk agar plate, while only two isolates out of the six proteolytic isolates displayed keratinolytic activity by degrading chicken feather in the feather meal broth. In another study, the isolation of a keratinolytic alkaliphilic *Bacillus* species from a novel habitat, limestone deposit site in Manipur, India was reported (Kshetri and Ningthoujam 2016). The strain was an efficient keratin degrader as it remarkably degraded chicken feather after 48 h of incubation at 30 °C. The strain also degraded other bird feathers apart from chicken feather as it degraded feathers of pigeon, broiler, kuroiler, duck, and turkey.

The isolation of keratin degrading strains of *Bacillus, Exiguobacterium, Deinococcus*, and *Micrococcus* from Patagonian Merino wool was recently reported (Iglesias et al. 2017). Out of the isolates, strains of *Bacillus* were predominant with the highest wool-keratinolytic activity being produced by *Bacillus* sp. G51. The

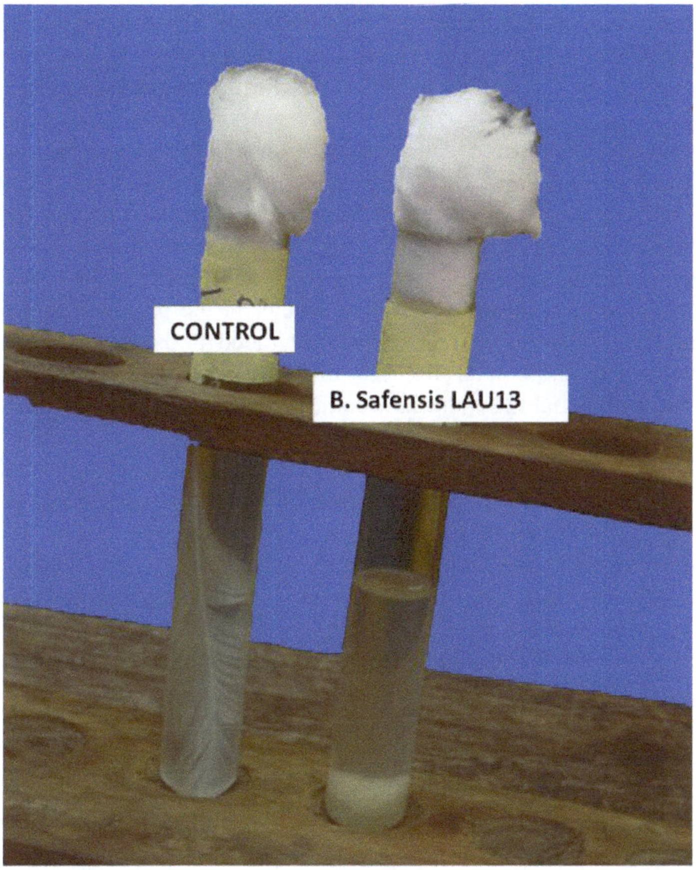

Fig. 1 Complete degradation of a native chicken feather by *B. safensis* LAU 13

strain secreted serine and metalloproteases in addition to other enzymes that could be involved in the reduction of keratin disulfide bonds. Hydrolysis of wool keratin by *Bacillus* sp. G51 caused a reduction in wool felting tendency without considerable weight loss. Iruolaje et al. (2016) had also reported the isolation of strains of *Bacillus licheniformis*, *Bacillus subtilis*, *Bacillus cereus,* and *Serratia marcescens* from feather dump sites. The rate of degrading powder and native feather samples by the isolates was studied and it was observed to be significantly higher than the rate of degradation of native feather by all the isolates. However, the isolates demonstrated synergistic effect on the degradation of native feather samples unlike that of powder samples. Similarly, Mohamad et al. (2017) isolated *Pseudomonas* sp. LM19 from a feather dump site to degrade chicken feathers. The rate of degradation was enhanced by 2.4-fold at optimized physical and nutritional conditions. The degradation led to production of free amino acids such as glycine, arginine, leucine, and serine. Also, the isolation of thermophilic strains of *Bacillus cytotoxicus* and *B. licheniformis* from geothermal areas in Patagonia, Argentina using feather enrichment medium was reported (Cavello et al. 2018). The isolates exhibited very great keratin degrading potential to achieve complete whole chicken feather degradation after 5 days of incubation.

Saieb et al. (2015) compared feather degradation potential of free and immobilized strain of *Bacillus licheniformis* isolated from chicken feather wastes. Cells immobilized by entrapment in alginate pellet demonstrated superior feather degradation than the free bacterial cells. The feather degradation potential of both the free and immobilized cells was improved by the inclusion of glucose in the basal medium. The immobilized cells degraded 53.0% of feather in the presence of glucose and 49.5% in its absence. However, the free cells degraded 44.0% of feather in the presence of glucose and 40.0% in its absence. Sobucki et al. (2017) investigated *Bacillus* sp. CL18 for its potential to synthesize keratinolytic protease through feather degradation. Feather broth was compounded such that feather was used as the sole source of carbon. It was observed that the degradation of feather was accompanied with the release of protease. Effect of peptone and NH_4Cl on protease production and feather degradation was further investigated using a central composite design. Supplementation with peptone and NH_4Cl increased protease production and hastened feather degradation. A sum of 842 spore-forming bacterial strains was isolated from 221 animal feces samples by Yue et al. (2017), out of which 196 strains were found capable of keratin degradation. Further investigation revealed *Bacillus* sp. 50-3, a contaminant of *Calotes versicolor* (an agamid lizard) feces as the highest keratinolytic strain. The bacterium completely degraded native feather under mild conditions after 36 h of cultivation with the maximum keratinase activity of 680 ± 25 U/mL.

Ferrareze et al. (2016) evaluated the keratin degrading potential of a probiotic strain of *Bacillus subtilis* using feather meal, white feather, melanized feather, and human hair as substrates, which was determined by measuring the amount of soluble protein released from each substrate. The organism efficiently degraded white feather, feather meal, and melanized feather, while human hair was weakly degraded as 98% of the substrate remained after cultivation. In an investigation carried out by de Oliveira et al. (2017) that reported the isolation of 15 bacterial strains from feather dump site, 7 strains out the 15 bacterial strains were positive for both proteolytic and keratinolytic activities. Among these, a strain identified as *Bacillus* sp. CL33A by 16S rRNA gene sequencing showed a high rate of degradation of feather in the range of 29–95% within 96–216 h of cultivation. The degradation was accompanied with the production of keratinase and increased soluble protein and alkaline pH of the medium. Moreover, *Chryseobacterium* sp. RBT obtained from a poultry waste disposable site showed remarkable sulfitolytic and keratinolytic activities (Gurav et al. 2016). The organism yielded maximum enzyme activity at 48 h (89.12 U/mL) with 83% of native black feather degradation. The high keratinolytic and sulfitolytic activities displayed by the organism facilitated the breakdown of protein and disulphide bonds in keratin accompanied with the release of melanin embedded in the feather matrix.

Maciel et al. (2017) investigated the feather-degrading ability of *Chryseobacterium* sp. kr6 and *Bacillus* sp. kr16 by cultivating them in mineral medium supplemented with different amount of chicken feather (10, 20, or 50 g/L). Efficient feather degradation by *Chryseobacterium* sp. kr6 and *Bacillus* sp. kr16 was achieved when cultivated in feather concentration of 50 g/L and 10 g/L, respectively. Biodegradation of feather by extremophilic bacterium, *Meiothermus taiwanensis* WR-220 isolated from Wu-rai Hot Spring in Northern Taiwan (Wu et al. 2017) showed that the strain demonstrated excellent feather degradation as it almost completely degraded native feather at 65 °C in two days. The feather degradation resulted in the production of novel heat-stable keratinase with optima activity obtained at 65 °C and pH of 10. The structural and phylogenetic analyses of the keratinase also led to the discovery of two novel keratinases from other extremophiles, *Thermus* and *Deinococcus* with remarkable keratin degrading ability. The work of Stiborova et al. (2016) compared hydrolysis of feather using whole cell microbial digestion, enzymatic, and chemical cleavage methods. Among all the isolates, a strain of *Pseudomonas* sp. was selected for microbial degradation, while its keratinase was chosen for enzymatic hydrolysis due to highest keratinolytic activity and efficiency to hydrolyze feather wastes demonstrated by the organism above other tested bacterial isolates. The three methods converted feather wastes into commercial value products including peptides and amino acids. Hydrolysates obtained from microbial digestion had up to 6.2 g/L of peptides and 301 mg/L of free amino acids. The enzymatic hydrolysis released hydrolysates containing 1191 mg/L of amino acids and 3.3 g/L of peptides, while chemical hydrolysis yielded 17.2 g/L of peptides but a significantly lesser quantity of amino acids. Although microbial transformation of feathers into soluble peptides and amino acids may be low, it is still very important with regard to its cost-effectiveness, simplicity, and eco-friendliness.

An attempt was made by Osman et al. (2017) to enhance feather degradation through the use of mixture of alkaliphilic strains isolated from Egypt. Four strains that belonged to genera *Lysinibacillus*, *Nocardiopsis,* and *Micrococcus* were selected for their remarkable feather-degrading abilities. The coculture setup was formulated using a two-level factorial design, while the central composite design was employed to optimize the conditions for the coculture. It was reported that 96% degradation of both white and black feathers were achieved in comparison with a maximum of 31% recorded in degradation by individual isolate. Similarly, there was a complete degradation of feather with very high keratinase production by *Bacillus pumilus* GRK cultivated in optimized conditions consisting 1% (w/w) feather substrate, 4% (v/v) inoculum size at pH 10, 200 rpm, and 37 °C. The hydrolysate was rich in essential amino acids, suggesting its potential to be used as feed supplement (Reddy et al. 2017). In a similar investigation as documented by Zaghloul et al. (2004), degradation of chicken feather by recombinant strain of *Bacillus subtilis* carrying keratinase gene was highly enhanced using optimization strategies including pretreatment of feather wastes with NaOH or two times autoclaving, and supplementation of culture medium with 0.1% yeast extract or 0.5% corn oil. The process was also optimized using 2% (w/v) feather waste, 5% inoculum size, with incubation at 45 °C for 3–4 days. Significant amounts of soluble proteins and amino acids released after the enzymatic hydrolysis indicated that feather wastes had been remarkably degraded. On the other hand, Fuke et al. (2017) recently reported tremendous keratin degradation and multiple proteolytic enzymes production by a strain of *Serratia marcescens* isolated from tannery wastewater treatment plant. The bacterium produced multiple proteolytic enzymes with the capability to hydrolyze both soluble (casein) and insoluble (feather and nails) proteins. This unique ability of the bacterium was justified through sequence and annotation of genes coding the proteolytic enzymes. The organism performed optimally at 30 °C and under alkaline condition (pH 8–10).

3.2 Degradation by Fungi

Keratinolytic fungi are among the ecologically important organisms as they contribute tremendously in the recycling of recalcitrant keratinous wastes. The extent of keratin degradation by microorganisms varies greatly as the process depends on factors such as culture conditions, sources of carbon, nitrogen, and energy, and the attribute of degrading organism. In fungi, morphological feature of filamentous fungi facilitates keratin degradation through the firm attachment of mycelia and penetration of keratin substrates by fungal hyphae. Keratin degradation by a wide range of fungi has been documented in quite a number of literature. Blyskal (2009) reported 299 species of keratinolytic fungi that are distributed among 100 genera. The most predominant genera were *Aspergillus, Penicillium, Chrysosporium, Fusarium, Microsporum, Trichophyton,* and *Acremonium*, while about 107 species were reported to be human pathogens. Other previously reported genera of keratinolytic fungi isolated from various sources like soil and poultry wastes include *Scopulariopsis, Myceliophthora, Candida, Cladosporium, Metarrhizium, Neurospora, Cunninghamella*, and *Westerdikella* (Vermelho et al. 2010; Eliades et al. 2010; Sharaf and Khalil 2011; Liang et al. 2011; More et al. 2013). Some members of these genera are grouped as dermatophytes and cannot be exploited for biotechnological applications (Gradisar et al. 2000).

In a study conducted by Saber et al. (2009), 82 fungal isolates obtained from poultry wastes were screened for keratinolytic activity, out of which only 27 isolates were found positive for keratinolytic activity. The most potent two keratinolytic isolates, *Alternaria tenuissima* K2, and *Aspergillus nidulans* K7 were selected for further studies. The incorporation of starch and maltose into the fermentation medium at pH 7.5, 35 °C and 7.5% inoculum enhanced both production of keratinase and degradation of feather by the two isolates. Also, among the keratinous wastes attempted for degradation, chicken, duck, and goose feathers were the most degraded by the two fungi. Their keratinases hydrolyzed 70% of chicken, duck, goose, and turkey feathers at pH 8.5, and at 40 °C after 24 h of incubation. Five soil-dwelling fungi were recently evaluated for keratinolytic activity by Bohacz (2017). The fermentation broth for the experiment was compounded such that native chicken feather was used as sole source of C, N, S, and energy, with incubation also held at 28 °C under static condition. Response surface methodology was used to optimize culture conditions for the feather degradation using Box–Behnken designs. Feather degradation and enzyme activities were determined periodically. The degradation of feather by the organisms led to production of keratinolytic enzyme, feather dry weight loss, and increase in ammonium ion concentration in the culture medium. The highest feather degradation and maximum keratinolytic activities occurred at pH 7.58, and temperature of 28.7 °C by isolate *Aphanoascus fulvescens* B21/4–5 identified by phenotypic traits and molecular characterization.

In a study conducted by Călin et al. (2017), keratin-degrading ability of keratinolytic fungal strains isolated from soil was compared using horsehair as substrate. The degree of degradation was determined by SEM, FTIR-ATR spectra, and TGA analyses. SEM revealed the interactions between microorganism and structural elements of hair strands where the mycelia was showing attaching to the hair surface and the hyphae was penetrating the cortex below cuticular scales. The FTIR spectra showed bands at 1035–1075 cm^{-1} assigned to sulfoxide bond, which was due to the breaking of S–S bond that resulted in the commencement of keratin biodegradation. The TGA analysis demonstrated the degree of thermal denaturation of the keratin substrate. Among the tested strains, *Fusarium* sp. 1A was identified to be the most active organism in the keratin degradation process as it displayed maximum denaturation of polypeptide chains.

Kumar et al. (2017) utilized *Chrysosporium tropicum* and *Malbranchea* sp. isolated from soil by hair baiting technique for degradation of feather wastes. The feather hydrolysates obtained after the degradation contained a significant quantity of proteins and amino acids. The hydrolysate was used as organic manure to improve soil nutritional quality which enhanced the germination of pea and rice seeds. In another study, Călin et al. (2016) evaluated the keratin-degrading ability of keratinolytic fungi strains obtained from geophilic and clinical samples, using horse hair strand as substrate. Keratin degradation was measured by determining the hair weight loss over 3 weeks of submerged fermentation. The morphological changes in hair sample were observed under light and electron microscopes. It was observed that a strain of *Trichophyton mentagrophytes* demonstrated highest keratin degradation among the considered isolates as it degraded 75% of hair strand under the 3 weeks of fermentation with serious morphological damage.

Hussein and Abdullah (2017) screened 100 samples collected from different residential houses and mosques in Duhok province, Iraq, during the year 2014–2015 for the presence of keratinolytic and opportunistic pathogenic fungi using hair baiting technique. The keratinolytic species detected among the 24 fungal species identified were *Gymnoascus ressii, Chrysosporium tropicum*, and *Arthroderma cuniculi*, while the opportunistic pathogens found were species of *Aspergillus, Arthrographis, Geomyces, Microascus, Scopulariopsis,* and *Neoscytalidium.* In addition, El-Ghonemy and Ali (2017) evaluated 15 keratinolytic fungi isolated from a poultry farm soil for keratinase production through feather hydrolysis. A novel strain of *Aspergillus* sp. displayed potent keratinolytic activity as it demonstrated maximum keratinolytic activity by producing highest yield of keratinase (199 ± 4.2 U/mL) after 4 days of incubation at pH 6.0, 120 rpm, and 30 °C using 2% of chicken feathers as substrate. The organism also degraded other keratinous wastes including goat hair, turkey feather, and sheep wool.

3.3 Degradation by Actinomycetes

Studies had shown that keratin degrading microbes are widespread in the microbial world, most importantly among bacteria, fungi, and actinomycetes. Some members of actinomycetes had been described as efficient keratin degraders due to their ability to hydrolyze a wide variety of keratin wastes such as feather, hair, and wool (Gupta and Ramnani 2006). The isolation of keratinolytic actinobacteria like *Actinoalloteichus* (Manivasagan et al. 2014) and *Actinomadura* (Habbeche et al. 2013) had been reported. Others are members of *Streptomyces*, viz: *S. fradiae*, and *S. gulbargensis* (Syed et al. 2009; Li et al. 2013). Saha et al. (2013) reported the isolation of 91 strains of actinobacteria from feather waste soil out of which a strain of *Nocardiopsis* exhibited significant keratinolytic activity. In a related study, Nayaka et al. (2013) reported the isolation of thermophilic *Streptomyces albus* from a hot soil zone of the Gulbarga and Bellary. The isolate was utilized for keratinase production through the hydrolysis of hair wastes. The quest for keratinolytic microorganisms led Chao et al. (2007) to screen over 200 strains of *Streptomyces*, *Aspergillus*, *Bacillus,* and *Vibrio* obtained from different sources for keratin degrading ability using native human foot skin as sole source of nitrogen. Of all the screened strains, a strain of *Streptomyces* sp. had the highest keratinolytic activity (110 KU/mL) with specific activity of 2870 KU/mg. The crude keratinase obtained from the strain was also found to possess wide substrate specificity as it degraded different substrates including keratin azure, human hair, cock feathers, and collagen.

Vasileva-Tonkova et al. (2009) successfully utilized a mixed culture of soil-dwelling thermoactinomycete strains for the degradation of feather wastes. It was observed that the feather degradation by the mixed culture was very rapid and efficient than single strains. Analysis of feather hydrolysates revealed the presence of essential amino acids, soluble proteins, enzymes, and other biotechnologically important products. Elhoul et al. (2016) reported the isolation of a novel thermostable keratinase-producing strain of *Actinomadura viridilutea* from Algerian fishing port. The isolate displayed substantial keratinase production from chicken feather hydrolysis after 96 h of incubation at 45 °C. Similarly, Ningthoujam et al. (2016) reported the isolation of keratinolytic strain of *Amycolatopsis* sp. from a limestone habitat. The strain demonstrated keratinolytic ability by degrading keratin azure and chicken feather substrates through its extracellular keratinase. Optimization by response surface methodology enhanced keratinase production by the strain as it produced maximum keratinase of 395 KU/mL in the presence of 0.5% (w/v) chicken feather, 0.13% (w/v) yeast extract, 1.1% (w/v) corn flour and 0.5% (w/v) soyabean meal, a 2.3-fold increase over the un-optimized conditions.

In a recent investigation, 22 keratinolytic actinobacteria isolated from diverse sources like feather wastes, agricultural and poultry farm soils were screened for the synthesis of bioactive secondary metabolite from feather hydrolysis (Omran 2017). The results of the evaluation of antimicrobial and cytotoxic activities of their secondary metabolites indicated *Nocardiopsis* sp. 28ROR as the best strain. It demonstrated the highest antimicrobial activities against some selected human pathogens and also exhibited considerable anticancer effect against breast cancer cell line MCF-7. The strain also displayed very potent keratinolytic activities by producing two varieties of keratinases with activities spread over a wide range of pH and hydrolyzed completely feather meal to amino acids and peptides after 14 days. It has also been demonstrated that microparticles from feather biomass displayed potent antioxidant and anticancer activities (Sharma et al. 2017). Verma et al. (2016) reported the isolation of a novel keratinolytic strain of *Thermoactinomyces* sp. from a rice mill waste dump site. The isolation was done using keratin supplemented medium, while the characterization was based on 16S rDNA and phylogenetic analyses. The strain completed degradation of native feather after 4 days at 60 °C; the degradation was accompanied with the production of keratinase enzyme, soluble proteins, and indole acetic acid. The keratinase was active over wide range of pH 6.0–12.0 and 60–80 °C with very high stability at pH 6.0–10.0 and 70 °C after 2 h of incubation. The enzyme showed high level of tolerance to organic solvents, while the presence of some heavy metals boosted its catalytic effect. Also produced alongside the keratinase during the feather hydrolysis by the strain was indole acetic acid, a very potent plant growth-promoting metabolite.

4 Mechanism of Keratin Degradation by Microorganisms

The exact mechanism of keratin degradation by microorganisms is not wholly understood, but researchers had proposed that two steps are involved in keratinolysis: sulfitolysis, i.e., the reduction of disulfide bonds and proteolysis which is hydrolysis of peptide bonds (Gupta and Ramnani 2006). It is assumed that sulfitolysis requires the presence of disulfide reductases (Ramnani et al. 2005) which act in collaboration with keratinases to ensure complete degradation of keratin. Some authors agreed that keratinases act synergistically with other microbial enzymes like disulfide reductases and cysteine dioxygenase for the effective degradation of keratinous wastes (Yamamura et al. 2002; Lange et al. 2016). Disulfide reductase and cysteine dioxygenase initiate denaturation of keratin structure by breaking the cross-linkages formed by disulfide bonds to make the peptide bonds accessible for hydrolysis by keratinase enzyme (Yamamura et al. 2002). Ramnani et al. (2005) reported that both keratinase and disulfilde reductase produced by *Bacillus licheniformis* RG1 acted synergistically to degrade feather wastes.

In a related study, Rahayu et al. (2012) compared the keratinolytic activities of each of purified keratinase and purified disulfide reductase produced by *Bacillus* sp. MTS with their combination. It was observed that keratinolytic activities of the combined enzymes were higher than each of the sole enzymes. Sometimes, keratinase and disulphide reductase ability are combined in a single protein, like in the case of *Bacillus cereus* Wu2 strain (Lo et al. 2012). However, several studies agreed that microbial degradation of keratin is mainly by their keratinases. Efficient degradation of keratin by microorganisms not only relies on enzymes, but on fermentation conditions like pH, temperature, rate of agitation, and sources of carbon, energy, and nitrogen. Keratin degradation is often efficient when keratin is used as the sole source of carbon, nitrogen and energy. In most cases, keratin serves as the inducer for keratinase production which is subsequently accompanied by its degradation (Gupta and Ramnani 2006). Keratinases sometimes attack peptide bonds of keratin at numerous sites, as they are not site specific in action, however, at times they show preference for cleaving bond involving aromatic and nonpolar amino acids. In addition, the synthesis of exopolysaccharides in some microorganisms during feather degradation plays a significant role in microbial adhesion to particles of feather thereby provoking the degradation of feather (Vasileva-Tonkova et al. 2009).

Studies have also shown that keratin biodegradation by some microorganisms like fungi involves enzymatic lysis combined with mechanical force (Kunert 2000). Fungal morphology assists in keratin degradation; they attach their mycelia firmly to keratin, perforate it with their hyphae as growth progresses. In-depth investigation to determine the mechanism of fungal keratin degradation using nonpathogenic fungus, *Onygena corvina* as a model organism was reported (Lange et al. 2014). The organism was selected owing to its physiology and preference to grow exclusively on keratinous materials. In these studies, molecular analyses including genomics and secretomics, and keratinolytic activity of different portions of the fractionated broth culture of *O. corvina* were considered. The outcome of these integrated studies implies that a combination of three fungal keratinases including an endoprotease (S8), exoprotease (M28), and an oligopeptidase/metalloprotease (M3) might have acted in synergistic manner to degrade the keratinous wastes. However, for bacterial keratinases, the M28 may be replaced by a bacterial exopeptidase with a similar function. Furthermore, Lange et al. (2016) proposed that together with the mixture of the three proteases, four additional components can still act in a synergistic manner to boost the keratin degradation process: they are *O. corvina* AA11/LPMOs (group of auxiliary proteins), disulfide reductase, cyteine dioxygenase, and sulfite. It can be concluded that efficient keratin degradation could be achieved if crude or partially purified keratinase enzyme is applied as several studies involving utilization of pure keratinase often result in weakly degraded keratin.

5 Keratinases

5.1 Overview on Production of Keratinase

Keratinases are very diverse in properties due to variations in the substrate and the producer organisms (Adelere and Lateef 2016a). They are serine or metallo proteases that can display activities over a broad range of pH and temperature with optimal activities at neutral to alkaline and 40–70 °C. Keratinases have wide substrate specificity, with the capability to degrade substrates like keratins, hemoglobin, fibrin, gelatin, and casein (Jeong et al. 2010a; Tiwary and Gupta 2010; Vermelho et al. 2010).

They are produced extracellularly by the majority of microorganisms when growing on keratinous substrates (Nam et al. 2002); only few microorganisms produced intracellular keratinases (Onifade et al. 1998). Keratinases are mostly inducible enzymes; nevertheless, a small number of constitutive keratinases are known (Gassesse et al. 2003; Manczinger et al. 2003). Mostly in the constitutive case, the nature of the enzyme is on the basis of caseinolytic activity rather than keratinolytic activity; therefore keratinolytic activity is generally described as been inducible (Gupta and Ramnani 2006). In most cases, keratin serves as the inducer, however, non-keratin substrates such as soy meal can also induce keratinase production (Gradisar et al. 2000). Simple sugars such as glucose can interfere with keratinase production due to catabolite repression (Daroit et al. 2011), whereas complex sugars like starch enhanced the enzyme production (Syed et al. 2009). Since keratin induces the production of keratinases, therefore, all fermentations that lead to the production of keratinases are accompanied by subsequent degradation of keratin substrates (Gupta and Ramnani 2006).

Several authors have documented the production of keratinases by different microorganisms under diverse fermentation conditions using varieties of substrates such as feathers, hair, horn, and sugarcane bagasse as inducers (Eliades et al. 2010; Awad et al. 2011; Lateef et al. 2015a). Sahoo et al. (2012) reported the optimal production of keratinase by *Bacillus weihenstephanensis* at the initial pH of 7.0, 2% (v/v) inoculum size, 40 °C and 120 rpm. Similarly, *Bacillus subtilis* SLC isolated from agro-industrial waste of a poultry farm in Brazil produced maximum keratinolytic activity of 280 U/mL in a feather medium with supplementation of 0.1% yeast extract (Cedrola et al. 2012). The production of extracellular caseinolytic and keratinolytic enzymes in basal media containing 10% feather at 30 °C and 160 rpm was also reported (Agrahari and Wadhwa 2012). Hmidet et al. (2010) gave the first account of the coproduction of α-amylase and keratinolytic protease by *B. licheniformis* NH1 using chicken feathers as inducer, while the use of human hair as inducer for keratinase production by *Bacillus subtilis* AMR was also the first of its kind (Mazotto et al. 2010). Saieb et al. (2015) successfully immobilized cells of *Bacillus licheniformis* (St. 24) on alginate beads for keratinase production from feather hydrolysis. The immobilized cells produced higher keratinase compared with free bacterial cells.

Furthermore, Vasileva-Tonkova et al. (2009) gave the first account of exploitation of mixed culture of different strains of thermoactinomycete to produce keratinolytic enzymes. However, *Aspergillus fumigatus* TKF 1 isolated from poultry waste soil produced keratinolytic enzyme in a basal medium with 0.5% supplementation of raw chicken feather at 30 °C and pH 6.0 (Paul et al. 2014b), while citrate and soybean meal were used as carbon and nitrogen supplements in feather meal medium to enhance keratinase production by alkaliphilic strain of *Bacillus* sp. (Kshetri and Ningthoujam 2016). It is difficult to compare the production condition for keratinase due to variety of organisms and the methods of cultivation. Recently, Chitturi and Lakshmi (2016) successfully utilized agricultural wastes including rice bran, black gram husk, green gram husk, and wheat bran each as a supplement in a feather-based medium to substantially increase keratinase production by strains of *Bacillus* sp. in semisolid state fermentation. However, the incorporation of 0.5% sucrose as a supplementary carbon source into the feather-based medium increased keratinase production by *Aspergillus* sp. DHE7 from 199 ± 4.2 to 226 ± 5.4 U/mL (El-Ghonemy and Ali 2017). In another study by Fellahi et al. (2016), a strain of *Bacillus pumilus* produced two different keratinases that degraded both α- and β-keratins. Both enzymes were of the same molecular mass 38 kDa. The two keratinase genes consisted of 1153 base pairs each with a gene similarity of 67%. They both belonged to subtilisin serine protease family according to their amino acids analysis.

5.2 Optimization of Production of Keratinase

The synthesis of keratinase during microbial degradation of keratin can be enhanced by different optimization strategies including Plackett–Burman design, Box–Behnken design, central composite design, one-variable-at-a-time (OVAT), response surface methodology (RSM), and statistical design (Kamarudin et al. 2017). Liang et al. (2010) successfully used Plackett–Burman factorial and Box–Behnken designs to increase keratinase production by a strain of thermophilic *Myceliophthora thermophila* by 6.4-fold, while Harde et al. (2011) improved keratinase production by *Bacillus subtilis* NCIM 2724 using OVAT, an L8 orthogonal array design, and RSM. According to Rai and Mukherjee (2011), the production of alkaline β-keratinase by a strain of *Brevibacillus* sp. after 48 h of incubation was boosted by response surface strategy. Furthermore, a feather-degrading strain of *Paenibacillus woosongensis* was reported to have produced maximum amount of keratinase under optimized conditions using OVAT protocol (Paul et al. 2013). In addition, the combination of three different strategies namely temperature-shift procedure, two-stage DO control, and fed-batch process in a fermenter induced a 62.2% improvement of keratinase yield from wool degradation by a strain of *Stenotrophomonas maltophilia* (Fang et al. 2013).

In another study, the production of thermostable keratinase by *Bacillus pumilus* KS 12 was improved by seven folds using statistical methods (Rajput and Gupta 2013). Moreover, Plackett–Burman method and RSM were used by Rajesh et al. (2016) to optimize keratinase production by a strain of *Bacillus cereus* isolated from a feather dump site from enzyme activity of 2.2–3.5 U/mL, whereas Ahmadpour et al. (2017) succeeded in using only RSM to optimized keratinase production by another strain of *B. cereus*. Ningthoujam et al. (2016) enhanced the production of keratinase by a feather-degrading strain of *Amycolatopsis* sp. isolated from a limestone deposit site using three different optimization strategies namely OVAT, Plackett–Burman method and RSM, respectively. According to Bohacz (2017), the keratinolytic activity of a strain of *Aphanoascus fulvescens* on feather substrate was significantly optimized using statistical tools and RSM.

5.3 Effect of Mutation on Keratinase-Producing Microorganisms

The production and properties of keratinases can be improved by the manipulation of the keratinolytic organisms genome using physical and chemical mutagens. Zhang (2012) reported that a strain of *Bacillus subtilis* mutated using UV irradiation and sodium nitrite solution demonstrated higher keratinolytic activity than the wild strain. Similarly, in our laboratory, the keratinolytic activity of a strain of *B. safensis* LAU 13 obtained from a feather dump site was improved by more than 100% after exposure to UV irradiation. The mutant keratinase displayed higher catalytic activities and improved thermo- and pH stabilities over the wild keratinase (Lateef et al. 2015a). The biocatalytic efficiency of natural enzymes can be improved using protein engineering technologies such as directed evolution, site-directed mutagenesis, truncation, and terminal fusion (Yang et al. 2014). Liu et al. (2013a) used the combination of computer algorithm with site-directed mutagenesis to improve the catalytic efficiency of a microbial keratinase. Similarly, single amino acid substitutions at positions 1 and 2 of Spf2 keratinase of *S. fradiae* var. k11 led to improved enzyme yield and specific activity according to Li et al. (2013). One of the mutants, L(-1)F demonstrated specific activity of 48935 U/mg, representing 9 times that of the wild-type Sfp2.

In a study carried out by Fang et al. (2016), the keratinolytic efficiency and heat stability of *Stenotrophomonas* sp. keratinase, KerSMD was appreciably enhanced by swapping its N/C-terminal domains with those from a keratinase-like protease, KerSMF while the swapping involving only the N-terminal domains increased the caseinolytic activity of the mutant enzyme by more than twofold. Recently, Fang et al. (2017a) further improved the keratinolytic activity, thermostability, and substrate specificity of KerSMD using site-directed mutagenesis and C-terminus fusion approaches.

5.4 Effect of DNA Recombinant Technology on Keratinase-Producing Microorganisms

The yield and characteristics of microbial keratinases can be improved using DNA recombinant technology. Isolation, alteration, cloning, and expression of keratinase genes are essential to boost the enzyme production and properties so as to increase their market value. The technology is equally important to enhance the keratin degrading ability of keratinolytic microorganisms to improve the valorization of keratin wastes. Studies have revealed many achievements in the cloning and expression of keratinase genes in diverse microorganisms. Radha and Gunasekaran (2009) reported that there was a remarkable increase in the keratinase production by the recombinant *Bacillus megaterium* and *Pichia pastoris* strains cloned with the keratinase gene obtained from a strain of *B. licheniformis*. A recombinant *Escherichia coli* HB101 carrying keratinase coding gene from *Pseudomonas aeruginosa* KS-1 was reported to degrade 2% chicken feather in LB-Amp medium in 24 h (Sharma and Gupta 2012). Furthermore, the transformation of two different keratinolytic strains of *B. cereus* with alkaline protease gene using a plasmid vector p5.2 increased their keratinolytic activities in feather medium by more than threefold (Ouled-Haddar et al. 2010). Fang et al. (2014) isolated two keratinolytic genes (KerSMD and KerSMF) from a strain of *Stenotrophomas maltophilia*, the two genes code for serine type of protease and were effectively expressed in *E. coli* cells. The recombinant KerSMD (48 kDa) was discovered to be the best as it displayed higher activity in hydrolyzing feather, thermostability, and substrate specificity. The enzyme also had a better tolerance for surfactants; SDS and triton X-100.

Most recently, Dong et al. (2017), reported a keratinase gene (KerP) obtained from *Bacillus polyfermenticus* B4 isolated from a poultry farm which was successfully cloned and overexpressed in *B. subtilis* PT5 using plasmid pDMT. The recombinant *B. subtilis* strain was better in activity as it produced higher yield of keratinase and completed feather degradation faster than *B. polyfermenticus* B4. The maximum keratinolytic activity of the recombinant KerP was achieved at pH 9.0 and 60 °C. Similar study by Gupta et al. (2017) reported the transformation of an *E. coli* strain by pSPORT expression vector carrying *Bacillus subtilis* RSE163 keratinase gene. The recombinant *E. coli* cells produced keratinase enzyme with significantly higher catalytic activity than the parental *Bacillus subtilis* RSE163 keratinase. The enzyme also displayed promising relevance in the drug delivery system.

5.5 Physicochemical Properties of Keratinases

Keratinases are extremely varied in properties on the basis of the nature of substrates and the producer organisms. They are robust enzymes that have displayed activities over a broad range of substrates, pH and temperature. Keratinases majorly have optimal pH in neutral to alkaline range (Bouacem et al. 2016; Elhoul et al. 2016; Su et al. 2017). However, a purified fungal keratinase obtained from *Cunninghamella echinulata* has two optimal pH of 4.5 and 10.0 (More et al. 2013). The optimum temperature of keratinolytic activities ranged from 30 to 85 °C (Ghasemi et al. 2012; Bose et al. 2014; Bouacem et al. 2016; Elhoul et al. 2016; Su et al. 2017). Furthermore, Nam et al. (2002) reported the optimum keratinolytic activity of keratinase from thermophilic *Fervidobacterium islandicum* AW-1 at 100 °C.

Keratinases are commonly stable and active over extensive range of pH and temperature. They are stable over the pH range of 4–13 (Radha and Gunasekaran 2009; Habbeche et al. 2013; Manivasagan et al. 2014; Sanghvi et al. 2016). Thermostability of keratinases have been reported in the range of 40–80 °C (Radha and Gunasekaran 2009; Kuo et al. 2012; Saha et al. 2013; Sanghvi et al. 2016). Most recently, Abdel-Naby et al. (2017) reported a purified *Bacillus pumilus* FH9 keratinase (50 kDa) with the appreciable thermostable property. Likewise, Wu et al. (2017) discovered novel thermostable keratinases from a keratinolytic and thermophilic bacterium, *Meiothermus taiwanensis* WR-220. The enzymes were active over a broad range of pH and temperature, and also retained its 100% activity after it was incubated at 70 °C for 1 h. Moreover, immobilization process has been reported to enhance the enzymatic properties and activities of keratinases (Rajput and Gupta 2013). A study was carried out by Abdel-Naby et al. (2016), whereby keratinase obtained from *Bacillus pumilus* FH9 was conjugated with activated pectin. The conjugated keratinase demonstrated better catalytic activity and improved kinetic and thermodynamic properties than the native keratinase. It was also more tolerant to some metal ions than the native keratinase.

Keratinases have been established to show the capability of hydrolyzing different types of soluble and insoluble proteinaceous substrates. The soluble protein substrates are bovine serum albumin, casein, hemoglobin, and gelatin, while the insoluble ones are feather, nail, hair, hoof, horn, wool, silk, collagen, elastin, stratum corneum, and azokeratin. The purified alkaline keratinase from *Scopulariopsis brevicaulis* was reported to have higher activity on chicken feathers, human nails and human hairs (Sharaf and Khalil 2011). Abdel-Naby et al. (2017) most recently reported a purified keratinase of *Bacillus pumilus* FH9 that demonstrated a broad range of substrate specificity whereby it hydrolyzed both soluble and insoluble proteins that included feather, horn, wool, bovine serum albumin, casein, and collagen.

Keratinases are stimulated, activated, and stabilized when exposed to various solvents, detergents, metallic ions, and reducing agents. They are largely serine type of proteases whereas very few are metalloproteases as determined by specific substrates and inhibitors. The keratinase produced by *B. subtilis* AMR of molecular weight 60 kDa is a serine type as it was fully inhibited by phenylmethanesulfonyl fluoride (PMSF) (Mazotto et al. 2010). Similarly, a serine-type dimeric 58 kDa keratinase obtained from *B. licheniformis* ER-15 was inhibited by thiol and mercaptoethanol (Tiwary and Gupta 2010). However, the keratinolytic activity of the 2.68-fold purified keratinase produced by a strain of *Acinetobacter* sp. was absolutely inhibited by EDTA indicating that the enzyme was a metallokeratinase (Zhang et al. 2016a). In another study, Prakash et al. (2010a) reported that the partially purified keratinase of *Bacillus halodurans* PPKS-2 was stable on exposure to NaCl, SDS, EDTA, H_2O_2, and detergent.

Furthermore, the two alkaline keratinase-I and II produced by *Bacillus halodurans* PPK-2 were purified by ammonium sulfate, DEAE-Sephadex, and Sephadex G-200 column chromatography; the purified enzymes were reportedly active and stable to surfactants, bleaching agents, commercial detergents, and SDS (Prakash et al. 2010b). Also, the purified keratinase of *Streptomyces albus* was characterized as a serine protease with 29–35 kDa (Nayaka et al. 2013), its catalytic activities were stoutly inhibited by PMSF and $HgCl_2$, while moderate inhibition was caused by pCMB, KCN, and 8-hydroxyquinoline. The enzyme displayed stability towards 2-mercaptoethanol and significantly activated by EDTA. Jeong et al. (2010a) cultivated *Bacillus* sp. SH-517 for the production of keratinolytic enzyme. The enzyme activity was significantly inhibited by Hg^{2+}, Ag^{2+}, and EDTA but slightly stimulated by K^+ and Na^+. In a similar study, a reducing agent 2-mercaptoethanol strongly activated the activity of keratinase from *Bacillus* sp. MKR 5. The enzyme activity was also moderately activated by Co^{2+}, Mg^{2+}, Triton X-100, Tween-80, and EDTA, while SDS inhibited the activity of the enzyme (Ghasemi et al. 2012).

The purified keratinase of *Meiothermus* sp. 140 was also discovered to be inhibited by PMSF, while the enzyme exhibited good stability in the presence of isopropanol, ethanol, dimethyl sulfoxide, and acetonitrile (Kuo et al. 2012). Similarly, *Cunninghamella echinulata* produced keratinolytic enzyme purified by acetone precipitation and lectin agarose affinity column. Its activity was activated by Mn^{2+} and inhibited by PMSF and by group specific reagents like TLCK, IAA, NEM, and NAI which indicate that the enzyme is a serine type (More et al. 2013). The catalytic activity of serine keratinase produced by *Stenotrophomonas maltophilia* K279a was highly increased on exposure to Ca^{2+}, while the strong inhibition was caused by $Cu^{2+}$and Hg^{2+}. The enzyme was tolerant to SDS and Triton X-100 but slightly inhibited by reducing agents β-mercaptoethanol and dithiothreitol (Shah and Vaidya 2017). The analysis of keratinase of 66 kDa produced by *Actinoalloteichus* sp. MA-32 was carried out by Manivasagan et al. (2014). The enzyme activity was inhibited by Hg^{2+}, Pb^{2+}, Tween-80, 1,10-phenanthroline, and EDTA, while Fe^{2+}, Mg^{2+}, Cu^{2+}, Ca^{2+}, Ni^{2+}, Mn^{2+}, SDS, ethoxylated octyphenol, DMSO, sodium sulfite, and *β*-mercaptoethanol enhanced the enzyme activity. The enzyme was also strongly stable to most of the commercial laundry detergents.

Paul et al. (2014c) characterized the crude keratinase of *Paenibacillus woosongensis* TKB2 as serine protease as it was strongly inhibited by PMSF, aprotinin, and leupeptin. The crude enzyme had good stability to EDTA and commercial detergents. Similarly, the extracellular keratinase secreted by *Caldicoprobacter algeriensis* was purified by ammonium sulfate precipitation, dialysis, and heat treatment (2 h at 50 °C) followed by UNO Q-6 FPLC anion exchange chromatography. The purified keratinase was analyzed by matrix-assisted laser desorption ionization time of flight mass spectrometry (MALDI TOF/MS) which showed that the enzyme was a monomer with a molecular weight of 33.246 KDa. The enzyme was suggested to be serine keratinase as it was totally inhibited by PMSF and diiodopropyl fluorophosphates (DFP) (Bouacem et al. 2016). The keratinase encoding gene (*ker* gene) of *B. licheniformis* BBE11-1 was successfully cloned and expressed in *B. subtilis* WB600. The recombinant keratinase showed stability against bleaching agent (H_2O_2) and activated by Mg^{2+} and Co^{2+} (Liu et al. 2013b). *Bacillus halodurans* JB 99 produced thermostable keratinolytic protease. Its N-terminal amino acid sequence showed homology with other thermostable alkaline proteases. It is a serine protease because it was inhibited by PMSF (Shrinivas and Naik 2011).

Keratinase Ahm 1 and Ahm 2 were purified from the culture supernatant of *Penicillium* spp. Morsy1 by precipitation using ammonium sulfate, DEAE-Sepharose, and gel filtration chromatography. Their activities were moderately inhibited by PMSF and iodoacetamide but strongly inhibited by EDTA and EGTA which suggest that they are both metalloproteases (El-Gendy 2010). Recently, Su et al. (2017) reported that the keratinolytic activity of a recombinant keratinase purified to homogeneity by nickel affinity chromatography was strongly inhibited by PMSF, while the enzyme displayed appreciable tolerance to surfactants such as Tween 40 and Triton X-114. The 13.18-fold purified *Brevibacillus parabrevis* CGMCC 10798 keratinase with an estimated molecular weight of 28 kDa was classified as serine-metallo protease on the basis of inhibition of its catalytic activity by both EDTA and PMSF. The enzyme activity was slightly enhanced by 5 mM Tween 40 and Triton X-100 (Zhang et al. 2016b). Similarly, Tork et al. (2016) characterized another serine metallokeratinase produced by *B. pumilus* NRC21, the catalytic efficiency of the 12-fold purified keratinase was stimulated by Na^+, K^+, and Mg^{2+}. The enzyme was classified as serine metallokeratinase because of its catalytic inhibition by serine and metalloproteinase inhibitors.

6 Applications of Keratinases

Keratin degradation by microorganisms is a safe and efficient alternative means of keratin wastes management and a process that produces products of high biotechnological value. Keratinolytic microorganisms convert keratin wastes into soluble proteins, peptides, and amino acids. The process is exclusively accompanied with the release of keratinases. Recently, research interest is rapidly growing on keratinases owing to their numerous present and promising applications in waste treatment, agro-industrial, pharmaceutical, and biomedical fields. Others include nanobiotechnology, leather, detergent, and cosmetic industries (Adelere and Lateef 2016a). Some of these biotechnological applications are briefly discussed.

6.1 Keratin Waste Management

A significant quantity of keratin in the form of feathers, nails, hairs, hooves, and horns are mostly produced as waste products of agro-industrial processing. These wastes have constituted a hindrance in the management of solid waste because of their recalcitrance to biodegradation. The conventional methods of treating feather wastes such as burning, landfilling, alkali hydrolysis, and steam pressure cooking are not eco-friendly, labor intensive, expensive, energy consuming and also capable of destroying certain essential amino acids. Attention has recently been focused on keratinases/keratinolytic microorganisms because of their ability to degrade keratinous materials like feather, hair, nail, and others and therefore become a part of solid waste management strategy as recycling of these wastes is difficult.

The hydrolysis of keratin wastes by microorganisms is a biotechnological alternative for their recycling and valorization (Călin et al. 2017). There exist several microbial strains that could be useful in the feather waste management due to their very remarkable feather-degrading ability (Lateef et al. 2010). Gareth et al. (2010) have used bioaugmentation strategies for the treatment of poultry wastes, while Vasileva-Tonkova et al. (2009) reported improved feather degradation by mixed culture of thermoactinomycete strains. Lateef et al. (2015a) utilized a novel *B. safensis* LAU 13 for effective degradation of chicken feather wastes (Fig. 1). Queiroga et al. (2012) isolated five wool-degrading bacterial strains from the wool of Portuguese Merino sheep. They hydrolyzed feathers, wool, human hair, and nails very remarkably. Relatedly, Fang et al. (2013) reported the degradation of wool waste by a keratinase-producing strain of *Stenotrophomonas maltophilia.* More interestingly, in a study carried out by Omran (2017), feather wastes were converted into anticancer and antimicrobial products after fermentation by a strain of *Nocardiopsis* sp.

The keratinolytic enzyme of *Bacillus licheniformis* strain has been reported to significantly degrade melanized feather without using reducing agents (Okoroma et al. 2012). Moreover, the protein-rich wastes such as keratin, elastin, collagen, and prion proteins generated in meat processing industry were efficiently degraded by the keratinolytic enzyme E77 (Zhao et al. 2012). The mixture of keratinase with disulfide reductase of *Bacillus* sp. MTS demonstrated synergy to effectively degrade feather and wool wastes (Rahayu et al. 2012). Recently, Łaba et al. (2017) successfully applied crude keratinase of *Bacillus cereus* PCM 2849 to degrade pretreated pig bristle waste to release free amino acids. Keratinases are therefore could be an indispensable tool in the formulation of novel biomaterials for the treatment of agro-industrial wastes in an eco-friendly manner.

6.2 Animal Feed Production

The ability of keratinases to hydrolyze keratins to produce soluble proteins and amino acids is of immense benefits in the development of high grade animal feeds. In a related study, Tiwary and Gupta (2012) prepared feather meal by soaking native chicken feather in water and allowed to boil for few minutes before adding the dimeric bacterial keratinase. The feather meal produced was rich in nitrogen and amino acids and showed very remarkable in vitro digestibility. Similarly, Vasileva-Tonkova et al. (2009) reported that the hydrolysate obtained after the fermentation of feather wastes by mixed microbial culture was rich in soluble proteins and essential amino acids including lysine, threonine, leucine, isoleucine, and valine. The authors therefore suggested that the hydrolysate could be used in the formulation of animal feed because of its high nutritional value.

Keratinases are commercially important due to their capability of transforming feather wastes into animal feeds (Harde et al. 2011). The supplementation of diet with keratinase had been reported to boost immune response, weight gain, nutrient digestibility, intestinal morphology, and ecology in growing and nursery pigs (Wang et al. 2011a, b). Fakhfakh et al. (2013) also reported the production of proteinous hydrolysates of high antioxidative potential from wool waste degradation by a keratinolytic strain of *B. subtilis* A1. Similarly, bioconversion of feather wastes by *Chryseobacterium* sp. kr6 and *Bacillus* sp. kr16 generated protein-rich hydrolyates that could be used as feed supplement in animal food production (Maciel et al. 2017). Most recently, Saarela et al. (2017) suggested the possibility of utilizing feather hydrolysate obtained after feather hydrolysis by nonpathogenic keratinolytic bacterial strains isolated from birds' nest for animal feed formulation due to its high amino acids composition. Similarly, Mazotto et al. (2017) applied the feather hydrolysate obtained after feather hydrolysis by *B. subtilis* AMR in a blend with cornmeal to produce an extrudate food. Therefore, biodegradation of keratins is an indispensable source of nutrient-rich feeds, nutraceuticals, and feed supplements with lots of promising applications in animal husbandry. The enhancement of digestibility and nutritional improvement in the supplementation of feeds with keratinases may be an indication of its importance as a key component of animal feeds. It is envisaged that use of keratinases as additives to form essential components of feeds, particularly processed pellets would expand in the future.

In addition, the oligopeptides and amino acids obtained from the hydrolysates can veritably serve the purpose of nitrogen supplementation in the formulation and compounding of growth media for the cultivation of microorganisms. Therefore, keratin hydrolysates as complex media components could be cheap sources of nutrients, not only for the growth of microorganisms but also to produce novel bio-products in fermentation processes. The reported abilities of keratinases to degrade prions will further enhance the safety of these products in manufacturing of substances intended for animal and human consumption.

6.3 Production of Nitrogen Fertilizer/Biofertilizer

Keratin hydrolysate obtained from keratin hydrolysis is a very rich source of ammonia and amino acids, hence, can be exploited as a slow release nitrogen fertilizer. For instance, Bose et al. (2014) proposed the use of feather hydrolysate as bioactive organic nitrogen fertilizer. In the study conducted by Paul et al. (2013), the feather hydrolysate obtained after feather hydrolysis by *Paenibacillus woosongensis* TKB2 was evaluated for plant growth-promoting ability, and it was observed that the seed germination, nodule formation, and seedling growth of Bengal gram (*Cicer arietinum*) were all enhanced by the feather hydrolysate treatment. The treatment improved soil fertility by increasing the C/N ratio and N, P, and K contents in the soil. The microbial activities in the soil were also boosted such that the incidences of free nitrogen fixers and phosphate solubilizers were increased by 2- and 5.8-folds, respectively. Similarly, Arasu et al. (2009) investigated the plant growth-promoting ability of two keratinolytic *Bacillus* strains, and found out that the feather hydrolysates obtained after their fermentations significantly increased the vigor index of rice seeds by 30%.

A keratinolytic strain of *B. subtilis* was reported to demonstrate plant growth-promoting ability as it produced indoleacetic acid (IAA) and broad-spectrum antimicrobial metabolites in the course of keratinase production (Jeong et al. 2010b). In addition, the culture supernatant obtained after the fermentation of whole chicken feather by a strain of *Stenotrophomonas maltophilia* appreciably enhanced plant growth by 82 and 66% using two- and fourfold dilutions, respectively (Cao et al. 2012). Nitrogen is gradually released to the soil by the decomposition of keratinous wastes through soil microorganisms, thereby improving soil fertility. Most recently,

Tamreihao et al. (2017) reported a significant improvement in the growth of rice plant cultivated on the soil treated by feather hydrolysate obtained after feather hydrolysis by *Amycolatopsis* sp. MBRL 40 over the control rice plant cultivated on the ordinary soil. Similarly, nitrogen-rich organic manure was formulated from feather hydrolysate produced after feather wastes hydrolysis by *Chrysosporium tropicum* and *Malbranchea* sp. The soil treated with the manure considerably enhanced the germination of pea and rice seed (Kumar et al. 2017). Likewise the soil treatment with feather hydrolysate obtained after feather hydrolysis by *Thermoactinomyces* sp. RM4 supported early seed germination of *Cicer arietinum* than the control (Verma et al. 2016). In this manner, biodegradation of keratin wastes can occupy a prime position in enriching the soil fertility through the concept of biofertilizers, enhancement of the growth of plant-growth-promoting rhizobacteria (PGPR), and improvement of the nutritional and bioactive richness of cultivated crops. These, undoubtedly, will promote human and soil health with a total benefit on the ecosystem.

6.4 Leather Processing Industry

Chemical dehairing by sodium sulfide, lime, and chromate in leather industries releases toxic and solid wastes to the ecosystem. Sodium sulfide attacks the skin and also damages the quality of hair fragments which in turn lowers their commercial values. These chemicals are highly responsible for increased biochemical oxygen demand (BOD) and chemical oxygen demand (COD) (Pillai and Archana 2008). Dehairing by enzymes, precisely keratinolytic proteases have proved to be the best alternative as they selectively remove hair from the skin without damaging the skin quality and hair fragments. The method is simple, cheap, and environmentally benign as it avoids the release of toxic effluent to the environment (Paul et al. 2016; Fang et al. 2017b). Several authors have reported dehairing of animal skins by microbial keratinases (Chaturvedi et al. 2014; Lateef et al. 2015a; Paul et al. 2016).

The protease blend preparations containing keratinase, collagenase, alkaline, neutral, and acidic proteases produced by *Trichoderma harzianum* MH-20 have demonstrated very remarkable dehairing capabilities (Ismail et al. 2012) without the addition of lime and sulfide. Tiwary and Gupta (2010) reported the dehairing of buffalo hide by a dimeric keratinase in the presence of 3% lime. In a related study, Chaturvedi et al. (2014) compared keratinase-mediated dehairing of goat skin with the chemical-based process using sodium sulfide and lime. While the enzyme completely dehaired the skin under a limited period, incomplete dehairing with serious skin and hair damage were observed in the chemical method. In our laboratory, the keratinase produced during feather degradation by a strain of *Bacillus safensis* LAU 13 completely dehaired goat skin within 12 h without damaging the skin quality, whereas there was incomplete dehairing with a very noticeable skin damage in the chemical-based method (Lateef et al. 2015a) (Fig. 2).

Similarly, Zhang et al. (2016a) reported another complete removal of hair from goat hide by the crude keratinase produced by *Acinetobacter* sp. R-1 under 21 h

Fig. 2 Complete dehairing of goat skin by crude keratinase from wild strain of *B. safensis* LAU 13 (**a**), and incomplete dehairing by sodium sulfide and lime (**b**); control (**c**)

of incubation at 37 °C without observable damage to the skin. In a study conducted recently by Isaac and Abu-Tahon (2016), considerable dehairing of cowhide by crude alkaline keratinase produced by *Cochliobolus hawaiiensis* AUMC 8606 was achieved after 16 h without observable damage to the skin quality while chemical dehairing lasted for 28 h with some damages to the skin quality. Also, Bouacem et al. (2016) completely dehaired goat, sheep, and bovine hides with only purified keratinase produced by a strain of *Caldicoprobacter algeriensis* after 12 h of incubation at 30 °C with no observable damage to the skins. Therefore, scope exists for the application of keratinases in eco-friendly processing of leather and skin products.

6.5 Detergent Industry

Proteases are essential ingredients in the detergent formulation because of the uniqueness of their properties. The keratinolytic proteases will add more to the value of proteases in detergent formulation due to their ability to degrade the recalcitrant and insoluble keratin and their ruggedness like stability at high temperature and pH, activity at wide range of temperature and pH, high tolerance for surfactants, oxidizing and bleaching agents, chelating agents, and compatibility with some commercial laundry detergent. Gupta and Ramnani (2006) suggested the use of keratinase in detergent formulation to remove keratinous dirts that are mostly encountered in the laundry, such as collars of shirts that may be difficult to remove by common proteases, and for cleaning solid surfaces, and drains clogged with keratinous wastes. Keratinolytic proteases are essential additives for the production modern laundry detergent powder and liquid to improve washing performance by removing different proteinaceous stains (Paul et al. 2016).

The cleansing power of a detergent was boosted after its combination with *Paenibacillus woosongensis* TKB2 keratinase; the combination safely and effectively removed blood, fruit juice, and turmeric stains from fabric (Paul et al. 2014d). Also,

Paul et al. (2014c) removed blood, egg yolk and chocolate stains from cloths by crude keratinase of *Paenibacillus woosongensis* TKB2 with the release of eco-friendly effluent. Manivasagan et al. (2014) reported the first reference to a marine actinobacterium *Actinoallteichus* sp. MA-32 as a producer of thermo- and detergent stable keratinase in a feather-based medium. The resultant activity of the enzyme and its stability to SDS, ethoxylated octylphenol, DMSO, sodium sulphite, and mercaptoethanol are indicative of its potential application in laundry detergent formulation. In our findings, we achieved remarkable removal of bloodstain from stained cloth within 2 h of incubation with the crude keratinase produced by *B. safensis* LAU 13 (Lateef et al. 2015a).

In a study conducted by Zhang et al. (2016c), the production of surfactants and commercial detergents stable keratinase by a strain of *Gibberella intermedia* was reported. The enzyme completely removed bloodstain from stained cloth when combined with commercial detergents. Most recently, Reddy et al. (2017) reported the potential application of keratinase produced by *B. pumilus* GRK as biodetergent additive due to its high compatibility with commercial detergents and efficient bloodstain removal from stained cloth by the keratinase-mediated detergent without damaging the cloth quality. Keratinases are therefore essential ingredients that could be utilized as additives in detergent formulations for efficient removal of keratinous stains in an eco-friendly manner.

6.6 Textile Industry

Keratinases have potential application in textile processing industries due to their ability to modify silk and wool. The treatment of wool and polyester-blended fabrics with crude keratinase obtained from a strain of *Pseudomonas* sp. improved their shrink resistance and tensile strength (Cai et al. 2011). Three hydrolytic enzymes, cutinase, keratinase, and protease were used synergistically to improve the wettability and anti-felting property of wool fabrics (Wang et al. 2011c) with resultant reduction of contact angle to 66°, area shrinkage of 5.2%, and acceptable strength loss of 14%.

Keratinase and protease have been used to eliminate wool fibers, prevent shrinkage and increase the anti-felting property of wool (Liu et al. 2013b). Similarly, Zhang et al. (2016a) reported that the surface of the wool obtained after enzymatic dehairing of goat hide by crude keratinase of *Acinetobacter* sp. R-1 was modified which in turn improved the quality and the anti-felting properties of the wool. In a similar study, three cooperative keratinolytic enzymes purified from *S. maltophilia* BBE11-1 have been reported to show remarkable removal of cuticle layers of wool without damaging internal fibers (Fang et al. 2013), with potential applications in textile and leather industries. In the processing of raw silk, degumming of fibrous proteins to improve the quality of fibers in terms of softness and luster can be achieved in an eco-friendly process by using keratinases instead of using soap, alkali, and oxidizing agents. Some proteases have been demonstrated for application in this sector (More et al. 2013). Most recently, Iglesias et al. (2017) treated wool top with keratinase-containing enzyme extract of *Bacillus* sp. G51, and it was observed that the treatment reduced the wool felting tendency without significant weight loss.

6.7 Cosmetic and Pharmaceutical Applications

Keratinases have promising application in pharmaceutical and cosmetic industries preferably the non-collagenolytic keratinases. They are important ingredients in the production of hair shaving and skin lightening agents (Yang 2012). Keratinases can be used instead of the conventional salicylic acid to remove thickened layer of dead skin found in toes and fingers (Gupta and Ramnani 2006). In another study, keratinases displayed potential application in the production of hair care products as it improved hair qualities such as weight, flexibility, brightness, softness, and strength (Cao et al. 2012). Keratinases can be used to remove acnes and scars on the skin to allow for tissue regeneration (Selvam and Vishnupriya 2012). In a study conducted recently by Sanghvi et al. (2016), it was suggested that the thioglycolate chemical commonly used in the production of hair removal cream could be substituted by microbial keratinase for more effectiveness and safety.

Keratinases have been reported to degrade Prions PrP^{Sc}, the infective protein molecules responsible for scrapie, bovine spongiform encephalopathy (BSE or "mad cow disease"), kuru, chronic wasting disease, and Creutzfeldt–Jakob disease (CJD) (Caughey 2001). Studies have revealed the capability of keratinolytic enzymes to degrade this abnormal molecule which suggests their application in drug formulation and as a disinfectant for medical instruments suspected to be contaminated by this protein. Mitsuiki et al. (2010) reported the remarkable degradation of PrP^{Sc} protein by keratinase of *Nocardiopsis* strain TOA-1. Also, the combined action of γ-glutamyl transpeptidase-glutathione and keratinase effectively hydrolyzed surrogate prion protein (Sharma and Gupta 2012).

Keratinases can be utilized to boost the efficiency of drug delivery system for topical application. Keratinase conjugation with drug-facilitated drug penetration through the nail plate which may be a barrier to penetration by other drugs (Mohorcic et al. 2007). They could also be used to disinfect medical equipment and laboratory apparatus owing to their prion protein degrading ability (Liang et al. 2010). Feather hydrolysate obtained after feather fermentation by *Bacillus subtilis* S1-4 demonstrated promising application in the formulation of antioxidative peptides (Wan et al. 2016). In addition, keratinase obtained from *Paecilomyces lilacinus* (LPS#876) was immobilized using a mixture enrofloxacin and PVA-pectin cryogels. It was discovered that the presence of keratinase led to the slow release of the antibiotic (6.9%) at investigated conditions (pH 5.5, 37 °C, 5 h), compared to 15.0% that was obtained in the absence of keratinase in the cryogels (Martínez et al. 2013). Thus, it is practicable to apply keratinase in pharmaceutical formulations for slow drug delivery.

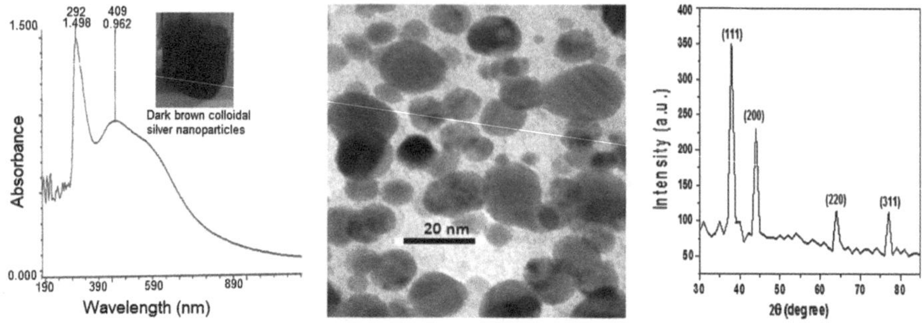

Fig. 3 UV–Vis spectrum, transmission electron micrograph and XRD spectrum of silver nanoparticles biosynthesized using crude keratinase of *B. safensis* LAU 13

6.8 Nanobiotechnology and Other Applications of Keratinases

Revathi et al. (2013) gave a first report on the green synthesis of silver nanoparticles (AgNPs) using microbial keratinase. Later in our laboratory, we investigated the crude extracellular keratinase obtained from *B. safensis* LAU 13 for its potential application in the green synthesis of AgNPs (Lateef et al. 2015c). The biosynthesized spherical-shaped particles (Fig. 3) were 5–30 nm in size with considerable antibacterial activity against clinical isolates of *E. coli*. There is increasing relevance of enzymes in the green synthesis of metallic nanoparticles (Lateef and Adeeyo 2015; Adelere and Lateef 2016b; Elegbede et al. 2018).

Keratinases are capable of modifying keratinous wastes to produce biodegradable films, coatings, and glues for compostable packaging, agricultural films, or edible film application (Gupta and Ramnani 2006; Sharma et al. 2018). The study of Brandelli et al. (2010) revealed the potential application of keratinases in energy generation and biofuel production through anaerobic digestion of biowastes. The solution of organic colloid obtained from the degradation of horns and hoofs have also been used as fire extinguisher (Datta 1993). Keratinases could be used as biocontrol molecules due to their pesticidal and insecticidal properties (Yue et al. 2011). The alkaline keratinase of *Bacillus* sp. 50-3 was reported to kill 98.5% juveniles of *Meloidogyne incognita* (a root-knot nematode) at concentration of 50 μg/mL (Yue et al. 2011).

Moreover, bioremediation of crude oil-contaminated environments is possible through keratinolytic fungi (Ulfig et al. 2003). Other promising applications of keratinases include the use of keratin hydrolysates to regulate heamostasis and nerve regeneration (Rouse and Van Dyke 2010), cleaning of contact lenses (Ray 2012), and recovery of silver from used X-ray films in an eco-friendly manner (Seid 2011). Keratinases can also find application in food processing industries due to their gelatinolytic activity. It has been suggested that keratinase may be useful in removing contaminants (mainly feather and plumage) in edible nests of swiftlets (Gupta et al. 2013). In addition, Zhang et al. (2010) demonstrated the effectiveness of keratinase to remove keratin impurities from pearls to improve their gem quality.

7 Conclusion

Degradation of keratins by keratinolytic microorganisms is an efficient alternative means of keratins waste management. The process has immense advantages as it reduces the environmental problem posed by these wastes in an eco-friendly manner, it is a cheap and simple method that often leads to the release of biotechnologically important products like keratinases, proteins, peptides, and amino acids. Keratinases are robust enzymes with very diverse applications. The enzymes are useful in processes related to the bioconversion of keratin wastes into feed and fertilizers. Other keratinase related applications include enzymatic dehairing for leather and cosmetic industry, medical and pharmaceutical applications, detergents, biocatalysts, cosmetics manufacturing, and others. Recently, studies had further extended the frontier of keratinase applications to the field of nanotechnology, where keratinases were used in the biosynthesis of versatile silver nanoparticles. All these underscored the relevance of keratinase as important enzyme in biocatalysis and white biotechnology to deliver novel biotechnological products to improve quality of life and safe environment.

References

Abdel-Naby MA, Ibrahim MHA, El-Refai HA (2016) Catalytic, kinetic and thermodynamic properties of Bacillus pumilus FH9 keratinase conjugated with activated pectin. Int J Biol Macromol 85:238–245

Abdel-Naby MA, El-Refai HA, Ibrahim MH (2017) Structural characterization, catalytic, kinetic and thermodynamic properties of keratinase from *Bacillus pumilus* FH9. Int J Biol Macromol 105:973–980

Adelere IA, Lateef A (2016a) Keratinases: emerging trends in production and applications as novel multifunctional biocatalysts. Kuwait J Sci 43:118–127

Adelere IA, Lateef A (2016b) A novel approach to the green synthesis of metallic nanoparticles: the use of agro-wastes, enzymes and pigments. Nanotechnol Rev 5:567–587

Agrahari S, Wadhwa N (2012) Isolation and characterization of feather degrading enzymes from *Bacillus megaterium* SN1 isolated from Ghazipur poultry waste site. Appl Biochem Microbiol 48:175–181

Ahmadpour F, Yakhchali B, Musavi MS (2017) Isolation and identification of a keratinolytic *Bacillus cereus* and optimization of keratinase production. J Appl Biotechnol Reports 3:507–512

Alibardi L, Dalla Valle L, Toffolo V, Toni M (2006) Scale keratin in lizard epidermis reveals amino acid regions homologous with avian and mammalian epidermal proteins. Anatom Record 288(7):734–752

Allure N, Madhusudhan DN, Agsar D (2015) Detection of keratinolytic Actinobacteria and evaluation of bioprocess for production of alkaline keratinase. Int J Curr Microbiol Appl Sci 4:907–918

Arasu VT, Sivakumar T, Ramasubramanian V, Nalini K, Kiruthiga R (2009) The potential application of keratinase from *Bacillus* sp. as plant growth promoters. J Pure Appl Microbiol 3(2):583–590
Awad GEA, Esawy MA, Salam WA, Salama BM, Abdelkader AF, El-diwany A (2011) Keratinase production by *Bacillus pumilus* GHD in solid-state fermentation using sugar cane bagasse: optimization of culture conditions using a Box-Behnken experimental design. Annals Microbiol 61:663–672
Blyskal B (2009) Fungi utilizing keratinous substrates. Int Biodeterior Biodegr 63(6):631–653
Bohacz J (2017) Biodegradation of feather waste keratin by a keratinolytic soil fungus of the genus *Chrysosporium* and statistical optimization of feather mass loss. World J Microbiol Biotechnol 33:13
Bose A, Pathan S, Pathak K, Keharia H (2014) Keratinolytic protease production by *Bacillus amyloliquefaciens* 6B using feather meal as substrate and application of feather hydrolysate as organic nitrogen input for agricultural soil. Waste Biomass Valor 5:595–605
Bouacem K, Bouanane-Darenfed A, Jaouadi NZ, Joseph M, Hacene H, Ollivier B, Fardeau ML, Bejar S, Jaouadi B (2016) Novel serine keratinase from *Caldicoprobacter algeriensis* exhibiting outstanding hide dehairing abilities. Int J Biol Macromol 86:321–328
Brandelli A, Daroit DJ, Riffel A (2010) Biochemical features of microbial keratinases and their production and applications. Appl Microbiol Biotechnol 85:1735–1750
Cai SB, Huang ZH, Zhang XQ, Cao ZJ, Zhou MH, Hong F (2011) Identification of a keratinase-producing bacterial strain and enzymatic study for its improvement on shrink resistance and tensile strength of wool- and polyester-blended fabric. Appl Biochem Biotechnol 163:112–126
Călin M, Drăcea O, Răut I, Vasilescu G, Doni MB, Arsene ML, Alexandrescu E, Constantinescu-Aruxandei D, Jecu L, Lazăr V (2016) In vitro biodegradation of keratinized substrates by keratinophilic fungi. Sci Bull Ser F Biotechnol 20:248–253
Călin M, Constantinescu-Aruxandei D, Alexandrescu E, Răut I, Doni MB, Arsene ML, Oancea F, Jecu L, Lazăr V (2017) Degradation of keratin substrates by keratinolytic fungi. Electronic J Biotechnol 28:101–112
Cao ZJ, Lu D, Luo LS, Deng YX, Bian YG, Zhang XQ, Zhou MH (2012) Composition analysis and application of degradation products of whole feathers through a large scale of fermentation. Environ Sci Pollut Res 19:2690–2696
Caughey B (2001) Interactions between prion protein isoforms: the kiss of death? Trends Biochem Sci 26:235–242
Cavello I, Urbieta MS, Segretin AB, Giaveno A, Cavalitto S, Donati ER (2018) Assessment of keratinase and other hydrolytic enzymes in thermophilic bacteria isolated from geothermal areas in Patagonia Argentina. Geomicrobiol J 35:156–165
Cedrola SML, de Melo ACN, Mazotto AM, Lins U, Zingali RB, Rosado AS, Peixoto RS, Beatriz A (2012) Keratinase and sulfide from *Bacillus subtilis* SLC to recycle feather waste. World J Microbiol Biotechnol 28:1259–1269
Chao YP, Xie FH, Jing Y, Lu JH, Qian SJ (2007) Screening for a new *Streptomyces* strain capable of efficient keratin degradation. J Environ Sci 19:1125–1128
Chaturvedi V, Bhange K, Bhatt R, Verma P (2014) Production of kertinases using chicken feathers as substrate by a novel multifunctional strain of *Pseudomonas stutzeri* and its dehairing application. Biocatalys Agric Biotechnol 3:167–174
Chitturi CMK, Lakshmi VV (2016) Development of semi-solid state fermentation of Keratinase and optimization of process by cheaper and alternative agricultural wastes. Eur J Biotechnol Biosc 4:1–4
Daroit DJ, Correa APF, Brandelli A (2011) Production of keratinolytic proteases through bioconversion of feather meal by the Amazonian bacterium *Bacillus* sp. P45. Int Biodeterior Biodegrad 65:45–51
Datta MS (1993) Role of keratin in fire fighting. J Ind Leather Technol Associat 43:297–299
de Oliveira CT, Rieger TJ, Daroit DJ (2017) Catalytic properties and thermal stability of a crude protease from the keratinolytic *Bacillus* sp. CL33A. Biocataly Agric Biotechnol 10:270–277

Dong YZ, Chang WS, Chen PT (2017) Characterization and overexpression of a novel keratinase from *Bacillus polyfermenticus* B4 in recombinant *Bacillus subtilis*. Bioresour Bioprocess 4:47

Elegbede JA, Lateef A, Azeez MA, Asafa TB, Yekeen TA, Oladipo IC, Adebayo EA, Beukes LS, Gueguim-Kana EB (2018) Fungal xylanases-mediated synthesis of silver nanoparticles for catalytic and biomedical applications. IET Nanobiotechnol 12:857–863

El-Gendy MMA (2010) Keratinase production by endophytic *Penicillium* spp. Morsy1 under solid-state fermentation using rice straw. Appl Biochem Biotechnol 162:780–794

El-Ghonemy DH, Ali TH (2017) Optimization of physico-chemical parameters for hyper keratinase production from a newly isolated *Aspergillus* sp. DHE7 using chicken feather as substrate-management of biowaste. J Appl Pharmaceut Sci 7:171–178

Elhoul MB, Jaouadi NZ, Rekik H, Benmrad MO, Mechri S, Moujehed E, Kourdali S, El Hattab M, Badis A, Bejar S, Jaouadi B (2016) Biochemical and molecular characterization of new keratinoytic protease from *Actinomadura viridilutea* DZ50. Int J Biol Macromol 92:299–315

Eliades L, Cabello M, Voget C, Galarza B, Saparrat M (2010) Screening for alkaline keratinolytic activity in fungi isolated from soils of the biosphere reserve "Parque Costero del Sur" (Argentina). World J Microbiol Biotechnol 26:2105–2111

Fakhfakh N, Ktari N, Siala R, Nasri M (2013) Wool-waste valorization: production of protein hydrolysate with high antioxidative potential by fermentation with a new keratinolytic bacterium, *Bacillus pumilus* A1. J Appl Microbiol 115:424–433

Fang Z, Zhang J, Liu B, Du G, Chen J (2013) Biodegradation of wool waste and keratinase production in scale-up fermenter with different strategies by *Stenotrophomonas maltophilia* BBE11-1. Bioresour Technol 140:286–291

Fang Z, Zhang J, Liu B, Jiang L, Du G, Chen J (2014) Cloning, heterologous expression and characterization of two keratinases from *Stenotrophomonas maltophilia* BBE11-1. Process Biochem 49:647–654

Fang Z, Zhang J, Liu B, Du G, Chen J (2016) Enhancement of the catalytic efficiency and thermostability of *Stenotrophomonas* sp. keratinase KerSMD by domain exchange with KerSMF. Microb Biotechnol 9:35–46

Fang Z, Yong YC, Zhang J, Du G, Chen J (2017a) Keratinolytic protease: a green biocatalyst for leather industry. Appl Microbiol Biotechnol 101:7771–7779

Fang Z, Zhang J, Du G, Chen J (2017b) Rational protein engineering approaches to further improve the keratinolytic activity and thermostability of engineered keratinase KerSMD. Biochem Eng J 127:147–153

Fellahi S, Chibani A, Feuk-Lagerstedt E, Taherzadeh MJ (2016) Identification of two new keratinolytic proteases from a *Bacillus pumilus* strain using protein analysis and gene sequencing. AMB Express 6:42

Ferrareze PAG, Correa APF, Brandelli A (2016) Purification and characterization of a keratinolytic protease produced by probiotic *Bacillus subtilis*. Biocataly Agric Biotechnol 7:102–109

Fraser RDB, Parry DA (2011) The structural basis of the filament-matrix texture in the avian/reptilian group of hard β-keratins. J Struct Biol 173:391–405

Fraser RDB, MacRae TP, Rogers GE (1972) Keratins: their composition, structure and biosynthesis. Charles C Thomas, Springfield, IL, p 304

Fuke P, Gujar VV, Khardenavis AA (2017) Genome annotation and validation of keratin-hydrolyzing proteolytic enzymes from *Serratia marcescens* EGD-HP20. Appl Biochem Biotechnol 184:970–986

Gareth LJ, Julian P, Callum T, Erik B (2010) Protease- and keratinase-producing microbial strains for compost bioaugmentation. Int Biodeterior Biodegr 64:574–580

Gassesse A, Hatti-Kaul R, Gashe BA, Mattiasson B (2003) Novel alkaline proteases from alkaliphilic bacteria grown on chicken feather. Enzyme Microb Technol 32:519–524

Ghasemi Y, Shahbazi M, Rasoul-Amini S, Kargar M, Safari A, Kazemi A, Montazeri-Najafabady N (2012) Identification and characterization of feather-degrading bacteria from keratin-rich wastes. Ann Microbiol 62:737–744

Gillespie JM (1990) The proteins of hair and other hard α-keratins. In: Goldman RD, Steinert PM (eds) Cellular and molecular biology of intermediate filaments. Springer, Boston, MA, pp 95–128
Gradisar H, Kern S, Friedrich J (2000) Keratinase of *Doratomyces microsporus*. Appl Microbiol Biotechnol 53:196–200
Gupta R, Ramnani P (2006) Microbial keratinases and their prospective applications: an overview. Appl Microbiol Biotechnol 70:21
Gupta R, Rajput R, Sharma R, Gupta N (2013) Biotechnological applications and prospective market of microbial keratinases. Appl Microbiol Biotechnol 97:9931–9940
Gupta S, Tewatia P, Misri J, Singh R (2017) Molecular modeling of cloned *Bacillus subtilis* keratinase and its insinuation in Psoriasis treatment using Docking studies. Indian J Microbiol 57:485–491
Gurav RG, Jadhav JP (2013) Biodegradation of keratinous waste by *Chryseobacterium* sp. RBT isolated from soil contaminated with poultry waste. J Basic Microbiol 53:128–135
Gurav RG, Tang J, Jadhav JP (2016) Sulfitolytic and keratinolytic potential of *Chryseobacterium* sp. RBT revealed hydrolysis of melanin containing feathers. 3 Biotech 6:145
Habbeche A, Saoudi B, Jaouadi B, Haberra S, Kerouaz B, Boudelaa M, Badis A, Ladjama A (2013) Purification and biochemical characterization of a detergent-stable keratinase from a newly thermophilic actinomycete *Actinomadura keratinilytica* strain Cpt29 isolated from poultry compost. J Biosci Bioeng 117:413–421
Harde SM, Bajaj IB, Singhal RS (2011) Optimization of fermentative production of keratinase from *Bacillus subtilis* NCIM 2724. Agric Food Anal Bacteriol 1:54–65
Hmidet N, Ali NE, Fakhfakh NZ, Haddar A, Nasri M, Kamoun AS (2010) Chicken feathers: a complex substrate for the co-production of α-amylase and proteases by *B. licheniformis* NH1. J Ind Microbiol Biotechnol 37:983–990
Holkar CR, Jain SS, Jadhav AJ, Pinjari DV (2017) Valorization of keratin based waste. Process Saf Environ Prot 115:85–98
Huang Y, Sun Y, Ma S, Chen L, Zhang H, Deng Y (2013) Isolation and characterization of *Keratinibaculum paraultunense* gen. nov., sp. nov., a novel thermophilic, anaerobic bacterium with keratinolytic activity. FEMS Microbiol Lett 345:56–63
Hussein HA, Abdullah SK (2017) Keratinolytic and opportunistic pathogenic fungi from carpet dust in mosques and residential houses in Duhok, Kurdistan Region, Iraq. Sci J Univ Zakho 5:16–19
Iglesias MS, Sequeiros C, Garcia S, Olivera NL (2017) Newly isolated *Bacillus* sp. G51 from Patagonian wool produces an enzyme combination suitable for felt-resist treatments of organic wool. Bioprocess Biosystems Eng 40:833–842
Iruolaje FO, Ogbeba J, Tula MY, Ijebor JA, Dogo BA (2016) Isolation and identification of keratinolytic bacteria that exhibit feather-degrading potentials. J Adv Biol Biotechnol 5(2):1–9
Isaac GS, Abu-Tahon MA (2016) Dehairing capability of alkaline keratinase produced by new isolated *Cochliobolus hawaiiensis* AUMC 8606 grown on chicken feather. Romanian Biotechnol Lett 22:12147–12154
Ismail AS, Housseiny MM, Abo-Elmagd HI, El-Sayed NH, Habib M (2012) Novel keratinase from *Trichoderma harzianum* MH-20 exhibiting remarkable dehairing capabilities. Int Biodeterior Biodegr 70:14–19
Jayalakshmi T, Krishnamoorthy P, Kumar GR, Sivamani P, Lakshmi CGA (2012) Application of pure keratinase on keratinous fibers to identify the keratinolytic activity. J Chem Pharmaceut Res 4:3229–3233
Jeong EJ, Rhee MS, Kim GP, Lim KH, Yi DH, Bang BH (2010a) Purification and characterization of a keratinase from a feather-degrading bacterium, *Bacillus* sp. SH-517. Jf Korean Soc Appl Biol Chem 53:43–49
Jeong JH, Jeon YD, Lee O, Kim JD, Lee NR, Park GT, Son HJ (2010b) Characterization of a multifunctional feather-degrading *Bacillussubtilis* isolated from forest soil. Biodegr 21:1029–1040

Kamarudin NB, Sharma S, Gupta A, Kee CG, Chik SMSBT, Gupta R (2017) Statistical investigation of extraction parameters of keratin from chicken feather using Design-Expert. 3 Biotech 7:127
Kannahi M, Ancy RJ (2012) Keratin degradation and enzyme producing ability of *Aspergillus flavus* and *Fusarium solani* from soil. J Chem Pharmaceut Res 4:3245–3248
Kim JM, Lim WJ, Suh HJ (2001) Feather-degrading *Bacillus* species from poultry waste. Process Biochem 37:287–291
Korniłłowicz-Kowalska T, Bohacz J (2011) Biodegradation of keratin waste: theory and practical aspects. Waste Manag 31:1689–1701
Kreplak L, Doucet J, Dumas P, Briki F (2004) New aspects of the α-helix to β-sheet transition in stretched hard α-keratin fibers. Biophys J 87:640–647
Kshetri P, Ningthoujam DS (2016) Keratinolytic activities of alkaliphilic *Bacillus* sp. MBRL 575 from a novel habitat, limestone deposit site in Manipur, India. SpringerPlus 5:595
Kumar M, Kumar R, Malik DK (2016) Keratin degradation by bacterial strain isolated from poultry farm soil. J Pharm Res 10:113–115
Kumar J, Sharma A, Kumar P, Kushwaha RKS (2017) Enhancement of soil nutrition using fermented feather and their efficacy on seed germination. Int J Pure Appl Biosci 5:92–98
Kunert J (2000) Physiology of keratinophilic fungi. Rev Iberoamericana de Micologia 1:77–85
Kuo JM, Yang JI, Chen WM, Pan MH, Tsai ML, Lai YJ, Hwang B, Pan BS, Lin CY (2012) Purification and characterization of a thermostable keratinase from *Meiothermus* sp. 140. Int Biodeterior Biodegr 70:111–116
Łaba W, Chorążyk D, Pudło A, Trojan-Piegza J, Piegza M, Kancelista A, Kurzawa A, Żuk I, Kopeć W (2017) Enzymatic degradation of pretreated Pig bristles with crude keratinase of *Bacillus cereus* PCM 2849. Waste Biomass Valoriz 8:527–537
Lange L, Busk PK, Huang Y (2014) Use of a microbial composition for the degradation of keratinaceous materials. Denmark Patent WO 169920, A2
Lange L, Huang Y, Busk PK (2016) Microbial decomposition of keratin in nature—a new hypothesis of industrial relevance. Appl Microbiol Biotechnol 100:2083–2096
Lateef A, Adeeyo AO (2015) Green synthesis and antibacterial activities of silver nanoparticles using extracellular laccase of *Lentinus edodes*. Not Sci Biol 7:405–411
Lateef A, Adelere IA, Gueguim-Kana EB (2015a) *Bacillus safensis* LAU 13: a new source of keratinase and its multi-functional biocatalytic applications. Biotechnol Biotechnol Equip 29:54–63
Lateef A, Adelere IA, Gueguim-Kana EB (2015b) The biology and potential biotechnological applications of *Bacillus safensis*. Biologia 70:411–419
Lateef A, Adelere IA, Gueguim-Kana EB, Asafa TB, Beukes LS (2015c) Green synthesis of silver nanoparticles using keratinase obtained from a strain of *Bacillus safensis* LAU 13. Int Nano Lett 5:29–35
Lateef A, Oloke JK, Gueguim-Kana EB, Sobowale BO, Ajao SO, Bello BY (2010) Keratinolytic activities of a new feather-degrading isolate of *Bacillus cereus* LAU 08 isolated from Nigerian soil. Int Biodeterior Biodegr 64:162–165
Li J, Chen D, Yu Z, Zhao L, Zhang R (2013) Improvement of expression level of keratinase Sfp2 from *Streptomyces fradiae* by site-directed mutagenesis of its N-terminal pro-sequence. Biotechnol Lett 35:743–749
Liang X, Bian Y, Tang XF, Xiao G, Tang B (2010) Enhancement of keratinolytic activity of a thermophilic subtilase by improving its autolysis resistance and thermostability under reducing conditions. Appl Microbiol Biotechnol 87:999–1006
Liang JD, Han YF, Zhang JW, Du W, Liang ZQ, Li ZZ (2011) Optimal culture conditions for keratinase production by a novel thermophilic *Myceliophthora* strain GZUIFR-H49-1. J Appl Microbiol 110:871–880
Lin X, Inglis GD, Yanke LJ, Cheng KJ (1999) Selection and characterization of feather degrading bacteria from canola meal compost. J Ind Microbiol Biotechnol 23:149–153
Liu B, Zhang J, Fang Z, Gu L, Liao X, Du G, Chen J (2013a) Enhanced thermostability of keratinase by computational design and empirical mutation. J Ind Microbiol Biotechnol 40:697–704

Liu B, Zhang J, Li B, Liao X, Du G, Chen J (2013b) Expression and characterization of extreme alkaline, oxidation-resistant keratinase from *Bacillus licheniformis* in recombinant *Bacillus subtilis* WB600 expression system and its application in wool fiber processing. World J Microbiol Biotechnol 29:825–832

Lo WH, Too JR, Wu JY (2012) Production of keratinolytic enzyme by an indigenous feather-degrading strain *Bacillus cereus* Wu2. J Biosci Bioeng 114:640–647

Maciel JL, Werlang PO, Daroit DJ, Brandelli A (2017) Characterization of protein-rich hydrolysates produced through microbial conversion of waste feathers. Waste Biomass Valoriz 8:1177–1186

Manczinger L, Rozs M, Vagvolgyi C, Kevei F (2003) Isolation and characterization of a new keratinolytic *Bacillus licheniformis* strain. World Microbiol and Biotechnol 19:35–39

Manivasagan P, Sivakumar K, Gnanam S, Venkatesan J, Kim SK (2014) Production, biochemical characterization and detergents application of keratinase from the marine *Actinobacterium actinoalloteichus* sp. MA-32. J Surfactants Detergents 17:669–682

Martínez YN, Cavello I, Hours R, Cavalitto S, Castro GR (2013) Immobilized keratinase and enrofloxacin loaded on pectin PVA cryogel patches for antimicrobial treatment. Bioresour Technol 145:280–284

Mazotto AM, Cedrola SML, Lins U, Rosado AS, Silva KT, Chaves JQ, Rabinovitch L, Zingali RB, Vermelho AB (2010) Keratinolytic activity of *Bacillus subtilis* AMR using human hair. Lett Appl Microbiol 50:89–96

Mazotto AM, Ascheri JLR, de Oliveira GRL, Damaso MCT, Couri S, Vermelho AB (2017) Production of feather protein hydrolyzed by *B. subtilis* AMR and its application in a blend with cornmeal by extrusion. LWT-Food Sci Technol 84:701–709

Mini KD, George SM, Mathew J (2015) Screening and selection of fungus for keratinase production by solid state fermentation and optimization of conditions of SSF and formulation of low cost medium for the production of keratinase by *Aspergillus flavus* S125. Int J Curr Microbiol App Sci 4(9):535–548

Mitsuiki S, Takasugi M, Moriyama Y, Futagami T, Goto M, Kanouchi H, Oka T (2010) Identification of an alkaliphilic actinomycetes that produces a PrP^{Sc}-degrading enzyme. Ann Microbiol 60:349–353

Mohamad N, Phang LY, Abd-Aziz S (2017) Optimization of metallo-keratinase production by *Pseudomonas* sp. LM19 as a potential enzyme for feather waste conversion. Biocatalys Biotransform 35:41–50

Mohorcic M, Torkar A, Friedrich J, Kristl J, Murdan S (2007) An investigation into keratinolytic enzymes to enhance ungula drug delivery. Int J Pharmaceut 332:196–201

More SS, Sridhar DL, Prakash SN, Vishwakarma J, Umashankar S (2013) Purification and properties of a novel fungal alkaline keratinase from *Cunninghamella echinulata*. Turkish J Biochem 38:68–74

Mortuza MF, Rahman MH, Rahman MH, Nahar A, Khan MRI, Hasan AM, Rahman M (2017) Isolation, biochemical and genetic characterization of extracellular protease producing cattle hide dehairing bacterium—a potential alternative to chemical dehairing. Ecol Gen Genomics 2:3–12

Nam GW, Lee DW, Lee HS, Lee NJ, Kim BJ, Choe EA, Hwang JK, Suhartono MT, Pyun YR (2002) Native feather degradation by *Fervidobacterium islandicum* AW-1, a newly isolating keratinase-producing thermophilic anaerobe. Arch Microbiol 178:538–547

Nayaka S, Gireesh BK, Vidyasagar GM (2013) Purification and characterization of keratinase from hair-degrading *Streptomyces albus*. Int J Bioassays 2:599–604

Ningthoujam DS, Devi LJ, Devi PJ, Kshetri P, Tamreihao K (2016) Optimization of keratinase production by *Amycolatopsis* sp. Strain MBRL 40 from a limestone habitat. J Bioprocess Biotechniq 6:1000282

Nwadiaro PO, Chuku A, Onyimba IA, Ogbonna AI, Nwaukwu IA, Adekojo DA (2015) Keratin degradation by *Penicillium purpurogenum* isolated from tannery soils in Jos, Nigeria. British Microbiol Res J 8:358–366

Okoroma EA, Garelick H, Abiola OO, Purchase D (2012) Identification characterization of a *Bacillus licheniformis* strain with profound keratinase activity for degradation of melanised feather. Int Biodeterior Biodegr 74:54–60
Omran R (2017) Production of keratinases from *Nocardiopsis* sp. 28ROR as a novel Iraqi strain. Asian J Pharmaceut Clin Res 10:160–168
Onifade AA, Al-Sane NA, Al-Musallam AA, Al-Zarban S (1998) A review: potentials for biotechnological applications of keratin-degrading microorganisms and their enzymes for nutritional improvement of feathers and other keratins as livestock feed resources. Bioresourc Technol 66:1–11
Osman Y, Elsayed A, Mowafy AM, Abdelrazak A, Fawzy M (2017) Bioprocess enhancement of feather degradation using alkaliphilic microbial mixture. Br Poult Sci 58:319–328
Ouled-Haddar H, Zaghloul TI, Saeed HM (2010) Expression of alkaline proteinase gene in two recombinant *Bacillus cereus* feather-degrading strains. Folia Microbiol 55:23–27
Parry DA, North ACT (1998) Hard α-Keratin intermediate filament chains: substructure of the N- and C-terminal domains and the predicted structure and function of the C-terminal domains of type I and type II chains. J Struct Biol 122:67–75
Paul T, Das A, Mandal A, Halder SK, DasMohapatra PK, Pati BR, Mondal KC (2014a) Valorization of chicken feather waste for concomitant production of keratinase, oligopeptides and essential amino acids under submerged fermentation by *Paenibacillus woosongensis* TKB2. Waste Biomass Valoriz 5:575–584
Paul T, Das A, Mandal A, Halder SK, DasMohapatra PK, Pati BR, Mondal KC (2014b) Production and purification of keratinase using chicken feather bioconversion by a newly isolated *Aspergillus fumigatus* TKF1: detection of valuable metabolites. Biomass Convers Biorefinery 4:137–148
Paul T, Das A, Mandal A, Halder SK, DasMohapatra PK, Pati BR, Mondal KC (2014c) Biochemical and structural characterization of a detergent stable alkaline serine keratinase from *Paenibacillus woosongensis* TKB2: a potential additive for laundry detergent. Waste Biomass Valoriz 5:563–574
Paul T, Das A, Mandal A, Halder SK, Jana A, Maity C, DasMohapatra PK, Pati BR, Mondal KC (2014d) An efficient cloth cleaning properties of a crude keratinase combined with detergent: towards industrial viewpoint. J Clean Prod 66:672–684
Paul T, Jana A, Mandal AK, Mandal A, Mohpatra PKD, Mondal KC (2016) Bacterial keratinolytic protease, imminent starter for NextGen leather and detergent industries. Sustain Chem Pharm 3:8–22
Paul T, Halder SK, Das A, Bera S, Maity C, Mandal A, Das PS, Mohapatra PKD, Pati BR, Mondal KC (2013) Exploitation of chicken feather waste as a plant growth promoting agent using keratinase producing novel isolate *Paenibacillus woosongensis* TKB2. Biocatal Agric Biotechnol 2:50–57
Pillai P, Archana G (2008) Hide depilation and feather disintegration studies with keratinolytic serine protease from a novel *Bacillus subtilis* isolate. Appl Microbiol Biotechnol 78:643–650
Poovendran P, Kalaigandhi V, Kanan VK, Rani EJ, Poongunran E (2011) A study of feather keratin degradation by *Bacillus licheniformis* and quantification of keratinase enzyme produced. J Microbiol Biotechnol Res 1:120–126
Prakash P, Jayalakshmi SK, Sreeramulu K (2010a) Production of keratinase by free and immobilized cells of *Bacillus halodurans* strain PPKS-2: partial characterization and its application in feather degradation and dehairing of the goat skin. Appl Biochem Biotechnol 160:1909–1920
Prakash P, Jayalakshmi SK, Sreeramulu K (2010b) Purification and characterization of extreme alkaline, thermostable keratinase, and keratin disulfide reductase produced by *Bacillus halodurans* PPKS-2. Appl Microbiol Biotechnol 87:625–633
Queiroga AC, Pintado ME, Malcata FX (2012) Potential use of wool-associated *Bacillus* species for biodegradation of keratinous materials. Int Biodeterior Biodegr 70:60–65
Radha S, Gunasekaran P (2009) Purification and characterization of keratinase from recombinant *Pichia* and *Bacillus* strains. Protein Expr Purif 64:24–31
Rahayu S, Syah D, Suhartono MT (2012) Degradation of keratin by keratinase and disulfide reductase from *Bacillus* sp. MTS of Indonesian origin. Biocatal Agric Biotechnol 1:152–158

Rai SK, Mukheriee AK (2011) Optimization of production of an oxidant and detergent-stable alkaline β-keratinase from *Brevibacillus* sp. strain AS-S10-II: application of enzyme in laundry detergent formulations and in leather industry. Biochem Eng J 54:47–56

Rajesh TP, Mathan RKH, Rajasekar S, Anandaraj B (2016) Computational approach for the optimization of keratinase enzyme production from *Bacillus cereus*. J Environ Biol 37:1473–1478

Rajput R, Gupta R (2013) Thermostable keratinase from Bacillus pumilus KS12: production, chitin crosslinking and degradation of Sup35NM aggregates. Bioresour Technol 133:118–126

Ramnani P, Singh R, Gupta R (2005) Keratinolytic potential of *Bacillus licheniformis* RG1: structural and biochemical mechanism of feather degradation. Can J Microbiol 51:191–196

Ray A (2012) Protease enzyme-potential industrial scope. Int J Technol 2:1–4

Reddy MR, Reddy KS, Chouhan YR, Bee H, Reddy G (2017) Effective feather degradation and keratinase production by *Bacillus pumilus* GRK for its application as bio-detergent additive. Bioresour Technol 243:254–263

Revathi K, Shaifali S, Mohd AK, Suneetha V (2013) A potential strain of keratinolytic bacteria VIT RSAS2 from Katpadi and its pharmacological benefits. Int J Pharmaceut Sci Res 20:89–92

Rieger TJ, De Oliveira CT, Pereira JQ, Brandelli A, Daroit DJ (2017) Proteolytic system of *Bacillus* sp. CL18 is capable of extensive feather degradation and hydrolysis of diverse protein substrates. Br Poult Sci, 58(3):329–335

Rouse JG, Van Dyke ME (2010) A review of keratin-based biomaterials for biomedical applications. Mater 3:999–1014

Saarela M, Berlin M, Nygren H, Lahtinen P, Honkapää K, Lantto R, Maukonen J (2017) Characterization of feather-degrading bacterial populations from birds' nests—potential strains for biomass production for animal feed. Int Biodeterior Biodegr 123:262–268

Saber WIA, El-Metwally MM, El-Hersh MS (2009) Keratinase production and biodegradation of some keratinous wastes by *Alternaria tenuissima* and *Aspergillus nidulans*. Res J Microbiol 5:21–35

Saha S, Dhanasekaran D, Shanmugapriya S, Latha S (2013) *Nocardiopsis* sp. SD5: a potent feather degrading rare actinobacterium isolated from feather waste in Tamil Nadu, India. J Basic Microbiol 53:608–616

Sahoo DK, Das A, Thatoi H, Mondal KC, Mohapatra PKD (2012) Keratinase production and biodegradation of whole chicken feather keratin by a newly isolated bacterium under submerged fermentation. Appl Biochem Biotechnol 167:1040–1051

Saieb FM, Boaker SA, El-komy HM, Issa A (2015) Production and estimation of keratinase by immobilized and free *Bacillus licheniformis* (St. 24). J Appl Environ Microbiol 3:119–122

Sanghvi G, Patel H, Vaishnav D, Oza T, Dave G, Kunjadia P, Sheth N (2016) A novel alkaline keratinase from *Bacillus subtilis* DP1 with potential utility in cosmetic formulation. Int J Biol Macromol 87:256–262

Sarkar AK, Rai V, Gupta AK (2014) Incidence of keratinophilic fungi in areas of Raipur city, Chhattisgarh region, India. Afr J Microbiol Res 8:264–269

Seid M (2011) Production, characterization, and potential applications of a keratinolytic alkaline protease produced by alkaliphilic *Vibrio* sp. A Master of Science (Biotechnology) thesis submitted to the School of Graduate Studies, Addis Ababa University, Addis Ababa, p 74

Sekar V, Kannan M, Ganesan R, Dheeba B, Sivakumar N, Kannan K (2016) Isolation and screening of keratinolytic bacteria from feather dumping soil in and around Cuddalore and Villupuram, Tamil Nadu. Proc Natl Acad Sci India Sect B: Biol Sci 86:567–575

Selvam K, Vishnupriya B (2012) Biochemical and molecular characterization of microbial keratinase and its remarkable applications. Int J Pharm Biol Arch 3:267–275

Shah M, Vaidya R (2017) Partial characterization of keratinase from *Stenotrophomonas maltophilia* K279a and study of its dehairing potential. Int J Biotechnol Biochem 13:95–110

Sharaf EF, Khalil NM (2011) Keratinolytic activity of purified alkaline keratinase produced by *Scopulariopsis brevicaulis* (Sacc.) and its amino acids profile. Saudi J Biol Sci 18:117–121

Sharma R, Devi S (2017) Versatility and commercial status of microbial keratinases: a review. Rev Environ Sci Bio/Technol 17:19–45

Sharma R, Gupta R (2012) Coupled action of γ-glutamyl transpeptidase-glutathione and keratinase effectively degrades feather keratin and surrogate prion protein, Sup 35NM. Bioresour Technol 120:314–317

Sharma S, Gupta A (2016) Sustainable management of keratin waste biomass: applications and future perspectives. Braz Arch Biol Technol 59:e16150684

Sharma S, Gupta A, Kumar A, Kee CG, Kamyab H, Saufi SM (2018) An efficient conversion of waste feather keratin into ecofriendly bioplastic film. Clean Technol Environ Policy. https://doi.org/10.1007/s10098-018-1498-2

Sharma S, Gupta A, Chik SMS, Kee CG, Mistry BM, Kim DH, Sharma G (2017) Characterization of keratin microparticles from feather biomass with potent antioxidant and anticancer activities. Int J Biol Macromol 104:189–196

Shrinivas D, Naik GR (2011) Characterization of alkaline thermostable keratinolytic protease from thermoalkalophilic *Bacillus halodurans* JB 99 exhibiting dehairing activity. Int Biodeterior Biodegr 65:29–35

Sobucki L, Ramos RF, Daroit DJ (2017) Protease production by the keratinolytic *Bacillus* sp. CL18 through feather bioprocessing. Environ Sci Pollut Res 24:23125–23132

Stiborova H, Branska B, Vesela T, Lovecka P, Stranska M, Hajslova J, Jiru M, Patakova P, Demnerova K (2016) Transformation of raw feather waste into digestible peptides and amino acids. J Chem Technol Biotechnol 91:1629–1637

Su C, Gong JS, Zhang RX, Tao LY, Dou WF, Zhang DD, Li H, Lu ZM, Xu ZH, Shi JS (2017) A novel alkaline surfactant-stable keratinase with superior feather-degrading potential based on library screening strategy. Int J Biol Macromol 95:404–411

Suh HJ, Lee HK (2001) Characterization of a keratinolytic serine protease from *Bacillus subtilis* KS-1. J Protein Chem 20:165–169

Syed DG, Lee JC, Li WJ, Kim CJ, Agasar D (2009) Production, characterization and application of keratinase from *Streptomyces gulbargensis*. Bioresour Technol 100:1868–1871

Tamreihao K, Devi LJ, Khunjamayum R, Mukherjee S, Ashem RS, Ningthoujam DS (2017) Biofertilizing potential of feather hydrolysate produced by indigenous keratinolytic *Amycolatopsis* sp. MBRL 40 for rice cultivation under field conditions. Biocatalys Agric Biotechnol 10:317–320

Tiwary E, Gupta R (2010) Medium optimization for a novel 58 kDa dimeric keratinase from *Bacillus licheniformis* ER-15: biochemical characterization and application in feather degradation and dehairing of hides. Bioresour Technol 101:6103–6110

Tiwary E, Gupta R (2012) Rapid conversion of chicken feather to feather meal using dimeric keratinase from Bacillus licheniformis ER-15. J Bioprocess Biotech 2:123 https://doi.org/10.4172/2155-9821.1000123

Tork SE, Shahein YE, El-Hakim AE, Abdel-Aty AM, Aly MM (2016) Purification and partial characterization of serine-metallokeratinase from a newly isolated *Bacillus pumilus* NRC21. Int J Biol Macromol 86:189–196

Ulfig K, Płaza G, Worsztynowicz A, Mańko T, Tien AJ, Brigmon RL (2003) Keratinolytic fungi as indicators of hydrocarbon contamination and bioremediation progress in a petroleum refinery. Pol J Environ Stud 12:245–250

Vasileva-Tonkova E, Gousterova A, Neshev G (2009) Ecologically safe method for improved feather wastes biodegradation. Int Biodeterior Biodegr 63:1008–1012

Veenayohini K, Sangeetha D (2016) Isolation and identification of keratinolytic bacteria from poultry waste and assessment of its keratinase activity on chicken feathers. Int J Appl Res 2:396–402

Verma A, Singh H, Anwar MS, Kumar S, Ansari MW, Agrawal S (2016) Production of thermostable organic solvent tolerant keratinolytic protease from *Thermoactinomyces* sp. RM4: IAA production and plant growth promotion. Frontiers Microbiol 7:1189

Vermelho AB, Mazotto AM, de Melo ACN, Vieira FHC, Duarte TR, Macrae A, Nishikawa MM, Bon EPS (2010) Identification of a *Candida parapsilosis* strain producing extracellular serine peptidase with keratinolytic activity. Mycopathol 169:57–65

Voet D, Voet JG (1995) Three-dimensional structure of proteins. In: Stiefel J (ed) Biochemistry, 2nd edn. Wiley, New York, pp 154–156

Wan MY, Dong G, Yang BQ, Feng H (2016) Identification and characterization of a novel antioxidant peptide from feather keratin hydrolysate. Biotechnol Lett 38:643–649
Wang B, Yang W, McKittrick J, Meyers MA (2016) Keratin: structure, mechanical properties, occurrence in biological organisms, and efforts at bioinspiration. Progr Mater Sci 76:229–318
Wang D, Piao XS, Zeng ZK, Lu T, Zhang Q, Li PF, Xue LF, Kim SW (2011a) Effects of keratinase on performance, nutrient utilization, intestinal morphology, intestinal ecology and inflammatory response of weaned piglets fed diets with different levels of crude protein. Asian–Austral J Anim Sci 24:1718–1728
Wang D, Zeng Z, Piao X, Li P, Xue L, Zhang Q, Han X, Zhang H, Dong B, Kim SW (2011b) Effects of keratinase supplementation of corn-soybean meal based diets on apparent ileal amino acid digestibility in growing pigs and serum amino acids, cytokines, immunoglobulin levels and loin muscle area in nursery pigs. Arch Anim Nutr 65:290–302
Wang P, Wang Q, Cui L, Gao M, Fan X (2011c) The combined use of cutinase, keratinase and protease treatments for wool bio-antifelting. Fibers Polym 12:760–764
Werlang PO, Brandelli A (2005) Characterization of a novel feather-degrading *Bacillus* sp. strain. Appl Biochem Biotechnol 120:71–79
Wu KC, Bryan JT, Morasso MI, Jang SI, Lee JH, Yang JM, Marekov LN, Parry DA, Steinert PM (2000) Coiled-coil trigger motifs in the 1B and 2B rod domain segments are required for the stability of keratin intermediate filaments. Mol Biol Cell 11:3539–3558
Wu WL, Chen MY, Tu IF, Lin YC, EswarKumar N, Chen MY, Ho MC, Wu SH (2017) The discovery of novel heat-stable keratinases from *Meiothermus taiwanensis* WR-220 and other extremophiles. Sci Reports 7:4658
Yamamura S, Morita Y, Hasan Q, Yokoyama K, Tamiya E (2002) Keratin degradation: a cooperative action of two enzymes from *Stenotrophomonas* sp. Biochem Biophys Res Comm 294:1138–1143
Yang Y (2012) Skin-whitening and freckle-dispelling essence and preparation method thereof. Patent: CN102614104
Yang H, Li J, Shin H, Du G, Liu L, Chen J (2014) Molecular engineering of industrial enzymes: recent advances and future prospects. Appl Microbiol Biotechnol 98:23–29
Yue XY, Zhang B, Jiang DD, Liu YJ, Niu TG (2011) Separation and purification of a keratinase as pesticide against root-knot nematodes. World J Microbiol Biotechnol 27:2147–2153
Yue X, Zhang B, Zou J, Chen W, Yang N (2017) Characterization of a new bacterium with high alkaline keratinase activity from *Calotes versicolor* feces. J Biotechnol Res 8:83
Zaghloul TI, Haroun MA, El-Gayar K, Abedalal A (2004) Recycling of keratin-containing materials (chicken feather) through genetically engineered bacteria. Polym-Plast Technol Eng 43:1589–1599
Zhang X (2012) Applying the mutation of *Bacillus subtilis* and the optimization of feather fermentation medium to improve keratinase activity. Adv Biol Chem 2:64–69
Zhang RX, Gong JS, Dou WF, Zhang DD, Zhang YX, Li H, Lu ZM, Shi JS, Xu ZH (2016a) Production and characterization of surfactant-stable fungal keratinase from *Gibberella intermedia* CA3-1 with application potential in detergent industry. Chem Pap 70:1460–1470
Zhang RX, Gong JS, Su C, Zhang DD, Tian H, Dou WF, Li H, Shi JS, Xu ZH (2016b) Biochemical characterization of a novel surfactant-stable serine keratinase with no collagenase activity from *Brevibacillus parabrevis* CGMCC 10798. Int J Biol Macromol 93:843–851
Zhang RX, Gong JS, Zhang DD, Su C, Hou YS, Li H, Shi JS, Xu ZH (2016c) A metallo-keratinase from a newly isolated *Acinetobacter* sp. R-1 with low collagenase activity and its biotechnological application potential in leather industry. Bioprocess Biosys Eng 39:193–204
Zhang S, Long L, Yin H, Xiao Z, Li Q, Zhang S, Tian X, Li C, Chen S, Yin T, Chen Y (2010) Pearl albefaction method mediated by keratinase and combined with redox. Patent: CN100579412
Zhao H, Mitsuiki S, Takasugi M, Sakai M, Goto M, Kanouchi H, Oka T (2012) Decomposition of insoluble and hard-to-degrade animal proteins by enzyme E77 and its potential application. Appl Biochem Biotechnol 166:1758–1768

Chapter 6
Keratin as a Biopolymer

Sarthak Saha, Muhammad Arshad, Muhammad Zubair and Aman Ullah

Abstract Keratin has recently gained a lot of limelight among various proteins. Keratin is a renewable, sustainable, biocompatible, and biodegradable bioresource, and these characteristics make it a promising candidate for diverse applications. It is a fibrous protein mainly found in feathers, hair, wool, animal claws, and fingernails. It is one of the key structural materials present in the outer layer of human skin, hence having a lot of applications in developing cosmetics. As an effective biopolymer, keratin containing a different sequence of amino acids have several functional properties and applications. One of the major applications is to be used as a biosorbent in wastewater treatment, which is attributed to expose functional groups present in the highly folded network of amino acids and their modification that improves the reactivity as adsorbent. Keratin-based nanocomposites are used in tissue engineering as it can impart characteristics like increased regeneration and hydration. Another great application in the field of biomedical engineering is drug delivery using keratin-based carriers. Thermoplastics developed from keratins can be used as films in biomedical application. This chapter will highlight the major applications of keratin proteins in detail.

Keywords Keratin · Extraction · Biosorbents · Biomedical · Wounds · Cosmetics Thermoplastics

---

S. Saha · M. Arshad · M. Zubair · A. Ullah (✉)
Department of Agricultural, Food and Nutritional Science, University of Alberta,
4-10 Agriculture/Forestry Centre, Edmonton, AB T6G2P5, Canada
e-mail: ullah2@ualberta.ca

S. Saha
Department of Chemical Engineering, Indian Institute of Technology,
Kharagpur 781021, India

S. Sharma and A. Kumar (eds.), *Keratin as a Protein Biopolymer*, Springer Series on Polymer and Composite Materials, https://doi.org/10.1007/978-3-030-02901-2_6

1 Introduction

Presently, a significant interest has been developed in bio-based materials due to their potential as a substitute for the traditional petroleum-derived polymers. The production of the synthetic polymeric material is heavily dependent on the petroleum resources. The excessive use of petroleum creates issues related to environmental security, economic stability, and sustainability. Furthermore, the petroleum-based resources are finite and depleting with every passing day. There is an urgent need to address the problems interconnected to the petroleum-based polymeric materials since the demand continues to rise due to ever-increasing world population, strict environmental legislations and public awareness.

In recent years, the focus of scientific community has been diverted toward renewable, biodegradable, and sustainable resources for the development of polymeric materials. The use of renewable resources offers global sustainability and reduction in environmental pollution. There has been much attention in exploiting naturally occurring biopolymer as an alternative for the crude oil based polymers. Biopolymers such as carbohydrates, proteins, and lipids are favorable choices over the petroleum-based polymers because they are abundantly available, renewable, biodegradable, and biocompatible.

Amongst renewable biopolymer resources, proteins are one of the favorite candidates, being used as a bioresource to produce bio-based materials. Proteins have different structural network due to the presence of different amino acid chains which provide several chemical functional groups as compared to other bio-based macromolecules such as carbohydrates and lipids. Moreover, protein's abundant availability, biodegradability, and non-eco-toxicity make them suitable for developing bio-based materials. Numerous proteins have been studied for the development of naturally bio-based materials such as keratin, collagen, albumin, gelatin, and fibroin (Khosa and Ullah 2013; Barone et al. 2005; Parenteau-Bareil et al. 2010).

Among proteins, keratin-derived materials have shown potential to transform the world of bio-based materials owing to their intrinsic biocompatibility, biodegradability, mechanical durability, and natural abundance (Balaji et al. 2012; Poole et al. 2008; Parenteau-Bareil et al. 2010). Likewise, keratin is the most abundant structural proteins in epithelial cells (Coulombe and Omary 2002) and a most important biopolymer in animals along with collagen (Wang et al. 2016). All the above-mentioned properties of keratin make it an interesting contender for the development of keratin-based materials with novel characteristics in a variety of applications.

Keratin does not have any harmful components and can be used directly in a variety of applications such as cosmetics (Matsunaga et al. 1983, 1984; Barba et al. 2007), pharmaceutical and biomedical (Kelly 2016; Sierpinski et al. 2008; Tachibana et al. 2002; Srinivasan et al. 2010). Various keratin-based biosorbents have also been developed for heavy metals removal from drinking water and industrial wastewaters (Khosa and Ullah 2014; Khosa et al. 2013; García-Sabido et al. 2016; Aluigi et al. 2009). Keratin-derived materials are also being used as a thermoplastic materials (Ullah and Wu 2013; Barone et al. 2005; Hong and Wool 2005; Arshad et al. 2016a;

Kaur et al. 2018). Additionally, keratin from feathers has been used to develop films, fibers (Xie et al. 2005; Barone and Schmidt 2005), hydrogels (Hill et al. 2011; Pace et al. 2013) and in the textile industry (Reddy et al. 2014; Zoccola et al. 2009).

This chapter will discuss the keratin structure and its various resources with a detailed explanation of keratin proteins extraction methods. The industrial usage of keratin-based materials will also be discussed with an emphasis on biomedical, cosmetic, thermoplastic, and biosorbent applications.

2 Structure and Characteristics

Keratin is a polypeptide consisting of amino acids having intermolecular bonding of cysteine and few intramolecular bonding of polar and nonpolar groups. The cysteine residues have thiol groups which produce strong disulfide bonds leading to the crosslinking of the matrix molecule. The model proposed by Astbury and Bell (1941) is the most widely accepted. Keratins generally have a folded configuration. The polypeptide backbone contains a number of side chains which allow the sequential folding of the keratin fibers. These structures under stress become straight but attain its original form when released (Astbury and Bell 1941). Keratins are normally distinguished into two ordered structure on the basis of X-ray diffraction α and β patterns (Fig. 1). Also, the internal structure of every keratin has α-helices and β-sheets. The elastic nature of keratin fiber is due to the interplay between α-helices and β-sheets configuration of the protein. The disulfide linkage between cysteine molecules present as an intrachain and interchain bonds responsible for its good stability and less solubility. It helps the formation of a stable 3D structure (Fig. 2b) which can provide good amount of resistance to enzymatic reactions. Ionic bonds also exist in the keratin structure between the ammonium cations and carboxylic anions (Fig. 2a). The ionic bonds are dependent on the pH as these are lowest in severe acidic or basic conditions or lower at the isoelectric point as they form zwitterions. The isoelectric point is observed to be somewhere around pH 4–5 (Shavandi et al. 2017).

Regarding methods of biosynthesis (Giroud and Leblond 1951) and the measure of sulfur cross connections, keratins can be assigned as delicate keratins (e.g., stratum corneum) generally pitifully united and with a lower measure of sulfur and lipids, and hard keratins found in hair, nails, paws, noses, plumes, which have a more rational structure and a higher measure of sulfur (Argyris 1973).

3 Sources of Keratin

There are two types of keratin, which can be differentiated by X-ray diffraction patterns α- and β-keratins. The former found highly in wool, hair, horn, nails, hooves (Fig. 3), and the outer most layer of the skin also known as the stratum corneum. The β-keratin is the real part of hard avian and reptilian tissues, for example, plumes,

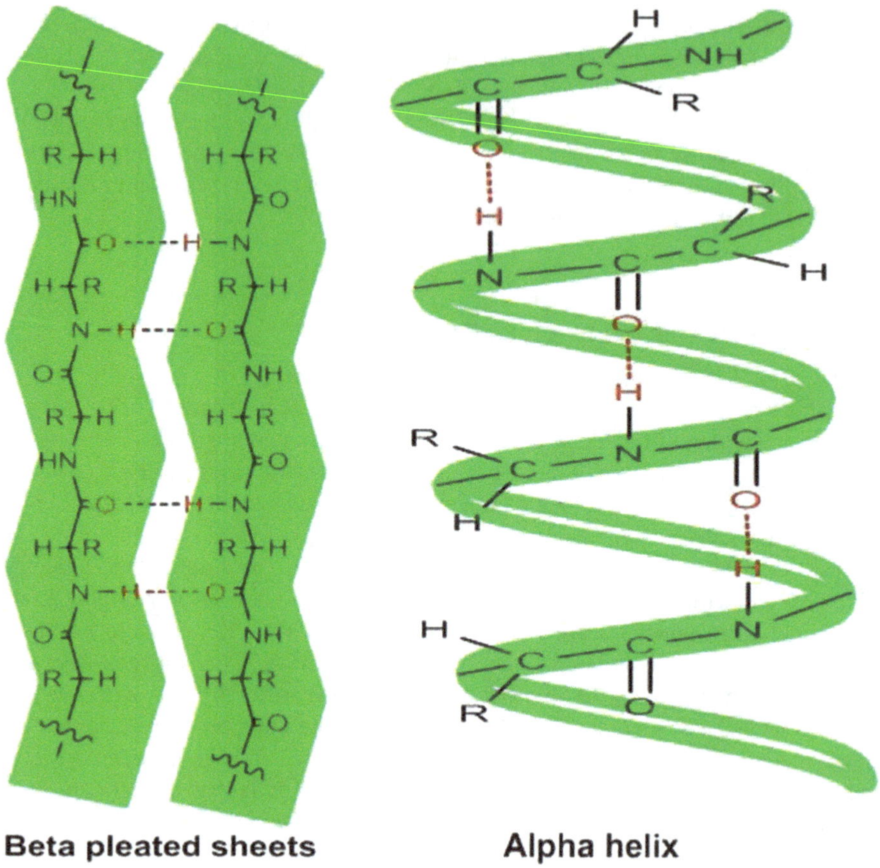

Fig. 1 Secondary structures of keratin protein (beta pleated sheets and alpha helix)

hooks, and mouths of flying creatures, and scales and paws of reptiles (Chen et al. 2012).

Stratum corneum is the outermost layer of mammalian skin, fills in as a dissemination boundary, barrier from outside assault and even cover from predators (Meyers et al. 2008; Tomlinson et al. 2018). It is made out of leveled cornified keratinocytes; these anucleated cells are implanted in a lipid-rich intercellular grid. The keratin fibers reach out all through the whole cytoplasm in a web-like structure, also coordinating cell–cell intersections, hence expanding the mechanical help (Coulombe and Wong 2004).

Wool is a critical case of the hard keratinous material. It is by a long shot the most vital creature fiber utilized as a part of material application, while the structure and mechanical conduct have been broadly contemplated. A clean wool fiber contains keratinous proteins up to 82% with a high centralization of cysteine. Around 17% is protein material of low cysteine content known as "non-keratinous material" found

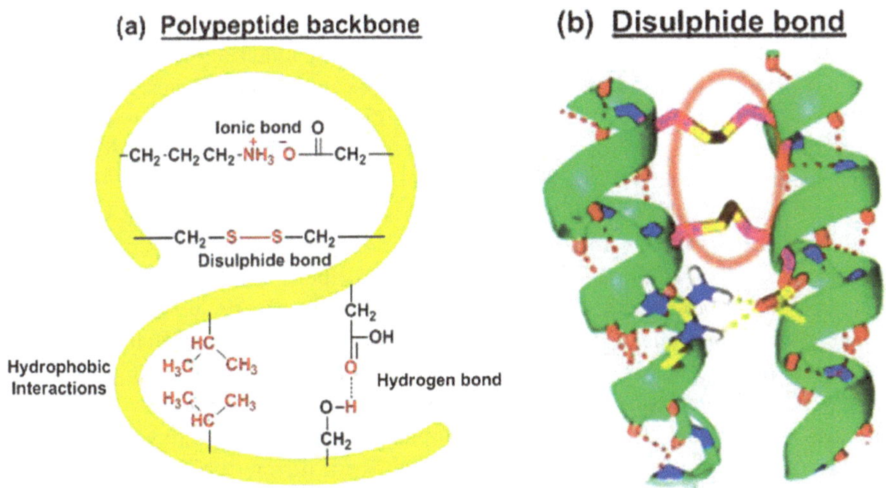

Fig. 2 Polypeptide backbone (**a**) and 3D structure of keratin protein (**b**) representing inter and intramolecular bonding including disulfide linkages between α-helices accountable for intensified strength and stability of keratin (Zhang et al. 2017) (**b** has been reproduced with permission from ACS publisher)

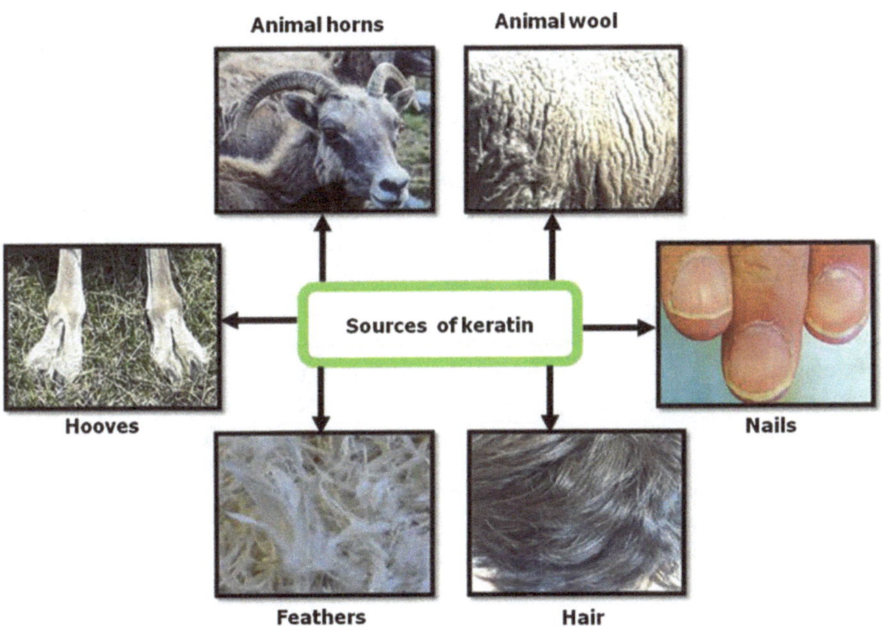

Fig. 3 Various sources of keratin biomass

basically in the plasmalemma complex, and around 1% of non-proteinaceous material contains waxy lipids with some traces of polysaccharide material. Keratin from wool fiber can be ordered into four distinctive molecular weight categories: a low sulfur content (LS) with a molecular weight (MW) of 45–60 kDa and a portion with a high sulfur content with a MW of 11–28 kDa, and divisions with a high glycine or high tyrosine content with a MW of 9–12 kDa.

Hooves are hard keratinous materials and have been the focal point of various examinations. The hoof wall adapts to an assorted variety of high ground-response powers and exchanges these to the hard skeleton, furthermore, any harms stay in the foot until the point when that part is worn off. Subsequently, the wall must be competent to withstand rehashed high burdens, and studies demonstrate that the stratum medium of the hoof wall is a standout amongst the most break safe organic materials known (Kasapi and Gosline 1997). Nails, covering the closures of fingers and toes in primates and a couple of warm-blooded animals, and bent paws from warm-blooded animals (e.g., feline) demonstrate an α-type X-beam diffraction pattern (Bear and Rugo 1951); both are gathered as nails in this segment. Nails fill in as a firm support to the delicate terminal cushions, keeping the skin from moving in reverse.

Among β-keratin feather is the most abundant source along with beaks and claws. In general, the design of a feather shaft has resemblance with a sandwich-organized composite: a thick keratin cortex encompasses an elastic keratin medulla, which amplifies quality and opposes flexure and burst while limiting weight. Chicken feathers are one of the abundant pool of keratin and waste of poultry industry contributing more than 65 million tons around the world (Poole et al. 2008). These account for almost 4–6% of the aggregate weight of a develop chicken (Rosa 2008). Poultry preparing industry of USA alone is producing an excess amount (4 billion pounds per year) of chicken feathers (Arshad et al. 2016a). They are extremely lightweight and have 91% keratin macromolecules with a normal sub-atomic weight of 10 kDa (Pedram Rad et al. 2012; Wang and Cao 2012). They have properties like high rigidity, water insolubility, basic durability, and stability over an extensive variety of pH (Rosa 2008). Poultry quills have limited uses and most are disposed into landfills expanding natural weights. Claws are bent, pointed extremities found toward the finish of digits in many amniotes (earthbound egg-laying creatures), and they contrast from nails which are leveled and do not have a sharp point. Claws of winged animals and reptiles demonstrate a β-type structure, while the claws of warm-blooded creatures demonstrate an α-type structure. They comprise of a shallow and profound layers of hard keratin, these two layers being created by the basal and terminal lattices individually (Thorndike 1968). The β-keratin fibers are arranged parallel to the heading of development in claws of different species and in between angle in claws of few primates (Baden 1970).

4 Extraction of Keratin

One of the major hurdles is to extract or isolate the keratin protein from keratin-rich materials. Here, we will briefly discuss few of the most extensively used methods.

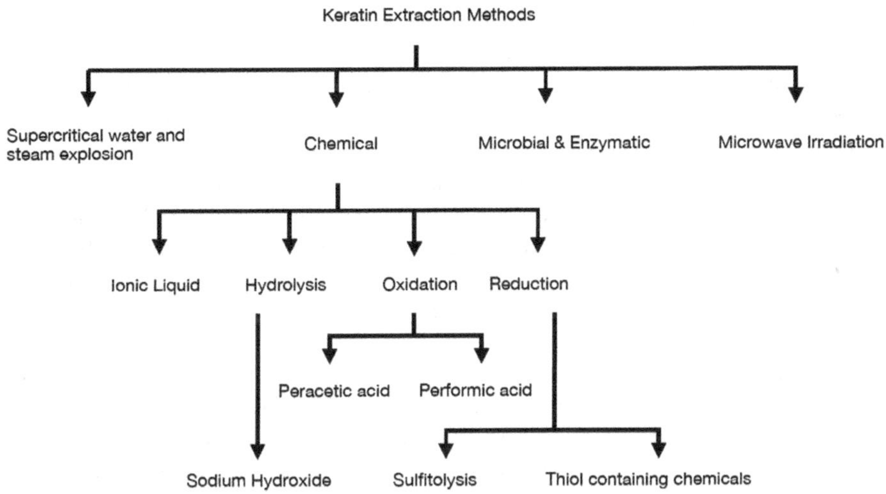

Fig. 4 Classification of keratin extraction methods used for different keratin-rich sources

There are mainly four types of chemical methods—reduction, oxidation, hydrolysis, and use of ionic liquid (Fig. 4). Other methods involve using microbes and enzymes, microwave irradiation, and steam explosion (Shavandi et al. 2017). This disulfide linkage can be diminished utilizing thiol containing chemicals. Various reducing agents under different preparing conditions, i.e., denaturing agents and different pH levels have been reported (Kamarudin et al. 2017; Goddard and Michaelis 1935).

It has been reported that treating wool with alkali causes the sulfur nucleus to split and degrade into cysteine residues (Blackburn and Lee 1956). High concentration of alkali separates the hydrogen from sulfate and carboxylic groups and encourages solubilization, in spite of the damage to the peptide chains (Zhang et al. 2015). The breakdown of these bonds can prompt the development of the basic sulfide scent amid the treatment procedure, which has an exceptionally bad odor. There are few drawbacks that inhibit the commercialization and scale-up of soluble base technique, these include mangling and splitting up of protein backbone, utilization of high quantity of alkali reagents and therefore, high measures of acid needed to counteract the effect of alkali and precipitate the protein.

A standard reduction method is the use of mercaptoethanol to reduce the disulfide bond. But it is a dangerous chemical and is not commercially viable or eco-friendly because of its high cost and issues identified with its upsetting odor and for being normally harmful to the surrounding. Hence, sodium sulfite is a viable alternative for reduction. This reduction method is known as sulfitolysis and creates S-sulfonated ($R–S–SO_3^-$) residue and cysteine thiol ($R–S^-$) (Shavandi et al. 2017) as shown in Eq. (1).

$$R - S - S - R + Na^{2+}SO_2^{3-} \rightleftharpoons R - S^- + R - S - SO_3^- \tag{1}$$

Earland and Knight (1955) and Buchanan (1977) were early researchers to report oxidation methods for extraction of keratin. They used peracetic acid and performic acid, respectively, to obtain a solubilized homogeneous solution. Often it has been reported that the entire wool keratin is not dissolved using this method as it is not a homogeneous material. Hence, often residues of insolubilized keratins are found, which are postulated to be β-keratin. The α-keratin which has a crystalline form is believed to be solubilized whereas the β part forms a jelly-like material (Earland and Knight 1955). Proteins extracted via oxidation (keratosis) are modified synthetically and the bisulfite bonds are changed over to sulfonic acids. In this way, these proteins may have diverse physiochemical properties than the keratins that are acquired through other preparing techniques.

Another technique used for extraction is steam flash explosion (SFE) that has been utilized for the generation of bio-based materials. In this procedure, the material is presented to high-temperature steam for a brief timeframe, which infiltrates in the tissues, the cells of the material, and after that a quick decompression and blast happen in a millisecond response. This method has an edge over other methods in terms of processing time, digestibility, potential application in low temperature, and higher dissolubility but lacks in one aspect, i.e., the small concentration of cysteine in the final extracted keratin (Tonin et al. 2006; Zhao et al. 2012).

Microwave heating is another method which was proposed by Chen et al. (2015). Microwaves are believed to reduce the activation energy due to the uniform heating as compared to the irregular and nonuniform heating of the traditional heating method. Though the exact reason behind lower activation energy is not known, it is postulated that the ester group which gets hydrolyzed due to the microwave radiation can be one of the possible reasons.

5 Application as Biosorbent

Sorbents are the substances which have the capability to remove toxic contaminants/pollutants by a phenomenon of adsorption or absorption and sometimes by a combination of both (Karan et al. 2011). Over the years, number of techniques such as photocatalytic degradation, aerobic degradation, coagulation, reverse osmosis, filtration, ozonation, flocculation, ion exchange, extraction, precipitation, and adsorption have been developed for contaminants removal from water (Ahmad et al. 2010; Ahmed and Mohammad 2014; Al-Zaben and Mekhamer 2017; Khosa et al. 2013; Kong et al. 2012; Pham and Inoue 2012). Adsorption process is the best among these techniques being an environment friendly, economical, and highly efficient for the removal of pollutants even in low concentration (Chen et al. 2011; Xi and Chen 2014; Yakout and Daifullah 2013). Adsorption is the process which involves the transfer of ions in between two phases: from the liquid phase to the surface of a solid (solid phase). The substance being attached/adsorbed onto the surface of solid is termed as adsorbate, while the one which is adsorbing named as adsorbent. The interaction among adsorbate and adsorbent could be physical or chemical mainly depending upon their nature or composition. When there is physical interaction between solid surface and adsorbed molecules such as van der Waals forces, London dispersion forces or hydrogen bonding, the process is called as physisorption. However, when the attractive forces between the adsorbent and adsorbate molecules occur as a result of chemical bonding, the process is named as chemisorption. When conditions (pH and temperature) are appropriate, both physical and chemical interactions can occur simultaneously (Coonery 1999).

Nevertheless, the most commonly used conventional adsorbent such as granular-activated charcoal has higher production cost making it economically less viable to be employed for large-scale water treatments (Ata et al. 2012). Recently, exploitation of low-cost adsorbents exhibiting highest metal binding capacities increased (Leung et al. 2000). Utilization of agricultural wastes and industrial byproducts for the development of adsorbents are a potential substitute of conventional sorbents in terms of low cost, environment friendliness, biodegradability, remarkably efficient for solutions with low concentration of pollutants and ease of design and operation (Mohammad 2013). Application of keratin as a biosorbent for the removal of pollutants from naturally contaminated and process water is promising due to its high surface area and number of reactive functional sites acquired after slight modification (Khosa and Ullah 2013; Al-Asheh et al. 2002; Al-Asheh et al. 2003; Banat et al. 2002; Gupta et al. 2006; Mittal 2006). Keratin materials having number of active functional groups such as $-NH_2$, –COOH, –OH, and –SH susceptible for metal binding. Most of these functional groups are entrapped or hidden in a highly cross-linked and folded network of protein mainly due to disulphide (–S–S–) linkages. The hidden functional groups can be exposed by unfolding the network of protein on treatment with chemical reagents; as a result, more reactive sites will emerge to interact with foreign species. Increase in surface area and improved reactivity of highly exposed functional groups demonstrated higher interaction with organic dyes and toxic metals for the purification of wastewater as reported in literature (Al-Asheh et al. 2002; Arai et al. 1983; Banat et al. 2002; Gupta et al. 2006; McGovern 2000; Mittal 2006).

One of the most abundant, cheap and renewable source of keratin are poultry feathers containing more than 90% of it. To date, most of the poultry feathers are disposed of in landfills with little usage as an animal feed, composite preparation, and other applications (Arshad et al. 2016a; Huda and Yang 2009; Kaur et al. 2018; Schrooyen et al. 2001; Ullah and Wu 2013; Esparza et al. 2018b; Yussef et al. 2017). Recently, an increased interest has been developed in the use of keratin biopolymer to treat wastewater for the sorption of dissolved contaminants (Sekhar et al. 1998). Gamze-Turan reported poultry litter, a mixture of excrement, bedding material, feed and keratin protein (chicken feathers) as an effective material for the removal of Zn(II) and Cu(II) from wastewater (Gupta et al. 1997; Stephenson et al. 1990). Modified chicken feathers (keratin) have also been used by Turan (2011) to treat wastewater for the sorption of chromium(VI) ions. They treated chicken feathers with sodium hydroxide and inferred that this treatment brought structural changes only on outer surface instead of internal network of keratin protein. As a result, low sorption of Cr(VI) ions (10–80 ppm) was observed by modified keratin biosorbent(Sun et al. 2009). Khosa et al. modified chicken feathers by treating individually with four different dopants including poly(N-isopropylacrylamide), poly(ethylene glycol), trisilanolcyclohexyl polyhedral oligomeric silsesquioxanes (POSS) and allyl alcohol and applied them for the soption of arsenic from water. They found trisilanolcyclohexyl POSS and allyl alcohol modified chicken feathers highly effective for the sorption of As(III) on treatment with arsenic contaminated water solution (Khosa and Ullah 2014). The same research group also reported chemical modification of chicken feathers with maleimide terminated poly(N-isopropylacrylamide) and methanol to enhance the reactivity/sorption capacity of the biosorbents. They concluded modified sorbents displayed considerable performance for the sorption of negatively charged arsenic species as compared to unmodified chicken feathers when treated with arsenic contaminated water (Khosa et al. 2013). The application of keratin protein (chicken feathers) for the removal of organic contaminants specifically for naphthenic acids from oil sand process-affected water (OSPW) has been investigated by Arshad et al. (2016b), where the keratinous structural reactivity or binding capacity of biopolymer was improved by treating with POSS nanocages and goethite dopant. Both of the modified biosorbents represented highest biosorption capacities (up to 4880) at maximum concentration of naphthenic acids in OSPW (Arshad et al. 2016b). Modified keratin biopolymers have also been found as an effective sorbent for oil–water separation (Radetić et al. 2003). These studies represent that when tertiary structure of keratin biopolymer is altered/unfolded to get more porous network and enhanced chemical reactivity of the functional groups (amides and amino acids) to interact with the pollutants makes keratin an acceptable alternative of conventional sorbents for the purification of wastewater.

6 Keratin in Biomedical Applications: Treatment of Wounds

Health has always been the primary concern of human beings. Humans and all other creatures are very much prone to injury and different civilizations have come up with their own way of dealing with this issue. Tissue engineering is a field in which scientists have indulged to find an appropriate solution. Scaffold fabrication is one key factor for the applications of tissue engineering as present research focuses on developing porous scaffold which are biodegradable and can be used to regenerate tissues. The modern concept of wound healing revolves around the concept of providing optimal conditions around the wound for better growth of epithelial cells and bioactivity alongside preventing microbial infection. Hence, efforts are made to develop porous materials which would impart antimicrobial characteristics clubbed with proper supply of bioactive molecules (Nayak and Gupta 2017). Keratin being biodegradable and biocompatible serves as an excellent material for scaffolds. They are well known for their ability to enhance the growth of fibroblast and keratinocyte which are kinds of mammalian cells. The extracellular matrix of epidermis are keratin based, keratinocytes is also a part of epidermis. Keratin lacks the water retention capacity and tends to dehydrate easily at room temperature. To curb this problem, composites using polysaccharides like agar, alginate, and gellan (Gupta and Nayak 2015; Nayak and Gupta 2015) can be prepared which have high water retention capacity due to their hydrophilic nature. Studies have shown up to 223% absorption capacity (Nayak and Gupta 2017) for such composites. Another interesting property of keratin is its antimicrobial property. The present research related to scaffolds revolves around the following properties (i) biodegradable and biocompatibility (ii) suitable pore sizes for gaseous exchange and cell infiltration (iii) mechanical strength (iv) allow proliferation by allowing cells to adhere to surface (v) antibacterial property to prevent wound infection (vi) should be able to mimic the extracellular matrix's function (Karageorgiou and Kaplan 2005; Ma 2008; Mandal and Kundu 2009; Pyun et al. 2015; Ulery et al. 2011).

Wool keratin is used as conventional hydrocolloids for wound dressing, but several studies had showed human hair based keratin to have better wound healing capacity. Most of the recent studies focus on keratin-based hydrogels because of hydrogel's high potential in tissue engineering related application (Esparza et al. 2017, 2018a). This is because of its cross-linked polymer networks which resembles natural extracellular matrix of the cell. Kakkar and Madhan (2016) reported a keratin silica hydrogel with good biocompatibility having considerable hardness was claimed to be very well suited for developing scaffolds. Keratin, in this case, was extracted from bovine hooves. Nakata et al. (2015) studied various keratin-based hydrogels, namely, carboxymethylated keratin, acetamidated keratin, and aminoethylated keratin. The latter two modifications were observed to promote good cell proliferation along with sustained drug release up to 3 days. Keratins, in each of the cases, were derived from wool. The high compatibility of hair keratin with keratinocyte differentiation and proliferation is a very beneficial attribute for a scaffold but cannot be widely used due to its poor tensile strength. This observation has been well explained by Park et al. (2015) They made a brilliant comparison between wool derived and human hair derived hydrogels. The latter was observed to have reduced wound sizes significantly as it could enhance the fibroblast proliferation and collagen synthesis. Although the tensile strength of human hair based keratins were less as compared to wool based keratins. The peak tensile strength was observed on day upon which the keratin started degenerating and hence low tensile strengths were observed on 14th day. Xu et al. (2013) observed the effect of human hair based keratin concentration in scaffolds created by freeze-drying method. The porosity was observed to be the highest (94%) at 2 wt% keratin, while highest water uptake capacity of 554% was observed at 8 wt% keratin. The smaller pore size distribution enhanced the water uptake capacity and hence was a tradeoff between porosity and water uptake capacity. The hydrophilic nature of keratin was responsible for the good cytocompatibility, as it enhances cell attachment. Very high proliferation was observed till 22nd day especially for lower keratin concentration due to the high porosity as the cell growth was facilitated in the larger pores. In one of the recent studies Kim et al. (2017) tried to develop a bilayer scaffold. The upper layer of which was composed with human hair based keratin and chitosan nanofibers which provided improved tensile strength and hence could be used for epidermis regeneration. This layer protects the wound from infection and external stimuli. The lower layer, on the other hand, was made of gelatin methacrylate which was photopolymerized to form the entire scaffold. The bottom layer mainly functions to facilitate the culture of fibroblasts. This study used electrospinning to create the keratin–chitosan nanofiber matrix which was cross-linked to glutaraldehyde in order to maintain the shape of matrix in aqueous conditions for 2 h. The tensile strength and elongation at break of composite nanofiber were 3.38 MPa and 6%, respectively, whereas for pure keratin it showed 0.87 MPa and 3% respectively, in dry conditions. The bilayer scaffold in a nutshell could facilitate the proliferation of human keratinocytes cells and normal human dermal fibroblasts and keratinocyte differentiation in the different layers. In another study, Wang et al. (2017) developed a keratin hydrogel scaffold derived from chicken feathers which showed characteristics similar to that of human hair based keratins. The composite was prepared by crosslinking the thiol group of keratins with H_2O_2. They showed improved biodegradability, regenerative properties, and compatibility with the tissues. The porosity of this scaffold ranged from 77% to 82% with varying concentration. Porosity was observed to be inversely dependent with the keratin concentration. The time period for complete re-epithelialization was observed to be 21 days while 90% of wound closure was attained in 10 days. Han et al. (2015) suggested new modification on human hair based keratins to control the erosion of hydrogel which could further affect the release of therapeutic agents. The tunable therapeutic agent's gel erosion and release rates were achieved by capping the cysteine thiol group using iodoacetamide alkylation which could result in controlling the amount of disulfide crosslinking. Though they did not prepare a scaffold, the results showed the modifications could be well integrated into a scaffold.

Few researchers have used chitosan blends to enhance several characteristics of scaffold (Tan et al. 2015). Chitosan is of semi-crystalline nature which gives properties like biocompatibility and biodegradability. Saravanan et al. (2013) reported a keratin chitosan composite scaffold which could enhance the above-mentioned characteristics. Keratin extracted from chicken feathers was transformed into nanoparticles using desolvation process. These were blended with chitosan and freeze-drying method was used to create the scaffold. SEM images confirmed pores ranging from 6 μm to 17 μm, these allowed higher cell penetration and nutrient diffusion. Kakkar et al. (2014) used keratin extracted from bovine hooves, blended with chitosan and gelatin to create a composite scaffold. The scaffold was created by freezing a homogenized solution having keratin, chitosan, and gelatin in the ratio of 1:1:2 by weight. The scaffold showed extremely high water uptake capacity $1796 \pm 23\%$ and porosity of 86.86%. They compared the results with collagen scaffold and found the biocompatibility to be similar after 3, 5 and 7 days. Tensile strength of the dry composite scaffold was also observed to be pretty good (95.69 MPa) as compared to collagen (107.33 MPa). Tanase and Spiridon (2014) reported a similar composite only by using polylactic acid (PLA) instead of gelatin, the material formed was hard with lower tensile strength of nearly 50 MPa and cell viability of 100% at the end of 1st and 2nd day.

Nayak and Gupta (2015) studied the compatibility of keratin agar based patches with myofibroblast as studies have shown it helps in re-epithelialization and enhances healing of wound (Moulin et al. 2000). The scaffold fabricated showed an apparent porosity of 94% along with water retention capacity of 160%. No further antimicrobial agents were added as keratin was observed to impart the required characteristic against both Gram-positive and negative bacteria. One of the salient features of any scaffold is the biodegradability to allow production of extracellular matrix by the cells. This system showed up to 50% decrease in weight after 4 weeks and hence had good biodegradable property. The scaffold was found biocompatible against myofibroblast cell line as it promoted growth of cells and their proliferation showing highest cellular activities for 7th day. Lv et al. (2016) reported a novel scaffold which exhibited excellent mechanical properties with improved oxygen generative ability at low cytotoxicity. Keratin for this study was also extracted from human hair and blended with silk fibrion at various ratios along with gelatin and calcium peroxide. One of the best results were found at silk fibron:keratin ratio of 60:40 with 2% gelatin blend and 20% calcium peroxide with a young modulus of 179 MPa. The porosity studies showed higher results of 82.53% after 28 days while 77.97% on day 0. The use of calcium peroxide as the oxygen-generating agent improved the cell viability to 92–94% to that of 88–93% over a period of 10 days. Although the water retaining capacity was affected in this experiment (30% after 28 days), it came at the cost of higher mechanical properties. Li et al. (2016) developed nanofibrous mats by electrospinning keratins in various ratio with poly(ε-caprolactone) (PCL). As evident from the earlier works, keratin here also improved the cytocompatibility and hydrophilic properties of the scaffold while PCL provided the tensile strength. PCL to keratin ratio of 7:3 showed the lowest contact angle of 29 at the first second which was reduced to 0 ° after 30 s. Lower clotting time was observed upon having higher keratin concentration. Singaravelu et al. (2016) used freeze-drying technique to develop a scaffold composite comprising of keratin fibrin and gelatin. Among various compositions, keratin:fibrin:gelatin ratio of 1:1:3 was found to have the maximum tensile strength and Young's modulus of 8.62 MPa and 1.75 MPa, respectively. The scaffold had a porosity of 81.7%. Mupirocin was incorporated to study the drug release characteristics and was observed to be 82.4% in a phosphate solution of pH 7.4 after 32 h. All these properties could enhance oxygen permeability alongside enhancing the growth of keratinocytes cells while effectively releasing the drug. Yao et al. (2017) developed a bilayer membrane using keratin/gelatin composite nanofibrous mats as the inner layer and polyurethane as the outer layer which acts as the first layer of defense against bacteria and external agents. The initial keratin/gelatin blend was electrospun in top of polyurethane, this inner mat's layer mimics the natural extracellular matrix of the cell. The bilayer membrane showed a very high water uptake capacity of 480% due to high hydrophilic property of gelatin and high surface to volume ratio and porosity of keratin/gelatin mats. This bilayer membrane could reduce the wound area by 40% and 97% at the end of 7th and 14th day respectively.

A wound can be differentiated on the basis of different categories like type of patient, stage of wound or the kind of wound, e.g., acute and chronic. A universal dressing has not yet been developed which can be employed across different types of wound. Nayak and Gupta (2017) tried to address one of the problems by providing various bioactive components which were required in different stages of the wound. Papain and glucose oxidase loaded with silver nanoparticles were used as an antioxidant and fibroblast growth factor. They created a homogeneous blend solution of keratin along with agar, alginate and gellan which were used to fabricate porous dermal patches using freeze extraction method (Deville 2010; Ho et al. 2004). The SEM images showed that the pores were interconnected and varied in size for different blends. Water adsorption capacity is one of the most important characteristics as it imparts the hydrophilic property required to provide moist environment for effective transport of bioactive molecule and accumulation of growth factor. The observed capacities were 223% for keratin/gellan, 160% for keratin/agar and 274% for keratin/alginate patches. Silver nanoparticles having concentration as low as 0.3 mM provided sufficient antimicrobial characteristics. They were tested for various pathogens and were found most effective against *Staphylococcus aureus*. High glucose concentration, presence of collagen and fibrin inhibits the healing process. Glucose oxidase and papain were incorporated to check the glucose concentration and as debridement agent for fibrin and collagen, respectively. Other bioactive molecules like trolox-based antioxidants and fibroblast growth factor were incorporated for enhanced healing rate.

It can be clearly observed that water retention capacity has an inverse relation with mechanical strength of the keratin-based scaffold and hence one must find the optimum design. Although water retention is essential property for a scaffold and lower pore sizes facilitate higher retention capacity but it has been observed, larger pore sizes enhances faster healing of the wound. So this is another design parameter researchers need to optimize. Freeze drying and electrospinning are the most common methods used for fabrication. Moreover, the use of hydrogel is highly preferred due to its resemblance with extracellular matrix. Lot of work have been done which gives a brief idea of the separate components that need to be used in a scaffold but very few works have been performed to give us in-depth knowledge of biocompatibility and effectiveness of a scaffold. Properties like mechanical, water retention, effective delivery of therapeutic ingredients to the wound need to be merged by optimizing the design parameter that could be effectively used for the welfare of the human kind.

7 Cosmetics and Thermoplastic

Skin aging is partitioned into intrinsic and extrinsic (Jenkins 2002). Intrinsic aging, also referred to as chronological aging, is caused normally by changes in skin versatility after some time, whereas extrinsic aging is essentially a consequence of outside incitement, and is generally simple to control contrast to intrinsic aging. Introduction to ultraviolet (UV) illumination is a main consideration in outward skin aging, which is likewise called photoaging. Ultraviolet B (UVB) light additionally initiates action and articulation of framework metalloproteinases (MMPs), which affects the skin by degrading the extracellular matrix of the cell (ECM) (Varani et al. 2002). MMPs are categorized based on the substrate, the ones containing collagenases (MMP-1 and

MMP-13) are targeted by the anti-aging cosmetics as the skin dermis mainly contains collagen. Most peptides utilized as components of cosmetics are engineered peptides or are acquired with low efficacy from a few protein assets, so they are restricted by huge cost. In any case, poultry feathers, which are made fundamentally out of keratin, are superb wellsprings of natural peptides, however, they are obstinate to proteolysis because of the nearness of a strong and fibrous framework which involves disulfide bonds.

Yeo et al. (2018) built up an anaerobic procedure to liquefy poultry feathers utilizing an extremophilic bacterium, *Fervidobacterium islandicum* AW-1, trailed by fractionation using ultrafiltration to produce feather hydrolysates of low molecular weight (LMW). The results suggested that the LMW keratins unmistakably smothered the expression of MMP-1 and MMP-13 that is induced by UVB, which constitutes a noteworthy reason for skin maturing.

Cosmetics comprises a wide assortment of added substances, for example, those for against wrinkle, nourishing, defensive, recovering, smoothing, and repairing characteristics, along with emollients, emulsifiers, and film formers. Regular substances containing proteins found in vegetables and animals are used as humectants. In light of the poor dissolvability of most proteins in their normal frame, proteins obtained from the animal are not appropriate in the generation of cosmetics. To consolidate a protein in cosmetics use, they must be water soluble. Hence, a typical practice is to effectively dissolve hydrosylates in water while preparing the cosmetic. In certain cases, it is dissolved in organic solvents (Lodén 2009).

Mokrejs et al. (2017) studied the effect of keratin hydrolysate infused in ointments on the moisturization of the skin. They created various concentrations of keratin hydrolysates (KH) whose properties were checked by measuring transepidermal water loss and hydration of the skin. It was observed that the concentration of 2% KH in ointment base had the optimal effect of increasing hydration by 14–23%, while 4% KH reduced trans epidermal water loss by 26–46%. Hence they concluded KH could act both as humectants and occlusive. Although LMW KH showed best characteristics as they diffuse very well into the cell membrane as evident from the previous studies also.

Another important use of keratin is thermoplastics. As mentioned in the previous section, keratin is widely used in adsorption, wound treatment due to its resemblance to ECM water retention capacity and tensile strength. Films developed from keratin-based thermoplastics impart similar characteristics as they have been used in scaffold development, adsorption of pollutant, and other applications.

Glycerol is one of the major components used to prepare the bioplastics. The usage of its various concentration leads to different characteristics, which have been studied and exploited by various researchers (Ramakrishnan et al. 2018; Sharma et al. 2018; Ullah and Wu 2013). Zhuang et al. (2014) used 40% of glycerol along with keratin derived from chicken feathers to test the adsorption of Cr(VI). The prepared film had an adsorbent capacity of 75.4 mg/g and a very high removal rate of 99.1%. In another experiment, Dou et al. (2016) once again studied the structural properties of bioplastics which were prepared by varying the content of glycerol in hydrolyzed keratin feather. The study implemented a low-cost alkali-based extraction of keratin.

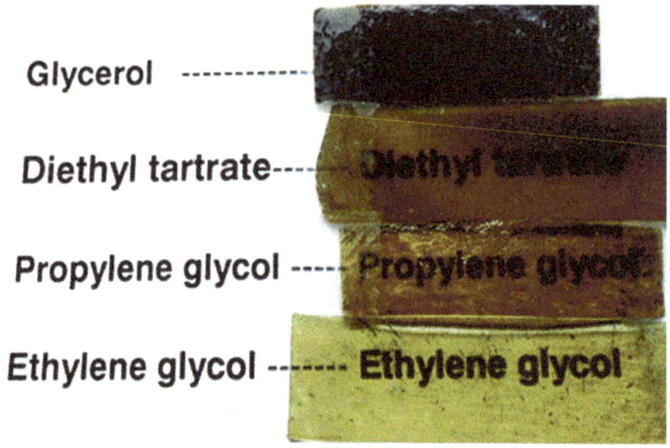

Fig. 5 Digital photographs of the films obtained after plasticizing extruded poultry feather quill keratin with different plasticizers (Ullah et al. 2011) (this figure has been reproduced with permission from ACS publisher)

These were observed to be compatible with glycerol at high temperature and pressure. This was concluded based on their glass transition temperature. Glycerol with 35% content was observed to provide maximum elongation at break of 63%. The incorporation of glycerol in the keratin film improved the water-resistant property and better microstructures were observed, hence they could be used in packaging applications. Ramakrishnan et al. (2018) varied the concentration of glycerol form 2–10% to conclude best tensile and thermal properties for 2% glycerol. Gokce et al. (2017) investigated keratin fiber reinforced with polyurethane and its potential use as scaffolds. The study suggested the new thermoplastic blend showed an 80% increase in both hydrophilicity and cell viability, both of which are essential characteristics of scaffold. Casadesús et al. (2018) tried to develop low-density polyethylene and polypropylene-based thermoplastics. Similar results of low tensile strength were obtained, but upon the addition of adhesion promoters, a significant increase of 30% was observed at 180 °C. In another experiment, addition of various plasticizers was studied. Ethylene glycol (EG), propylene glycol (PG) and glycerol were observed to be most efficient. EG was found to be the most compatible due to strong hydrogen bonding and exhibited best mechanical properties, transparency, processability, and flowability. The transparency of films obtained using different plasticizers is shown in Fig. 5 (Ullah and Wu 2013; Ullah et al. 2011). All the studies suggest that thermoplastics developed from keratin has a wide range of applicability from scaffolds to anti-pollutant, biodegradable films, and numerous potential applications.

8 Conclusions

Keratin is one of the hardest natural materials, filling in as a successful defensive surface layer, despite the fact that it is simply made out of proteins. It demonstrates extraordinary potential for novel bioinspired models, in which an intensive comprehension of the natural chemistry, structure, and mechanical properties are available. Current chemical methods used to extract keratin are efficient but are expensive, as results these methods becomes industrially less viable. Hence, there is an open door to find out new strategies for the feasible segregation of keratin.

Biosorption studies represent that keratins upon being altered/unfolded have high chemical reactivity due to the porous network, which allows pollutant to bind with the functional groups. In biomedical application, wound healing is one of the major fields which have a diverse application of keratin protein due to effective mechanical properties as a scaffold, water retention capacity and effective delivery of required nutrients. Although, there is a need for deeper investigation because most of the studies have worked on a few sets of properties at a time. The need of the hour is to develop an effective keratin-based scaffold, which has optimized design parameters. These parameters include biocompatibility, water retention capacity and effective degradation.

Thermoplastic films made of keratin exhibit similar characteristics, which can be used to remove pollutant and serve as an effective scaffold material. It can also be used as an effective packaging material due to its water resistant trait. Keratin fits very aptly in the present day trend to shifting towards renewable resources from petroleum products. In the process to curb pollution, keratins have come as a good replacement due to its excellent biocompatible properties. For use of keratin on large scale, it must be extracted from feathers because they serve as an excellent low-cost source. The assorted hierarchal structures and properties of keratinous materials provide valuable motivation to develop new bioinspired structures. The better comprehension of keratin properties and how they impact the biomaterial properties will permit a more extensive commercialization of keratin materials.

References

Ahmed SM, Mohammad SG (2014) Egyptian apricot stone (*Prunus armeniaca*) as a low cost and eco-friendly biosorbent for oxamyl removal from aqueous solutions. Am J Exp Agri 4(3):302–321

Ahmad T, Rafatullah M, Ghazali A, Sulaiman O, Hashim R, Ahmad A (2010) Removal of pesticides from water and wastewater by different adsorbents: a review. J Environ Sci Health C 28(4):231–271. https://doi.org/10.1080/10590501.2010.525782

Al-Asheh S, Banat F, Al-Rousan D (2002) Adsorption of copper, zinc and nickel ions from single and binary metal ion mixtures on to chicken feathers. Adsorpt Sci Technol 20(9):849–864. https://doi.org/10.1260/02636170260555778

Al-Asheh S, Banat F, Al-Rousan D (2003) Beneficial reuse of chicken feathers in removal of heavy metals from wastewater. J Clean Prod 11(3):321–326. https://doi.org/10.1016/S0959-6526(02)00045-8

Al-Zaben MI, Mekhamer WK (2017) Removal of 4-chloro-2-methyl phenoxy acetic acid pesticide using coffee wastes from aqueous solution. Arab J Chem 10:S1523–S1529. https://doi.org/10.1016/j.arabjc.2013.05.003

Aluigi A, Vineis C, Tonin C, Tonetti C, Varesano A, Mazzuchetti G (2009) Wool keratin-based nanofibres for active filtration of air and water. J Biobased Mater Bioenergy 3(3):311–319

Arai KM, Takahashi R, Yokote Y, Akahane K (1983) Amino-acid sequence of feather keratin from fowl. Eur J Biochem 132(3):501–507. https://doi.org/10.1111/j.1432-1033.1983.tb07389.x

Argyris TS (1973) Keratins. Their composition, structure and biosynthesis. R. D. B. Fraser, T. P. MacRae, G. E. Rogers. Q Rev Biol 48(2):378–379. https://doi.org/10.1086/407656

Arshad M, Kaur M, Ullah A (2016a) Green biocomposites from nanoengineered hybrid natural fiber and biopolymer. ACS Sustain Chem Eng 4(3):1785–1793

Arshad M, Khosa MA, Siddique T, Ullah A (2016b) Modified biopolymers as sorbents for the removal of naphthenic acids from oil sands process affected water (OSPW). Chemosphere 163:334–341. https://doi.org/10.1016/j.chemosphere.2016.08.015

Astbury WT, Bell FO (1941) Nature of the intramolecular fold in alpha-keratin and alpha-myosin. Nature 147:696–699. https://doi.org/10.1038/147696a0

Ata A, Nalcaci OO, Ovez B (2012) Macro algae Gracilaria verrucosa as a biosorbent: a study of sorption mechanisms. Algal Res 1(2):194–204. https://doi.org/10.1016/j.algal.2012.07.001

Baden HP (1970) The physical properties of nail. J Invest Dermatol 55(2):115–122. doi:https://doi.org/10.1111/1523-1747.ep12291628

Balaji S, Kumar R, Sripriya R, Rao U, Mandal A, Kakkar P, Reddy PN, Sehgal PK (2012) Characterization of keratin–collagen 3D scaffold for biomedical applications. Polym Adv Technol 23(3):500–507

Banat F, Al-Asheh S, Al-Rousan D (2002) Comparison between different keratin-composed biosorbents for the removal of heavy metal ions from aqueous solutions. Adsorp Sci and Technology 20(4):393–416. https://doi.org/10.1260/02636170260295579

Barba C, Mendez S, Roddick-Lanzilotta A, Kelly R, Parra J, Coderch L (2007) Wool peptide derivatives for hand care. J Cosmet Sci 58(2):99–107

Barone JR, Schmidt WF (2005) Polyethylene reinforced with keratin fibers obtained from chicken feathers. Compos Sci Techn 65(2):173–181

Barone JR, Schmidt WF, Liebner CF (2005) Thermally processed keratin films. J Appl Poly Sci 97(4):1644–1651

Bear S, Rugo J (1951) The results of X-ray diffraction studies on keratin fibers. Ann N Y Acad Sci 53(3):627–648. https://doi.org/10.1111/j.1749-6632.1951.tb31964.x

Blackburn S, Lee GR (1956) The reaction of wool keratin with alkali. Biochim Biophys Acta 19:505–512. https://doi.org/10.1016/0006-3002(56)90474-7

Buchanan JH (1977) A cystine-rich protein fraction from oxidized alpha-keratin. Biochem J 167(2):489–491. https://doi.org/10.1042/bj1670489

Casadesús M, Macanás J, Colom X, Cañavate J, Álvarez MD, Garrido N, Molins G, Carrillo F (2018) Effect of chemical treatments and additives on properties of chicken feathers thermoplastic biocomposites. J Compos Mater. https://doi.org/10.1177/0021998318766652

Chen B, Yuan M, Liu H (2011) Removal of polycyclic aromatic hydrocarbons from aqueous solution using plant residue materials as a biosorbent. J Hazar Mater 188(1–3):436–442. https://doi.org/10.1016/j.jhazmat.2011.01.114

Chen P-Y, McKittrick J, Meyers MA (2012) Biological materials: functional adaptations and bioinspired designs. Progress Mater Sci 57(8):1492–1704. https://doi.org/10.1016/j.pmatsci.2012.03.001

Chen J, Ding S, Ji Y, Ding J, Yang X, Zou M, Li Z (2015) Microwave-enhanced hydrolysis of poultry feather to produce amino acid. Chem Eng Process 87:104–109. https://doi.org/10.1016/j.cep.2014.11.017

Coonery DO (1999) Adsorption design for wastewater treatment. Lewis Publishers, USA, p 182

Coulombe PA, Omary MB (2002) 'Hard' and 'soft' principles defining the structure, function and regulation of keratin intermediate filaments. Curr Opin Cell Biol 14(1):110–122

Coulombe PA, Wong P (2004) Cytoplasmic intermediate filaments revealed as dynamic and multipurpose scaffolds. Nat Cell Biol 6:699–706
Deville S (2010) Freeze-casting of porous biomaterials: structure, properties and opportunities. Materials 3(3):1913
Dou Y, Zhang B, He M, Yin G, Cui Y (2016) The structure, tensile properties and water resistance of hydrolyzed feather keratin-based bioplastics. Chinese J Chem Eng 24(3):415–420. https://doi.org/10.1016/j.cjche.2015.11.007
Earland C, Knight CS (1955) Studies on the structure of keratin: I. The analysis of fractions isolated from wool oxidized with peracetic acid. Biochim Biophys Acta 17:457–461. https://doi.org/10.1016/0006-3002(55)90406-6
Esparza Y, Bandara N, Ullah A, Wu J (2018a) Hydrogels from feather keratin show higher viscoelastic properties and cell proliferation than those from hair and wool keratins. Mater Sci Eng, C 90:446–453. https://doi.org/10.1016/j.msec.2018.04.067
Esparza Y, Ullah A, Wu J (2018b) Molecular mechanism and characterization of self-assembly of feather keratin gelation. Int J Biol Macromol 107:290–296. https://doi.org/10.1016/j.ijbiomac.2017.08.168
Esparza Y, Ullah A, Boluk Y, Wu J (2017) Preparation and characterization of thermally crosslinked poly(vinyl alcohol)/feather keratin nanofiber scaffolds. Mater Des 133:1–9. https://doi.org/10.1016/j.matdes.2017.07.052
García-Sabido D, López-Mesas M, Carrillo-Navarrete F (2016) Chicken feather fibres waste as a low-cost biosorbent of acid blue 80 dye. Desalin Water Treat 57(8):3732–3740
Giroud A, Leblond CP (1951) The keratinization of epidermis and its derivatives, especially the hair, as shown by X-ray diffraction and histochemical studies. Ann N Y Acad Sci 53(3):613–626. https://doi.org/10.1111/j.1749-6632.1951.tb31963.x
Goddard DR, Michaelis L (1935) Derivatives of keratin. J Bio Chem 112(1):361–371
Gokce O, Kasap M, Akpinar G, Ozkoc G (2017) Preparation, characterization, and in vitro evaluation of chicken feather fiber-thermoplastic polyurethane composites. J Appl Poly Sci 134(45):45338. https://doi.org/10.1002/app.45338
Gupta G, Borowiec J, Okoh J (1997) Toxicity identification of poultry litter aqueous leachate. Poult Sci 76(10):1364–1367. https://doi.org/10.1093/ps/76.10.1364
Gupta P, Nayak KK (2015) Compatibility study of alginate/keratin blend for biopolymer development. J Appl Biomater Funct Mater 13(4):332–339. https://doi.org/10.5301/jabfm.5000242
Gupta VK, Mittal A, Kurup L, Mittal J (2006) Adsorption of a hazardous dye, erythrosine, over hen feathers. J Colloid Interface Sci 304(1):52–57. https://doi.org/10.1016/j.jcis.2006.08.032
Hill PS, Apel PJ, Barnwell J, Smith T, Koman LA, Atala A, Van Dyke M (2011) Repair of peripheral nerve defects in rabbits using keratin hydrogel scaffolds. Tissue Eng Part A 17(11–12):1499–1505
Ho M-H, Kuo P-Y, Hsieh H-J, Hsien T-Y, Hou L-T, Lai J-Y, Wang D-M (2004) Preparation of porous scaffolds by using freeze-extraction and freeze-gelation methods. Biomaterials 25(1):129–138. https://doi.org/10.1016/S0142-9612(03)00483-6
Hong CK, Wool RP (2005) Development of a bio-based composite material from soybean oil and keratin fibers. J Appl Poly Sci 95(6):1524–1538
Huda S, Yang Y (2009) Feather fiber reinforced light-weight composites with good acoustic properties. J Poly Environ 17(2):131. https://doi.org/10.1007/s10924-009-0130-2
Jenkins G (2002) Molecular mechanisms of skin ageing. Mech Ageing Dev 123(7):801–810. https://doi.org/10.1016/S0047-6374(01)00425-0
Kakkar P, Madhan B (2016) Fabrication of keratin-silica hydrogel for biomedical applications. Mater Sci Eng, C 66:178–184. https://doi.org/10.1016/j.msec.2016.04.067
Kakkar P, Verma S, Manjubala I, Madhan B (2014) Development of keratin-chitosan-gelatin composite scaffold for soft tissue engineering. Mater Sci Eng, C 45:343–347. https://doi.org/10.1016/j.msec.2014.09.021
Kamarudin NB, Sharma S, Gupta A, Kee CG, Chik SMSBT, Gupta R (2017) Statistical investigation of extraction parameters of keratin from chicken feather using design-expert. 3 Biotech 7(2):127. https://doi.org/10.1007/s13205-017-0767-9

Karageorgiou V, Kaplan D (2005) Porosity of 3D biomaterial scaffolds and osteogenesis. Biomaterials 26(27):5474–5491. https://doi.org/10.1016/j.biomaterials.2005.02.002
Karan CP, Rengasamy RS, Das D (2011) Oil spill cleanup by structured fibre assembly. Indian J Fibre Text Res 36(2):190–200
Kasapi MA, Gosline JM (1997) Design complexity and fracture control in the equine hoof wall. J Exp Biol 200(11):1639–1659
Kaur M, Arshad M, Ullah A (2018) In-situ nanoreinforced green bionanomaterials from natural keratin and montmorillonite (MMT)/cellulose nanocrystals (CNC). ACS Sustain Chem Eng 6(2):1977–1987. https://doi.org/10.1021/acssuschemeng.7b03380
Kelly R (2016) Keratins in wound healing. In: Ågren MS (ed) Wound healing biomaterials. Elsevier, pp 353–365
Khosa MA, Ullah A (2013) A sustainable role of keratin biopolymer in green chemistry: a review. J Food Process Beverage 1:1–8
Khosa MA, Ullah A (2014) In-situ modification, regeneration, and application of keratin biopolymer for arsenic removal. J Hazar Mater 278:360–371. https://doi.org/10.1016/j.jhazmat.2014.06.023
Khosa MA, Wu J, Ullah A (2013) Chemical modification, characterization, and application of chicken feathers as novel biosorbents. Rsc Adv 3(43):20800–20810
Kim JW, Kim MJ, Ki CS, Kim HJ, Park YH (2017) Fabrication of bi-layer scaffold of keratin nanofiber and gelatin-methacrylate hydrogel: implications for skin graft. Int J Biol Macromol 105:541–548. https://doi.org/10.1016/j.ijbiomac.2017.07.067
Kong H, He J, Wu H, Wu H, Gao Y (2012) Phenanthrene removal from aqueous solution on sesame stalk-based carbon. Clean: Oil, Air, Water 40(7):752–759. https://doi.org/10.1002/clen.201100322
Leung WC, Wong MF, Chua H, Lo W, Yu PH, Leung CK (2000) Removal and recovery of heavy metals by bacteria isolated from activated sludge treating industrial effluents and municipal wastewater. Water Sci Tech, vol 41
Li Y, Wang Y, Ye J, Yuan J, Xiao Y (2016) Fabrication of poly(ϵ-caprolactone)/keratin nanofibrous mats as a potential scaffold for vascular tissue engineering. Mater Sci Eng, C 68:177–183. https://doi.org/10.1016/j.msec.2016.05.117
Lodén M (2009) Hydrating substances. In: Barel AO, Paye M, Maibach HI (eds) vol 36, 3rd edn. Informa Healtcare, New York, pp 107–119. https://doi.org/10.1111/j.1524-4725.2009.01444.x
Lv XG, Li Z, Chen SY, Xie MK, Huang JW, Peng XF, Yang RX, Wang HP, Xu YM, Feng C (2016) Structural and functional evaluation of oxygenating keratin/silk fibroin scaffold and initial assessment of their potential for urethral tissue engineering. Biomaterials 84:99–110. https://doi.org/10.1016/j.biomaterials.2016.01.032
Ma PX (2008) Biomimetic materials for tissue engineering. Adv Drug Deliv Rev 60(2):184–198. https://doi.org/10.1016/j.addr.2007.08.041
Mandal BB, Kundu SC (2009) Cell proliferation and migration in silk fibroin 3D scaffolds. Biomaterials 30(15):2956–2965. https://doi.org/10.1016/j.biomaterials.2009.02.006
Matsunaga K, Okumura T, Tsushima R (1983) Hair treatment cosmetics containing cationic keratin derivatives. Google Patents
Matsunaga K, Okumura T, Naito S, Tsushima R (1984) Shampoo composition. Google Patents
McGovern V (2000) Recycling poultry feathers: more bang for the cluck. Environ Health Perspect 108(8):A366–A369
Meyers MA, Chen P-Y, Lin AY-M, Seki Y (2008) Biological materials: structure and mechanical properties. Prog Mater Sci 53(1):1–206. https://doi.org/10.1016/j.pmatsci.2007.05.002
Mittal A (2006) Adsorption kinetics of removal of a toxic dye, malachite green, from wastewater by using hen feathers. J Hazar Mater 133(1):196–202. https://doi.org/10.1016/j.jhazmat.2005.10.017
Mohammad SG (2013) Biosorption of pesticide onto a low cost carbon produced from apricot stone (*Prunus armeniaca*): equilibrium, kinetic and thermodynamic studies. J Appl Sci Res 9:6459–6469

Mokrejs P, Hutta M, Pavlackova J, Egner P, Benicek L (2017) The cosmetic and dermatological potential of keratin hydrolysate. J Cosmet Dermatol 16(4):e21–e27. https://doi.org/10.1111/jocd.12319

Moulin V, Auger FA, Garrel D, Germain L (2000) Role of wound healing myofibroblasts on re-epithelialization of human skin. Burns 26(1):3–12. https://doi.org/10.1016/s0305-4179(99)00091-1

Nakata R, Osumi Y, Miyagawa S, Tachibana A, Tanabe T (2015) Preparation of keratin and chemically modified keratin hydrogels and their evaluation as cell substrate with drug releasing ability. J Biosci Bioen 120(1):111–116. https://doi.org/10.1016/j.jbiosc.2014.12.005

Nayak KK, Gupta P (2015) In vitro biocompatibility study of keratin/agar scaffold for tissue engineering. Int J Biol Macromol 81:1–10. https://doi.org/10.1016/j.ijbiomac.2015.07.025

Nayak KK, Gupta P (2017) Study of the keratin-based therapeutic dermal patches for the delivery of bioactive molecules for wound treatment. Mater Sci Eng, C 77(Suppl C):1088–1097. doi:https://doi.org/10.1016/j.msec.2017.04.042

Pace LA, Plate JF, Smith TL, Van Dyke ME (2013) The effect of human hair keratin hydrogel on early cellular response to sciatic nerve injury in a rat model. Biomaterials 34(24):5907–5914

Parenteau-Bareil R, Gauvin R, Berthod F (2010) Collagen-based biomaterials for tissue engineering applications. Materials 3(3):1863–1887

Park M, Shin HK, Kim BS, Kim MJ, Kim IS, Park BY, Kim HY (2015) Effect of discarded keratin-based biocomposite hydrogels on the wound healing process in vivo. Mater Sci Eng, C 55:88–94. https://doi.org/10.1016/j.msec.2015.03.033

Pedram Rad Z, Tavanai H, Moradi AR (2012) Production of feather keratin nanopowder through electrospraying. J Aero Sci 51:49–56. https://doi.org/10.1016/j.jaerosci.2012.04.007

Pham TL, Inoue K (2012) Minimal generalization for conjunctive queries. In: Lecture notes in computer science (including subseries lecture notes in artificial intelligence and lecture notes in bioinformatics). LNAI, vol 7694. https://doi.org/10.1007/978-3-642-35455-7_23

Poole AJ, Church JS, Huson MG (2008) Environmentally sustainable fibers from regenerated protein. Biomacromolecules 10(1):1–8

Pyun DG, Yoon HS, Chung HY, Choi HJ, Thambi T, Kim BS, Lee DS (2015) Evaluation of AgHAP-containing polyurethane foam dressing for wound healing: synthesis, characterization, in vitro and in vivo studies. J Mater Chem B 3(39):7752–7763. https://doi.org/10.1039/c5tb00995b

Radetić MM, Jocić DM, Jovančić PM, Petrović ZL, Thomas HF (2003) Recycled wool-based nonwoven material as an oil sorbent. Environ Sci Techn 37(5):1008–1012. https://doi.org/10.1021/es0201303

Ramakrishnan N, Sharma S, Gupta A, Alashwal BY (2018) Keratin based bioplastic film from chicken feathers and its characterization. Int J Biol Macromol 111:352–358. https://doi.org/10.1016/j.ijbiomac.2018.01.037

Reddy N, Chen L, Zhang Y, Yang Y (2014) Reducing environmental pollution of the textile industry using keratin as alternative sizing agent to poly(vinyl alcohol). J Clean Prod 65:561–567

Rosa Gdl (2008) Recycling poultry feathers for Pb removal from wastewater: kinetic and equilibrium studies. Int J Chem, Molec, Nucl, Mater Metallur Eng 2337–346

Saravanan S, Sameera DK, Moorthi A, Selvamurugan N (2013) Chitosan scaffolds containing chicken feather keratin nanoparticles for bone tissue engineering. Int J Biol Macromol 62:481–486. https://doi.org/10.1016/j.ijbiomac.2013.09.034

Schrooyen PMM, Dijkstra PJ, Oberthür RC, Bantjes A, Feijen J (2001) Partially carboxymethylated feather keratins. 2. Thermal and mechanical properties of films. J Agric and Food Chem 49(1):221–230. https://doi.org/10.1021/jf0004154

Sekhar KC, Subramanian S, Modak JM, Natarajan KA (1998) Removal of metal ions using an industrial biomass with reference to environmental control. Int J Mineral Process 53(1):107–120. https://doi.org/10.1016/S0301-7516(97)00061-6

Sharma S, Gupta A, Kumar A, Kee CG, Kamyab H, Saufi SM (2018) An efficient conversion of waste feather keratin into ecofriendly bioplastic film. Clean Technol Environ Policy. https://doi.org/10.1007/s10098-018-1498-2

Shavandi A, Silva TH, Bekhit AA, Bekhit AE-dA (2017) Keratin: dissolution, extraction and biomedical application. Biomater Sci 5(9):1699–1735. https://doi.org/10.1039/c7bm00411g
Sierpinski P, Garrett J, Ma J, Apel P, Klorig D, Smith T, Koman LA, Atala A, Van Dyke M (2008) The use of keratin biomaterials derived from human hair for the promotion of rapid regeneration of peripheral nerves. Biomaterials 29(1):118–128
Singaravelu S, Ramanathan G, Raja MD, Nagiah N, Padmapriya P, Kaveri K, Tiruchirapalli U (2016) Biomimetic interconnected porous keratin–fibrin–gelatin 3D sponge for tissue engineering application. Int J Biol Macromol 86:810–819. https://doi.org/10.1016/j.ijbiomac.2016.02.021
Srinivasan B, Kumar R, Shanmugam K, Sivagnam UT, Reddy NP, Sehgal PK (2010) Porous keratin scaffold-promising biomaterial for tissue engineering and drug delivery. J Biomed Mater Res Part B Appl Biomater 92(1):5–12
Stephenson AH, McCaskey TA, Ruffin BG (1990) A survey of broiler litter composition and potential value as a nutrient resource. Biol Wastes 34(1):1–9. https://doi.org/10.1016/0269-7483(90)90139-J
Sun P, Liu Z-T, Liu Z-W (2009) Chemically modified chicken feather as sorbent for removing toxic chromium(VI) ions. Indust Eng Chem Resear 48(14):6882–6889. https://doi.org/10.1021/ie900106h
Tachibana A, Furuta Y, Takeshima H, Tanabe T, Yamauchi K (2002) Fabrication of wool keratin sponge scaffolds for long-term cell cultivation. J Biotech 93(2):165–170
Tan HB, Wang FY, Ding W, Zhang Y, Ding J, Cai DX, Yu KF, Yang J, Yang L, Xu YQ (2015) Fabrication and evaluation of porous keratin/chitosan (KCS) scaffolds for effectively accelerating wound healing. Biomed Environ Sci 28(3):178–189. https://doi.org/10.3967/bes2015.024
Tanase CE, Spiridon I (2014) PLA/chitosan/keratin composites for biomedical applications. Mater Sci Eng, C 40:242–247. https://doi.org/10.1016/j.msec.2014.03.054
Thorndike EE (1968) A microscopic study of the marmoset claw and nail. Am J Physic Anthropo 28(3):247–261. https://doi.org/10.1002/ajpa.1330280314
Tomlinson DJ, Mülling CH, Fakler TM (2018) Invited reviews: formation of keratins in the bovine claw: roles of hormones, minerals, and vitamins in functional claw integrity. J Dairy Sci 87(4):797–809. https://doi.org/10.3168/jds.s0022-0302(04)73223-3
Tonin C, Zoccola M, Aluigi A, Varesano A, Montarsolo A, Vineis C, Zimbardi F (2006) Study on the conversion of wool keratin by steam explosion. Biomacromolecules 7(12):3499–3504. https://doi.org/10.1021/bm060597w
Turan NG (2011) Metal uptake from aqueous leachate of poultry litter by natural zeolite. Environ Prog Sustain Ener 30(2):152–159. https://doi.org/10.1002/ep.10456
Ulery BD, Nair LS, Laurencin CT (2011) Biomedical applications of biodegradable polymers. J Polym Sci Part B: Polym Phys 49(12):832–864. https://doi.org/10.1002/polb.22259
Ullah A, Vasanthan T, Bressler D, Elias AL, Wu J (2011) Bioplastics from feather quill. Biomacromolecules 12(10):3826–3832. https://doi.org/10.1021/bm201112n
Ullah A, Wu J (2013) Feather fiber-based thermoplastics: effects of different plasticizers on material properties. Macromol Mater Eng 298(2):153–162
Varani J, Perone P, Fligiel SEG, Fisher GJ, Voorhees JJ (2002) Inhibition of type I procollagen production in photodamage: correlation between presence of high molecular weight collagen fragments and reduced procollagen synthesis. J Invest Dermat 119(1):122–129. https://doi.org/10.1046/j.1523-1747.2002.01810.x
Wang Y-X, Cao X-J (2012) Extracting keratin from chicken feathers by using a hydrophobic ionic liquid. Process Biochem 47(5):896–899. https://doi.org/10.1016/j.procbio.2012.02.013
Wang B, Yang W, McKittrick J, Meyers MA (2016) Keratin: structure, mechanical properties, occurrence in biological organisms, and efforts at bioinspiration. Progress Mater Sci 76:229–318
Wang J, Hao S, Luo T, Cheng Z, Li W, Gao F, Guo T, Gong Y, Wang B (2017) Feather keratin hydrogel for wound repair: preparation, healing effect and biocompatibility evaluation. Colloids Surf B Biointerfaces 149:341–350. https://doi.org/10.1016/j.colsurfb.2016.10.038

Xi Z, Chen B (2014) Removal of polycyclic aromatic hydrocarbons from aqueous solution by raw and modified plant residue materials as biosorbents. J Environ Sci (China) 26(4):737–748. https://doi.org/10.1016/s1001-0742(13)60501-x

Xie H, Li S, Zhang S (2005) Ionic liquids as novel solvents for the dissolution and blending of wool keratin fibers. Green Chem 7(8):606–608

Xu S, Sang L, Zhang Y, Wang X, Li X (2013) Biological evaluation of human hair keratin scaffolds for skin wound repair and regeneration. Mater Sci Eng, C 33(2):648–655. https://doi.org/10.1016/j.msec.2012.10.011

Yakout SM, Daifullah AAM (2013) Removal of selected polycyclic aromatic hydrocarbons from aqueous solution onto various adsorbent materials. Desalin Water Treat 51(34–36):6711–6718. https://doi.org/10.1080/19443994.2013.769916

Yao C-H, Lee C-Y, Huang C-H, Chen Y-S, Chen K-Y (2017) Novel bilayer wound dressing based on electrospun gelatin/keratin nano fibrous mats for skin wound repair. Mater Sci Eng, C 79:533–540. https://doi.org/10.1016/j.msec.2017.05.076

Yeo I, Lee YJ, Song K, Jin HS, Lee JE, Kim D, Lee DW, Kang NJ (2018) Low-molecular weight keratins with anti-skin aging activity produced by anaerobic digestion of poultry feathers with Fervidobacterium islandicum AW-1. J Biotech 271:17–25. https://doi.org/10.1016/j.jbiotec.2018.02.003

Yussef E, Aman U, Jianping W (2017) Preparation and characterization of graphite oxide nano-reinforced biocomposites from chicken feather keratin. J Chem Technol Biotechnol 92(8):2023–2031. https://doi.org/10.1002/jctb.5196

Zhang Y, Zhao W, Yang R (2015) Steam flash explosion assisted dissolution of keratin from feathers. ACS Sustain Chem Eng 3(9):2036–2042. https://doi.org/10.1021/acssuschemeng.5b00310

Zhang Z, Nie Y, Zhang Q, Liu X, Tu W, Zhang X, Zhang S (2017) Quantitative change in disulfide bonds and microstructure variation of regenerated wool keratin from various Ionic Liquids. ACS Sustainable Chem. Eng. 5(3):2614–2622

Zhao W, Yang R, Zhang Y, Wu L (2012) Sustainable and practical utilization of feather keratin by an innovative physicochemical pretreatment: high density steam flash-explosion. Green Chem 14(12):3352–3360. https://doi.org/10.1039/c2gc36243k

Zhuang Y, Zhao XX, Zhou MH (2014) Cr(VI) adsorption on a thermoplastic feather keratin film. Desalin Water Treat 52(13–15):2786–2791. https://doi.org/10.1080/19443994.2014.889439

Zoccola M, Aluigi A, Tonin C (2009) Characterisation of keratin biomass from butchery and wool industry wastes. J Molecul Struct 938(1–3):35–40

Chapter 7
Keratin-Based Biofilms, Hydrogels, and Biofibers

James McLellan, Starla G. Thornhill, Spencer Shelton and Manish Kumar

Abstract Keratin is a nearly ubiquitous, filamentous protein with high amounts of sulfur molecules. The sulfur content of keratin, resulting from a high proportion of cystine residues, is significant to the form and function of keratin in naturally occurring structures and chemically synthesized biomaterials. The intrinsic physical and chemical properties of keratin have led to the development of a variety of biotechnical and materials science applications. The exploitation of the biochemical and physical properties of this abundant and naturally occurring protein has led to numerous novel uses including drug delivery, dental implants, wound dressings, and food packaging. This chapter describes the intrinsic properties of keratin, as well as the production and use of keratin-based biofilms and biofibers.

Keywords Keratin film · Keratin hydrogel · Keratin biofibers
Keratin-based biomaterials

1 Introduction

Keratins are a class of proteins commonly found in the epithelial cells of a wide range of vertebrates, and are the main structural element of wool, fur, feathers, quills, spines, horns, hooves, and reptile scales (Wang et al. 2016). The structural organization and hierarchical assembly of keratin are highly variable between different organisms, resulting in a variety of structures and features with important functions including movement, weatherproofing, impact protection, and structural integrity (Table 1) (McKittrick et al. 2012). The structural characteristics of keratin result from a high proportion of cystine residues that form both inter- and intramolecular disulfide bonds resulting in a linked matrix, especially in α-keratin (McKittrick et al. 2012). The

J. McLellan and S. G. Thornhill—Equal contributers.

J. McLellan · S. G. Thornhill · S. Shelton · M. Kumar (✉)
Department of Biology, Texas State University, San Marcos, TX 78666, USA
e-mail: m_k135@txstate.edu

S. Sharma and A. Kumar (eds.), *Keratin as a Protein Biopolymer*, Springer Series on Polymer and Composite Materials, https://doi.org/10.1007/978-3-030-02901-2_7

Table 1 Amino acid composition by percentage of total amino acid residues of keratin from different taxa (Sinkiewicz et al. 2018). This table has been reproduced with permission from Blackwell Publishing, Inc. Hultin H.O

Amino acid	Buffalo horn and hoof (Noda et al. 1996)	Cow hair (Coward-Kelly et al. 2006)	Feathers (Moore et al. 2006)	Wool (Cardamone 2010)
Alanine	6.3	4.5	3.6	5.8
Arginine	6.8	11.0	5.4	7.8
Aspartic acid	6.7	6.6	4.7	4.1
Cysteine	3.7	nd	7.7	6.1
Glutamic acid	12.6	14.5	7.7	11.4
Glycine	12.3	5.5	6.2	2.9
Histidine	0.6	1.3	–	–
Isoleucine	3.0	4.2	4.3	3.9
Leucine	8.2	9.8	7.0	11.9
Lysine	2.7	5.5	0.6	2.9
Methionine	0.6	0.7	1.3	0.2
Phenylalanine	2.9	3.1	4.2	1.9
Proline	6.8	7.7	8.7	4.1
Serine	10.8	8.9	9.3	8.3
Threonine	5.6	7.5	3.5	5.6
Tyrosine	5.9	2.4	2.0	2.4
Valine	4.1	6.8	6.9	6.1

nd, not determined

resilience of keratin occurs without the addition of any minerals to provide strength to the polymer, setting keratin apart from bones or teeth (Johnson et al. 2017). Similar to other polymers, water content influences the mechanical and physical properties of keratin, likely through disruption of the disulfide bonding between the filament matrix and individual fibers (Johnson et al. 2017). The wide range of mechanical properties, unique phenotypes, and differences in cornification have made classifying keratin challenging in the past, yet keratin is divisible based on differing characteristics. All genes that result in keratins are part of a large gene superfamily known as intermediate filaments (IFs).

2 Intermediate Filaments

IF proteins are the structural building blocks of animal cells and are termed "intermediate" because their size is larger than actin (~6 nm) and smaller than the largest microtubules (Herrmann and Aebi 2004; Fuchs and Weber 1994). Initial investigation of wool revealed two classes based on amino acid composition. Type I keratins are more acidic (pKi = 4−6) and type II keratins are more basic (pKi = 6−8) (Fuchs and Weber 1994). However, the current paradigm of classification divides keratins into two groups based on the primary motif as either an α-helix or a β-pleated sheet (Wang et al. 2016).

3 Keratin Structure

The molecular structure and associated filament types are found in a wide variety of keratin forms, which can be classified as either α-keratin or β-keratin based on the hydrogen bonding of the primary protein structure (Wang et al. 2016; Chou and Buehler 2012). The majority of mammalian keratin is composed of α-helixes, while bird feathers, beaks, and reptilian scales are mainly composed of the β-keratin variant common among sauropsids (Wang et al. 2016). Interestingly, both α-keratin and β-keratin have been observed in a single species, a small insectivore inhabiting the rain forests of Asia and Africa called the pangolin (Wang et al. 2016). While the form and function of keratin proteins are different among taxa, the peptide composition and order are also variants. These changes affect the solubility, hydrolysis, and enzymatic degradation of keratin (Sinkiewicz et al. 2018). Regardless of the associated organism or phenotype, three aspects of keratin remain constant: (i) ability to withstand mechanical shearing, (ii) the biosynthesis regime, and (iii) the observed formation of a filament matrix.

3.1 α-Keratin

α-Keratin is classified by the formation of an α-helix motif approximately 35 residues long, which is organized hierarchically into dimers, protofilaments, protofibrils, and intermediate filaments (Chou and Buehler 2012). The primary α-helix associates with another α-helix in antiparallel, forming a coiled-coil dimer stabilized by ionic interactions of the relatively high proportion of cystine residues. The coiled-coil scaffold allows for inter-rope packing, or filament formation (Parry et al. 1977). More current, atomistic models of the structure further explain the mechanical properties based on the proposed geometric association of the coiled-coil monomers (Chou and Buehler 2012). A fully developed understanding of the atomistic structure may lead to a better understanding of the mechanical and physical properties of keratin, allowing for specific and tunable applications.

3.2 β-Keratin

The structural framework of β-keratin is the β-pleated sheet, as identified by X-ray diffraction patterns (Fraser et al. 1971). These proteins form 3–4 nm diameter filaments in both avian and reptilian species, yet the sequence homology between the two species is observed to be relatively low (~25%) (Fraser and Parry 1996). While keratinization of epithelial cells is the same between the avian and reptilian variants, there are developmental differences which produce different structures. Avian feather production begins with a pin feather, which differentiates into down feathers, while the basal core regresses to produce rachis (Wang et al. 2016; Mousquès and Dullaart 2012). This differentiation allows for complex fibrous structures and macro-shapes, which translate to impressive mechanical responses and material specificities. For instance, the shape and structure of the feather shaft are different from distal to proximal end, which contributes to strength, flexibility, and overall lightness (Wang and Sullivan 2017). The β-keratin genetic components are not fully characterized, but studies are ongoing to describe novel genes and the role they have in the evolved structural integrity of extant bird feathers, including microscopic hooks that are integral to the barbule linkages in feathers (Kowata et al. 2014). Such studies have the potential to inspire new and exciting designs and materials.

4 Biosynthesis of Keratin and Keratinous Materials

Keratin synthesis occurs in highly specialized epidermal cells that produce only keratin, called keratinocytes (Wang et al. 2016). During specialization of an epidermal cell to a keratinocyte, other cellular processes are halted and the nucleus is disassembled, resulting in cell death (Wang et al. 2016). This process occurs slowly as the cell moves outward to the outside layers of the epidermis. When the keratinized cell reaches the stratum granulosum, it is a fully differentiated corneocyte (McKittrick et al. 2012). The specific differentiation pathway and physical location of the keratinocytes control the types of hard α-keratin hierarchal structure, each pathway resulting in unique associated physical and mechanical properties (Johnson et al. 2017; Wang and Sullivan 2017). For instance, IF-producing bundled cells sheathed in a series of cuticles and a cortex form a strand of hair (Fig. 1a) while three layers of keratinized epithelial cells overlap to form a fingernail (Fig. 1b).

4.1 Solubilization of Keratin

Keratin use in biotech requires the degradation and isolation of keratinous fibers or keratinous hydrolysates, which often requires processing steps (Sinkiewicz et al. 2018). Because the abundance of disulfide bonds renders keratin insoluble, the degradation of keratin is challenging, and strong polar solvents are used to disrupt the disulfide bonds by either reduction or oxidization (Zainal-Abidin et al. 2017).

Keratin processing through the reduction of di-sulfide bonds commonly results in non-polar, less soluble products which are more stable (Sinkiewicz et al. 2018). Common reduction reagents include urea, EDTA, 2-mercaptoethanol, and hydroxyl methyl aminomethane, which must be and dialyzed after keratin solubilization (Sinkiewicz et al. 2018). Spontaneous formation of the keratin scaffold is prevented by sodium dodecylsulfate treatment (Pedram Rad et al. 2012). Keratin extraction via oxidation produces water-soluble derivatives of

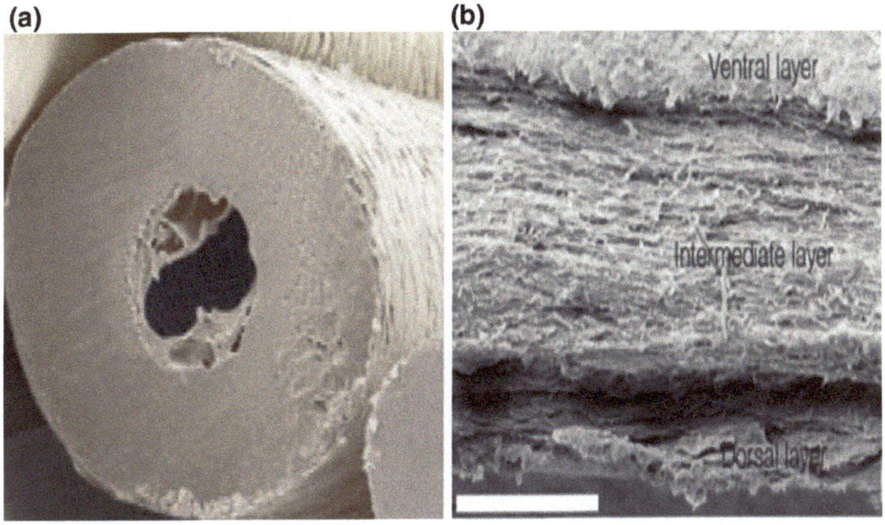

Fig. 1 SEM images of keratinized structures. **a** Cross-sectional SEM image of a strand of polar bear fur. **b** SEM image of a cross section of a primate nail (McKittrick et al. 2012) (This figure has been reproduced with permission from Springer Nature Publisher)

cysteic acid called keratoses (Sinkiewicz et al. 2018). Strong oxidants capable of producing keratoses include peracetic and performic acid, commonly dissolved in ammonia (Mousquès and Dullaart 2012).

More recently, highly polar deep eutectic solvents (DES) such as choline chloride:urea are used to extract keratin. Though the mechanism of keratin solubilization by DES is unknown, there are two proposed mechanisms that are widely accepted. DES is implicated in the disruption of the ionic binding in keratin, and urea interacts with the protein backbone causing denaturation (Zainal-Abidin et al. 2017). Alternatively, hydrophobic interactions may be weakened by the solvent, allowing the fibers to dissolve more easily (Zainal-Abidin et al. 2017). Regardless of the mechanism, these ionic liquids have great potential in the extraction of a wide range of biological compounds.

As an alternative to chemical extraction, keratin can be extracted via enzymatic hydrolysis using enzymes derived from microorganisms and fungi isolated from keratin-rich environments (Rajabinejad et al. 2016). Enzymatic hydrolysis is being used to produce keratin hydrolysates with very little damage to amino acids (Rajabinejad et al. 2016). Currently, the rate of extraction by enzymatic hydrolysis is too low to be commercially viable (Sharma et al. 2017). See Table 2 for a list describing keratinases by species (Sinkiewicz et al. 2018).

Extracted keratin can be allowed to spontaneously self-assemble to form a poorly defined scaffolding network (Kakkar et al. 2016) that is fragile in comparison to chemically cross-linked scaffolds (Tanabe et al. 2004). The addition of plasticizer (Maria Martelli et al. 2006), chemical crosslinkers (Tanabe et al. 2004), calcium

Table 2 Keratinase enzymes by organism (Sinkiewicz et al. 2018). Table 1 has been reproduced with permission from Blackwell Publishing, Inc. Hultin H.O

Source of keratinase	Molecular mass (kDa)	Optimum pH	Optimum temperature (°C)	References
Aspergillus fumigatus TKF1	24	6.0	50	Paul et al. (2014)
Aspergillus parasiticus	36	7.0	50	Anitha and Palanivelu (2013)
Bacillus licheniformis PWD-1	33	7.5	50	Lin et al. (1992)
Bacillus pumilus A1	–	9.0	55–60	Fakhfakh-Zouari et al. (2010)
Bacillus subtilis S14	–	8.0	50	Silva et al. (2014)
Brevibacillus sp.	83.2	12.5–13.0	45	Rai and Mukherjee (2011)
Chryseobacterium indologenes A22	–	7.5	45	Bach et al. (2011)
Chryseobacterium sp. kr6	64	8.5	50	Riffel et al. (2007)
Fervidobacterium islandiucum AW-1	>200	9.0	100	Nam et al. (2002)
Microsporum canis	33	8.0	35–45	Descamps et al. (2003)
Microsporum gypseum	33	8.0	35	Raju et al. (2007)
Stenotrophomonas sp. D-l	40	7.0	30	Yamamura et al. (2002)
Streptomyces fradiae	24	8.0	50	Galas and Kałuźewska (1991)
Streptomyces gulbargensis	46	9.0	45	Syed et al. (2009)
Streptomyces thermoviolaceus SD8	40	8.0	55	Chitte et al. (1999)
Trichophyton mentagrophytes	38	5.5	55	Muhsin and Aubaid (2001)

phosphate particles (Tachibana et al. 2005), and other polymers (Furth et al. 2007) have allowed chemical-based tuning of keratin scaffolds for a variety of medical and industrial applications. Visible light crosslinking improves the stability and strength of keratin scaffolding and allows for tuning the physical properties of the scaffolding, such as filament size (Yue et al. 2018).

5 Applications of Keratin-Based Biomaterials

The unique physical and chemical properties of keratin have lead researchers to develop keratin-based thin films and hydrogels, which can be useful in medical and industrial applications. Keratin-based biomaterials are biocompatible, reducing the likelihood of inducing allergic reactions, and are biodegradable, allowing implantation without the need for later removal (Kornillowicz-Kowalska and Bohacz 2011). Keratin films are thin sheets of keratin mixed with a plasticizer (Lee et al. 2015), and keratin hydrogels are three-dimensional gels of keratin biopolymers and water (Peralta Ramos et al. 2017). Keratin films and biopolymers have applications in medicine as a mechanism of tunable drug delivery (Han et al. 2015), regenerative medicine (Han et al. 2015), burn treatment (Poranki et al. 2016; Roy et al. 2016), and wound healing (Veerasubramanian et al. 2018; Wang et al. 2017).

5.1 Hydrogels and Thin Films

Keratin hydrogels allow tunable release of drugs including antibiotics, growth factors, and cancer treatment drugs (Table 3) directly to the application site. Localized drug delivery provides patients with many benefits: (i) the ability to deliver a high dose while avoiding toxicity (Eccleston et al. 1996), (ii) protection of the gastrointestinal microflora from dysbiosis typically caused by oral or intravenous administration of antibiotics, and (iii) promoting immune cell recruitment to an injury site (Fearing and Van Dyke 2014).

The method of preparing keratin hydrogel can be modified to alter the amount of disulfide bonds formed in the gel, which allows tuning of the release rate of growth factors, antibiotics, and keratins from the hydrogel, as well as the rate of biodegradation of the keratin hydrogel to promote healing at wound sites (Han et al. 2015). Since tissue regeneration is optimized when the rate of scaffolding material degradation is inversely proportional to the rate of replacement tissue growth (Alsberg et al. 2003), tuning keratin scaffolding degradation to match the expected regeneration rate of a specific injury can promote healing (Han et al. 2015). Oxidatively extracted keratose forms hydrogels that degrade in weeks, and reductively extracted keratin forms hydrogels that degrade in months (Han et al. 2015). The change in degradation time is due to the amount of free cysteine residues available for disulfide binding after extraction. Oxidative extraction of keratose caps cysteine residues with sulfonic acid groups, reducing the number of disulfide cross-links that occur during hydrogel formation (Han et al. 2015; Shavandi et al. 2017). Keratin

Table 3 Summary of some drugs tested for slow release delivery in keratin hydrogels

Drug	Application	References
Growth factors		
rhBMP-2	Wound healing	Han et al. (2015)[a]
rhIGF-1	Wound healing	Han et al. (2015)[a]
bFGF	Wound healing	Tomblyn et al. (2016)[a]
IGF-1	Wound healing	Tomblyn et al. (2016)[a]
VEGF	Wound healing	Tomblyn et al. (2016)[a]
Antibiotics		
Ciprofloxacin	Wound healing	Saul et al. (2011), Han et al. (2015)[a], Roy et al. (2016)
Minocyclin hydrochloride	Periodontitis treatment	Lee et al. (2015)[a]
Cancer treatment		
Doxorubicin hydrochloride	Cancer treatment	Sun et al. (2016)[a]
Plant extracts		
Avena sativa	Wound healing	Veerasubramanian et al. (2018)

[a]Denotes in vitro only studies

hydrogel degradation can be modulated by using alkylation agents like iodoacetamide to more finely tune drug delivery and hydrogel degradation (Han et al. 2015). Keratin hydrogels are also useful in tissue engineering and regenerative medicine because the keratin scaffolding contains cell adhesion domains which support cell proliferation (Sierpinski et al. 2008; Bordeleau et al. 2008). Of interest, diabetic wound healing is more rapid when *Avena sativa* (common oat) extract is delivered by keratin hydrogel applied directly to the wound site (Fig. 2) (Veerasubramanian et al. 2018). Burn wounds can be treated with keratin hydrogels or films by slowing wound progression (Poranki et al. 2016; Roy et al. 2016). Keratin hydrogels also promote nerve regeneration by providing scaffolding for new nerve cells (Pace et al. 2014).

Keratin hydrogels and films also have uses in the industry. Keratin films can be used to strengthen plastic (Oladayo et al. 2016) or as environmentally friendly food packaging when mixed with citric acid (Fig. 3) (Ramirez et al. 2017).

5.2 Keratin Biofibers

There are a number of traits that make keratin fibers useful for a variety of applications: keratin (i) is insoluble and relatively chemically unreactive, (ii) is naturally occurring and widely available as a secondary product from textile and poultry industry, (iii) has the ability to provide elasticity and mechanical resistance to shearing based upon its specific composition (Montes-Zarazúa et al. 2015; Sinkiewicz et al. 2018). The amino (NH_2), carboxyl (COO^-), and sulfhydryl

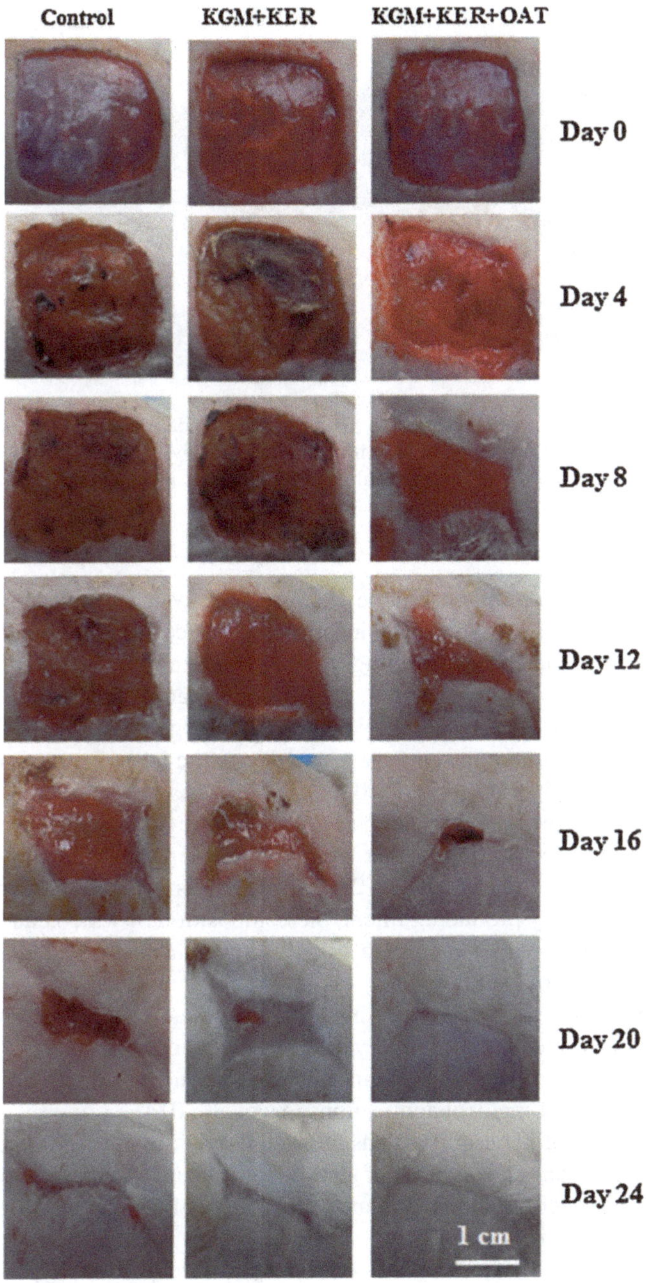

Fig. 2 Mock diabetic wound treatment by keratin scaffolding with *Avena sativa* extract in Wistar rats with induced hyperglycemia. Treatment with the keratin (KER) and konjac glucomannan (KGM) hydrogels, as well as the same infused with *A. sativa* extract (OAT) increased wound healing compared to standard dressing controls (Veerasubramanian et al. 2018) (This figure has been reproduced with permission from Elsevier BV)

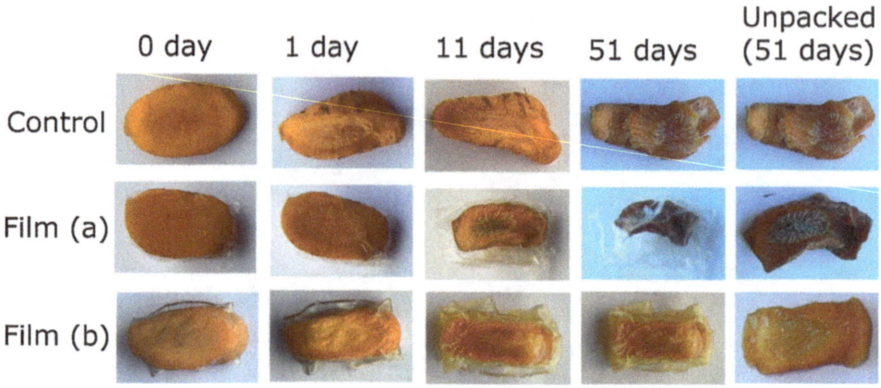

Fig. 3 Slices of carrots without preservation (control), with standard food storage film (Film (**a**)), and with keratin and citric acid food storage film (Film (**b**)), showing increased preservation over long amounts of time when wrapped with film (**b**) (Ramirez et al. 2017) (This figure has been reproduced with permission from Elsevier)

components of keratin biofibers can be chemically activated for use in a variety of applications including heavy metal biosorption (Saucedo-Rivalcoba et al. 2011). All hard keratinous materials, including α- and β-keratins, can be classified by four main morphologies: waterproof layers, solid blocks, filamentous forms, and porous materials enclosed in shells (McKittrick et al. 2012). Regardless of morphology, the polymer composite of keratin remains strong due to the ionic interactions between the matrix and the individual fiber itself.

Nanoscale diameter biofibers can be produced when a keratin polymer solution is released into an electric field and is stretched and deposited in mats via electric grounding (Mousquès and Dullaart 2012). This process is termed electrospinning and is capable of efficiently producing filters, the composition of which can be adjusted to contain other biopolymers including silk fibroin (Mousquès and Dullaart 2012).

The application of keratin fibers is seemingly unlimited, as the cystine bonds can be modified to activate the substrate to accomplish a chosen goal. One example of this occurs when the cystine bonds are reduced and reacted with [2-(acryloyloxy)ethyl]trimethylammonium chloride, producing an antimicrobial substrate with a variety of medical applications (Yu et al. 2014).

The low density of keratin fibers has allowed them to be incorporated into aerogels used in insulation. The addition of keratin fibers decreases thermal conductivity while improving mechanical performance of the aerogel (Sun et al. 2018). Furthermore, environmental remediation and soil stabilization using keratin feather fibers are advantageous due to the low density, low cost, biodegradability, and bioavailability of the fibers (Mousquès and Dullaart 2012; Montes-Zarazúa et al. 2015). One study concluded that the integration of feather fibers to in situ plots of soil mitigated soil expansion by reducing the swelling tendency of the soil (Montes-Zarazúa et al. 2015). In another study, processed chicken feathers were shown to absorb crude oil and diesel spills from soil (Ifelebuegu and Chinonyere 2016).

6 Conclusion

The availability and unique properties of keratin, including the ability to spontaneously self-assemble, increased biocompatibility which facilitates cell adhesion and proliferation, and biodegradation provide decided advantages for application in industry and medicine. Recent research has demonstrated that the extraction and scaffold formation processes used to generate keratin-based biomaterials can be tuned for specific applications, such as drug delivery or food storage. Future keratin-based biomaterial developments promise improvements to tunability and stability of keratin polymers, broadening the potential applications.

References

Alsberg E, Kong HJ, Hirano Y, Smith MK, Albeiruti A, Mooney DJ (2003) Regulating bone formation via controlled scaffold degradation. J Dent Res 82(11):903–908. https://doi.org/10.1177/154405910308201111

Anitha TS, Palanivelu P (2013) Purification and characterization of an extracellular keratinolytic protease from a new isolate of Aspergillus parasiticus. Protein Expr Purif 88(2):214–220. https://doi.org/10.1016/j.pep.2013.01.007

Bach E, Daroit DJ, Corrêa APF, Brandelli A (2011) Production and properties of keratinolytic proteases from three novel Gram-negative feather-degrading bacteria isolated from Brazilian soils. Biodegradation 22(6):1191. https://doi.org/10.1007/s10532-011-9474-0

Bordeleau F, Bessard J, Sheng Y, Marceau N (2008) Keratin contribution to cellular mechanical stress response at focal adhesions as assayed by laser tweezers. Biochem Cell Biol 86(4):352–359. https://doi.org/10.1139/o08-076

Cardamone JM (2010) Investigating the microstructure of keratin extracted from wool: peptide sequence (MALDI-TOF/TOF) and protein conformation (FTIR). J Mol Struct 969(1):97–105. https://doi.org/10.1016/j.molstruc.2010.01.048

Chitte RR, Nalawade VK, Dey S (1999) Keratinolytic activity from the broth of a feather-degrading thermophilic Streptomyces thermoviolaceus strain SD8. Lett Appl Microbiol 28(2):131–136. https://doi.org/10.1046/j.1365-2672.1999.00484.x

Chou CC, Buehler MJ (2012) Structure and mechanical properties of human trichocyte keratin intermediate filament protein. Biomacromolecules 13(11):3522–3532. https://doi.org/10.1021/bm301254u

Descamps F, Brouta F, Vermout S, Monod M, Losson B, Mignon B (2003) Recombinant expression and antigenic properties of a 31.5-kDa keratinolytic subtilisin-like serine protease from Microsporum canis. FEMS Immunol Med Microbiol 38(1):29–34. https://doi.org/10.1016/S0928-8244(03)00101-9

Eccleston DS, Horrigan MC, Ellis SG (1996) Rationale for local drug delivery. Semin Interv Cardiol 1(1):8–16

Fakhfakh-Zouari N, Haddar A, Hmidet N, Frikha F, Nasri M (2010) Application of statistical experimental design for optimization of keratinases production by Bacillus pumilus A1 grown on chicken feather and some biochemical properties. Process Biochem 45(5):617–626. https://doi.org/10.1016/j.procbio.2009.12.007

Fearing BV, Van Dyke ME (2014) In vitro response of macrophage polarization to a keratin biomaterial. Acta Biomater 10(7):3136–3144. https://doi.org/10.1016/j.actbio.2014.04.003

Fraser RD, Parry DA (1996) The molecular structure of reptilian keratin. Int J Biol Macromol 19(3):207–211. https://doi.org/10.1016/0141-8130(96)01129-4

Fraser R, Macrae T, Parry D, Suzuki E (1971) The structure of feather keratin. Polymer 12(1):35–56. https://doi.org/10.1016/0032-3861(71)90011-5

Fuchs E, Weber K (1994) Intermediate filaments: structure, dynamics, function, and disease. Annu Rev Biochem 63(1):345–382. https://doi.org/10.1146/annurev.bi.63.070194.002021
Furth ME, Atala A, Van Dyke ME (2007) Smart biomaterials design for tissue engineering and regenerative medicine. Biomaterials 28(34):5068–5073. https://doi.org/10.1016/j.biomaterials.2007.07.042
Galas E, Kałużewska M (1991) Proteinases of Streptomyces fradiae. III. Catalytic and some physicochemical properties of keratinolytic proteinase. Acta Microbiol Pol 41(3–4):169–177
Han S, Ham TR, Haque S, Sparks JL, Saul JM (2015) Alkylation of human hair keratin for tunable hydrogel erosion and drug delivery in tissue engineering applications. Acta Biomater 23:201–213. https://doi.org/10.1016/j.actbio.2015.05.013
Herrmann H, Aebi U (2004) Intermediate filaments: molecular structure, assembly mechanism, and integration into functionally distinct intracellular Scaffolds. Annu Rev Biochem 73:749–789. https://doi.org/10.1146/annurev.biochem.73.011303.073823
Ifelebuegu AO, Chinonyere P (2016) Oil spill clean-up from sea water using waste chicken feathers. In: Proceedings of the 4th international conference on advances in applied science and environmental technology (ASET'16), Bangkok, Thailand. IRED, pp 61–64
Johnson KL, Trim MW, Francis DK, Whittington WR, Miller JA, Bennett CE, Horstemeyer MF (2017) Moisture, anisotropy, stress state, and strain rate effects on bighorn sheep horn keratin mechanical properties. Acta Biomater 48:300–308. https://doi.org/10.1016/j.actbio.2016.10.033
Kakkar P, Balaraman M, Shanmugam G (2016) Transient structures of keratins from hoof and horn influence their self association and supramolecular assemblies. Int J Biol Macromol 93(Pt A):172–178. https://doi.org/10.1016/j.ijbiomac.2016.08.077
Kornillowicz-Kowalska T, Bohacz J (2011) Biodegradation of keratin waste: theory and practical aspects. Waste Manag 31(8):1689–1701. https://doi.org/10.1016/j.wasman.2011.03.024
Kowata K, Nakaoka M, Nishio K, Fukao A, Satoh A, Ogoshi M, Takahashi S, Tsudzuki M, Takeuchi S (2014) Identification of a feather β-keratin gene exclusively expressed in pennaceous barbule cells of contour feathers in chicken. Gene 542(1):23–28. https://doi.org/10.1016/j.gene.2014.03.027
Lee H, Hwang Y, Lee H, Choi S, Kim SY, Moon J, Kim JH, Kim KC, Han D, Park H, Bae H (2015) Human hair keratin-based biofilm for potent application to periodontal tissue regeneration. Macromol Res 23(3):300–308. https://doi.org/10.1007/s13233-015-3036-y
Lin X, Lee CG, Casale ES, Shih JC (1992) Purification and characterization of a keratinase from a feather-degrading Bacillus licheniformis strain. Appl Environ Microbiol 58(10):3271–3275
Maria Martelli S, Moore G, Silva Paes S, Gandolfo C, Laurindo JB (2006) Influence of plasticizers on the water sorption isotherms and water vapor permeability of chicken feather keratin films. LWT Food Sci Tech 39(3):292–301. https://doi.org/10.1016/j.lwt.2004.12.014
McKittrick J, Chen PY, Bodde SG, Yang W, Novitskaya EE, Meyers MA (2012) The structure, functions, and mechanical properties of keratin. JOM 64(4):449–468. https://doi.org/10.1007/s11837-012-0302-8
Montes-Zarazúa E, Colín-Cruz A, Pérez-Rea M, de Icaza M, Velasco-Santos C, Martínez-Hernández AL (2015) Effect of keratin structures from chicken feathers on expansive soil remediation. Adv Mat Sci Eng 2015:1–10. https://doi.org/10.1155/2015/907567
Moore GR, Martelli SM, Gandolfo CA, Pires AT, Laurindo JB (2006) Queratina de penas de frango: extração, caracterização e obtenção de filmes. Ciênc Tecnol Aliment 26(2):421. https://doi.org/10.1590/S0101-20612006000200027
Mousquès J, Dullaart R (2012) Keratin: structure, properties, and applications. Protein biochemistry, synthesis, structure, and cellular functions. Nova Science Publishers, Inc., Hauppauge, NY
Muhsin TM, Aubaid AH (2001) Partial purification and some biochemical characteristics of exocellular keratinase from Trichophyton mentagrophytes var. erinacei. Mycopathologia 150(3):121–125. https://doi.org/10.1023/A:1010900403141
Nam GW, Lee DW, Lee HS, Lee NJ, Kim BC, Choe EA, Pyun YR (2002) Native-feather degradation by Fervidobacterium islandicum AW-1, a newly isolated keratinase-producing thermophilic anaerobe. Arch Microbiol 178(6):538–547. https://doi.org/10.1007/s00203-002-0489-0

Noda J, Imai T, Kida K, Otagiri M (1996) The physicochemical and biopharmaceutical properties of fragmented keratin as a new drug carrier. Biol Pharm Bull 19(3):466–473. https://doi.org/10.1248/bpb.19.466
Oladayo OO, Ulmunna QC, Oluwasegun W, Joseph OS (2016) Physicochemical properties of cassava starch and starch-keratin prepared biofilm. Songklanakarin J Sci Tech 38(4):349–355
Pace LA, Plate JF, Mannava S, Barnwell JC, Koman LA, Li Z, Smith TL, Van Dyke M (2014) A human hair keratin hydrogel scaffold enhances median nerve regeneration in nonhuman primates: an electrophysiological and histological study. Tissue Eng Part A 20(3–4):507–517. https://doi.org/10.1089/ten.TEA.2013.0084
Parry DA, Crewther WG, Fraser RD, MacRae TP (1977) Structure of α-keratin: structural implication of the amino acid sequences of the type I and type II chain segments. J Mol Biol 113(2):449–454. https://doi.org/10.1016/0022-2836(77)90153-X
Paul T, Das A, Mandal A, Halder SK, DasMohapatra PK, Pati BR, Mondal KC (2014) Production and purification of keratinase using chicken feather bioconversion by a newly isolated Aspergillus fumigatus TKF1: detection of valuable metabolites. Biomass Convers Biorefin 4(2):137–148. https://doi.org/10.1007/s13399-013-0090-6
Pedram Rad Z, Tavanai H, Moradi AR (2012) Production of feather keratin nanopowder through electrospraying. J Aerosol Sci 51:49–56. https://doi.org/10.1016/j.jaerosci.2012.04.007
Peralta Ramos ML, Gonzalez JA, Fabian L, Perez CJ, Villanueva ME, Copello GJ (2017) Sustainable and smart keratin hydrogel with pH-sensitive swelling and enhanced mechanical properties. Mater Sci Eng C Mater Biol Appl 78:619–626. https://doi.org/10.1016/j.msec.2017.04.120
Poranki D, Goodwin C, Van Dyke M (2016) Assessment of deep partial thickness burn treatment with keratin biomaterial hydrogels in a swine model. Biomed Res Int 2016:1803912. https://doi.org/10.1155/2016/1803912
Rai SK, Mukherjee AK (2011) Optimization of production of an oxidant and detergent-stable alkaline β-keratinase from Brevibacillus sp. strain AS-S10-II: Application of enzyme in laundry detergent formulations and in leather industry. Biochem Eng J 54(1):47–56. https://doi.org/10.1016/j.bej.2011.01.007
Rajabinejad H, Bucişcanu II, Maier SS (2016) Practical ways of extracting keratin from keratinous wastes and by-products: a review. Env Eng Manag J 15(5):1131–1147
Raju KC, Neogi U, Saumya R, Goud NR (2007) Studies on extra cellular enzyme keratinase from dermatophyte Microsporum gypseum. Int J Biol Chem 1(3):174–178. https://doi.org/10.3923/ijbc.2007.174.178
Ramirez DOS, Carletto RA, Tonetti C, Giachet FT, Varesano A, Vineis C (2017) Wool keratin film plasticized by citric acid for food packaging. Food Packag Shelf Life 12:100–106. https://doi.org/10.1016/j.fpsl.2017.04.004
Riffel A, Brandelli A, Bellato CDM, Souza GH, Eberlin MN, Tavares FC (2007) Purification and characterization of a keratinolytic metalloprotease from Chryseobacterium sp. kr6. J Biotechnol 128(3):693–703. https://doi.org/10.1016/j.jbiotec.2006.11.007
Roy DC, Tomblyn S, Isaac KM, Kowalczewski CJ, Burmeister DM, Burnett LR, Christy RJ (2016) Ciprofloxacin-loaded keratin hydrogels reduce infection and support healing in a porcine partial-thickness thermal burn. Wound Repair Regen 24(4):657–668. https://doi.org/10.1111/wrr.12449
Saucedo-Rivalcoba V, Martinez-Hernandez AL, Martinez-Barrera G, Velasco-Santos C, Rivera-Armenta JL, Castano VM (2011) Removal of hexavalent chromium from water by polyurethane-keratin hybrid membranes. Water Air Soil Poll 218(1–4):557–571. https://doi.org/10.1007/s11270-010-0668-6
Saul JM, Ellenburg MD, de Guzman RC, Van Dyke M (2011) Keratin hydrogels support the sustained release of bioactive ciprofloxacin. J Biomed Mater Res A 98(4):544–553. https://doi.org/10.1002/jbm.a.33147
Sharma S, Gupta A, Chik S, Kee CG, Mistry BM, Kim DH, Sharma G (2017) Characterization of keratin microparticles from feather biomass with potent antioxidant and anticancer activities. Int J Biol Macromol 104(Pt A):189–196. https://doi.org/10.1016/j.ijbiomac.2017.06.015

Shavandi A, Silva TH, Bekhit AA, Bekhit AEA (2017) Keratin: dissolution, extraction and biomedical application. Biomater Sci 5(9):1699–1735. https://doi.org/10.1039/c7bm00411g
Sierpinski P, Garrett J, Ma J, Apel P, Klorig D, Smith T, Koman LA, Atala A, Van Dyke M (2008) The use of keratin biomaterials derived from human hair for the promotion of rapid regeneration of peripheral nerves. Biomaterials 29(1):118–128. https://doi.org/10.1016/j.biomaterials.2007.08.023
Silva LAD, Macedo AJ, Termignoni C (2014) Production of keratinase by Bacillus subtilis S14. Ann Microbiol 64(4):1725–1733. https://doi.org/10.1007/s13213-014-0816-0
Sinkiewicz I, Staroszczyk H, Sliwinska A (2018) Solubilization of keratins and functional properties of their isolates and hydrolysates. J Food Biochem 42(2). https://doi.org/10.1111/jfbc.12494
Sun K, Guo J, He Y, Song P, Xiong Y, Wang RM (2016) Fabrication of dual-sensitive keratin-based polymer hydrogels and their controllable release behaviors. J Biomater Sci Polym Ed 1–15. https://doi.org/10.1080/09205063.2016.1239955
Sun MZ, Sun H, Hostler S, Schiraldi DA (2018) Effects of feather-fiber reinforcement on poly(vinyl alcohol)/clay aerogels: structure, property and applications. Polymer 137:201–208. https://doi.org/10.1016/j.polymer.2018.01.008
Syed DG, Lee JC, Li WJ, Kim CJ, Agasar D (2009) Production, characterization and application of keratinase from Streptomyces gulbargensis. Biores Tech 100(5):1868–1871. https://doi.org/10.1016/j.biortech.2008.09.047
Tachibana A, Kaneko S, Tanabe T, Yamauchi K (2005) Rapid fabrication of keratin-hydroxyapatite hybrid sponges toward osteoblast cultivation and differentiation. Biomaterials 26(3):297–302. https://doi.org/10.1016/j.biomaterials.2004.02.032
Tanabe T, Okitsu N, Yamauchi K (2004) Fabrication and characterization of chemically crosslinked keratin films. Mater Sci Eng: C 24(3):441–446. https://doi.org/10.1016/j.msec.2003.11.004
Tomblyn S, Pettit Kneller EL, Walker SJ, Ellenburg MD, Kowalczewski CJ, Van Dyke M, Burnett L, Saul JM (2016) Keratin hydrogel carrier system for simultaneous delivery of exogenous growth factors and muscle progenitor cells. J Biomed Mater Res B Appl Biomater 104(5):864–879. https://doi.org/10.1002/jbm.b.33438
Veerasubramanian PK, Thangavel P, Kannan R, Chakraborty S, Ramachandran B, Suguna L, Muthuvijayan V (2018) An investigation of konjac glucomannan-keratin hydrogel scaffold loaded with Avena sativa extracts for diabetic wound healing. Colloids Surf B Biointerfaces 165:92–102. https://doi.org/10.1016/j.colsurfb.2018.02.022
Wang B, Sullivan TN (2017) A review of terrestrial, aerial and aquatic keratins: the structure and mechanical properties of pangolin scales, feather shafts and baleen plates. J Mech Behav Biomed Mater 76:4–20. https://doi.org/10.1016/j.jmbbm.2017.05.015
Wang B, Yang W, McKittrick J, Meyers MA (2016) Keratin: structure, mechanical properties, occurrence in biological organisms, and efforts at bioinspiration. Prog Mater Sci 76:229–318. https://doi.org/10.1016/j.pmatsci.2015.06.001
Wang J, Hao S, Luo T, Cheng Z, Li W, Gao F, Guo T, Gong Y, Wang B (2017) Feather keratin hydrogel for wound repair: preparation, healing effect and biocompatibility evaluation. Colloids Surf B Biointerfaces 149:341–350. https://doi.org/10.1016/j.colsurfb.2016.10.038
Yamamura S, Morita Y, Hasan Q, Yokoyama K, Tamiya E (2002) Keratin degradation: a cooperative action of two enzymes from Stenotrophomonas sp. Biochem Biophys Res Comm 294(5):1138–1143. https://doi.org/10.1016/S0006-291X(02)00580-6
Yu D, Cai JY, Liu X, Church JS, Wang L (2014) Novel immobilization of a quaternary ammonium moiety on keratin fibers for medical applications. Int J Biol Macromol 70:236–240. https://doi.org/10.1016/j.ijbiomac.2014.06.025
Yue K, Liu Y, Byambaa B, Singh V, Liu W, Li X, Sun Y, Zhang YS, Tamayol A, Zhang P, Ng KW, Annabi N, Khademhosseini A (2018) Visible light crosslinkable human hair keratin hydrogels. Bioeng Transl Med 3(1):37–48. https://doi.org/10.1002/btm2.10077
Zainal-Abidin MH, Hayyan M, Hayyan A, Jayakumar NS (2017) New horizons in the extraction of bioactive compounds using deep eutectic solvents: a review. Anal Chim Acta 979:1–23. https://doi.org/10.1016/j.aca.2017.05.012

Chapter 8
Keratin-Based Biotechnological Applications

Kush Kumar Nayak, Piyush Parkhey and Bidyut Mazumdar

Abstract The present chapter is aimed towards giving an overview of the applications of keratin as a potential biomaterial substitute in the field of biotechnology. Keratin is a fibrous protein and considered as biomaterial due to its biocompatible and biodegradable characteristics. Its use as biopolymer has been the subject of intense investigation over the past few years. Wool, feather, horn and hooves, etc., are strong in terms of mechanical strength due to the presence of keratin. But once keratin is extracted from natural sources then, it becomes poor in mechanical properties. So, the blending of keratin with other biopolymer can improve the material properties like strength, flexibility, and water vapor permeability. The malleable nature of keratin proves its biotechnological applications such as tissue engineering scaffold, green composites, green cement, bioplastic, etc. The functional groups and chemical structures of keratin govern its properties and morphology, which gives an opportunity to control the design of desired molecular structure for various applications, and varies from industrial to the biotechnological field. Recently, the biodegradable keratin-based biopolymers have gained considerable importance in the medical field as they avoid additional surgery to remove the implants and lack of medical waste burden. Thus, much attention needs to be undertaken on the development of composite biomaterial derived from keratin.

Keywords Keratin · Biomaterial · Tissue engineering · Wound · Packaging
Bio-cement · Bio-ink

K. K. Nayak (✉) · P. Parkhey
Amity Institute of Biotechnology, Amity University Chhattisgarh, Raipur 493225, India
e-mail: nayak23bt@gmail.com

B. Mazumdar
Department of Chemical Engineering, National Institute of Technology, Raipur 492010, India

S. Sharma and A. Kumar (eds.), *Keratin as a Protein Biopolymer*, Springer Series on Polymer and Composite Materials, https://doi.org/10.1007/978-3-030-02901-2_8

1 Introduction

In the past few decades, intensive studies were done to replace synthetic materials with the help of biopolymer-based composite materials. Many of the research activities are still going around the globe for the development of high performance "green products", which is environmental-friendly and nontoxic for use (Petrulyte 2008; Shi and Dumont 2014). In addition, the acceptance of renewable to nonrenewable materials is the center of interest for industrial researchers around the world because of the environmental legislation of the various states. The traditional synthetic petroleum-based materials are considered critical, due to restricted/limited disposal approach, environmental concerns, and minimization of the soil fertility (Calkins 2009). Hence, the focus has been diverted towards the study of biocompatible and biodegradable materials.

Various polymers have been widely used to develop biomaterial for the fabrication of medical devices and scaffolds (O'Brien 2011). In the past, significant efforts have been focused on developing scaffold from metallic- and ceramic-based materials that could bioactively interact with live tissues. For example, a coating of bioactive materials onto various dense alloys and ceramics has been observed to be an effective approach in the development of biomaterials (Lu et al. 2012; Yu et al. 2015). Ceramic material like hydroxyapatite and tri-calcium phosphate has been used for bone regeneration and is known for the hard brittle surface, high Young's modulus, and very low elasticity (Ma 2008). It exhibits excellent biocompatibility because of chemical and structural similarity to the bone's mineral phase, which makes it suitable for bone tissue engineering but for soft tissue engineering, it is insignificant (Bhardwaj et al. 2015). Typically, synthetic and natural polymers are being used in the designing of scaffolds for tissue engineering (O'Brien 2011). Synthetic polymers such as polylactic acid (PLA) (Jung et al. 2005), polyglycolic acid (PGA) (Wang et al. 2003), poly-e-caprolactone (PCA) (Ciardelli et al. 2005), polyortho esters (Harrison et al. 2004), polystyrene (PS) (El Fray et al. 2006) and polyvinyl alcohol (PVA) (Yang et al. 2014) etc., have been extensively used to construct biomaterials for medical applications. These materials have distinct advantages such as fabricability with a tailored architecture and controlled degradation characteristics by varying the composition of the individual polymer (Lu et al. 2000; Oh 2003; Rowlands et al. 2007). However, degradation of synthetic biomaterials like PLA and PGA produce CO_2 and therefore lowers the pH of application site, which causes cell and tissue necrosis (Liu et al. 2006). The other demerits are then poor hydrogel nature, as it is reported that synthetic materials such as polyolefin, polyethylene (PE), and polypropylene (PP) are hydrophobic in nature and they do not degrade in vivo (Maitz 2015), hence are not suitable to be used as temporary implants in tissue engineering. Synthetic biomaterials also have the risk of rejection due to then reduced bioactivity (Dreifke et al. 2013), which restricts biotechnological and biomedical applications. The problems of the synthetic material as mentioned above have been trying to be addressed by developing semisynthetic biomaterials like poly-*l*-lactic acid (PLLA) and poly-*d*-lactic acid (PDLA) (Maitz 2015), which can degrade faster than synthetic material. However, cytotoxicity still exists in these biomaterials, which again needs to be addressed for tissue engineering applications.

To overcome the limitations of synthetic and semisynthetic biomaterial, the use of natural polymers is currently undergoing the resurgence in the field of biotechnology. The environmental impact of biopolymer is negligible which favors its suitability over synthetic polymer. The chemical stability and biocompatibility of natural polymers greatly account for their utilization in biomedical devices (Dickinson and Gerecht 2016). It is reported that natural polymer helps host cells to replace degraded scaffold by producing their own extracellular matrix (ECM) (Cziple and Marques 2008). In tissue engineering, it is crucial for the scaffold to positively interact with surrounding tissues and facilitate the natural regeneration of cells (He and Lu 2016). To achieve this goal, several biopolymers have been used to develop biomaterial by various researchers (Hemsri et al. 2011; Isarankura Na Ayutthaya et al. 2015; Gupta and Nayak 2015a). Recently, number of biopolymer have been introduced such as collagen (Levingstone et al. 2014), chitosan (Ahmed and Ikram 2016), starch (Xiong et al. 2008), polyhydroxybutyrate (PHB) (Zhijiang et al. 2012), hyaluronic acid (HA) esters (An et al. 2016), polyhydroxyalkanoates (PHA) (Panchal et al. 2013) and copolymers of the above, etc., due to their biocompatibility, biodegradability, plentiful availability, and owing structures similar to the human tissue they intend to replace.

Keratin is a biopolymer, which supports the growth of mammalian cells like keratinocytes and fibroblast due to its excellent biocompatibility (Xing et al. 2011; Vasconcelos and Cavaco-Paulo 2013). It is a fibrous protein that has various characteristics, ranges from a structurally robust and impact-resistant material to a simple waterproof layer (Feughelman 1959). Keratin is found in several forms like skin, fur, wool, claws, hair, nails, feathers, hooves, beaks, scales and horns, etc. (Wrzesniewska-Tosik and Adamiec 2007; Reichl 2009; Wang and Cao 2012; Xu et al. 2014). The main difference in the keratin obtained from various sources is its sulfur content. The presence of many cysteine disulphide crosslink causes the rigid structure of keratin as in horns and nails (Reichl 2009). Fewer disulphide crosslink is available in wool and skin, which gives them flexible characteristics (Balaji et al. 2012). Keratin in its extracted form is fragile in nature with poor hydrogel property, which restricts its use in biomaterial. Blends of keratin with natural and synthetic polymers such as keratin–chitosan (Tanabe et al. 2002), keratin–polypropylene, keratin–cellulose–polypropylene (Bullions et al. 2003, 2004; Schuster 2003), and keratin–polyethylene, Barone and Schmidt (2005) etc., were studied by several researchers and found that the blending improves the material properties such as strength, flexibility and water vapor permeability.

The candidature of keratin as a biopolymer is predominant for the preparation of keratin-based materials with characteristics of several therapeutic and nontherapeutic sectors. Due to its natural abundance, structural feature, non-toxicity, biocompatibility, renewability, biological activity, and behavior as biopolymer, keratin secures its place in modern material sectors (McKittrick et al. 2012; Nayak and Gupta 2017). With this knowledge, application of native keratin for the development of the advanced material with value-added features is the core concept that provides extensive opportunities for multidisciplinary scientific experiment and research like a study of scaffolds, films, hydrogels and degradable plastic, etc. (Vasconcelos and Cavaco-Paulo 2013). The applications of such biomaterials have been redirecting in different sectors including pharmaceuticals, cosmetics, and biomedical/biotechnological applications (Bainbridge 2013; Bhardwaj et al. 2015; Gupta and Nayak 2015b). Figure 1 demonstrates an outline of the potential application of keratin-based biopolymers for various fields that include both biotechnological and biomedical sectors.

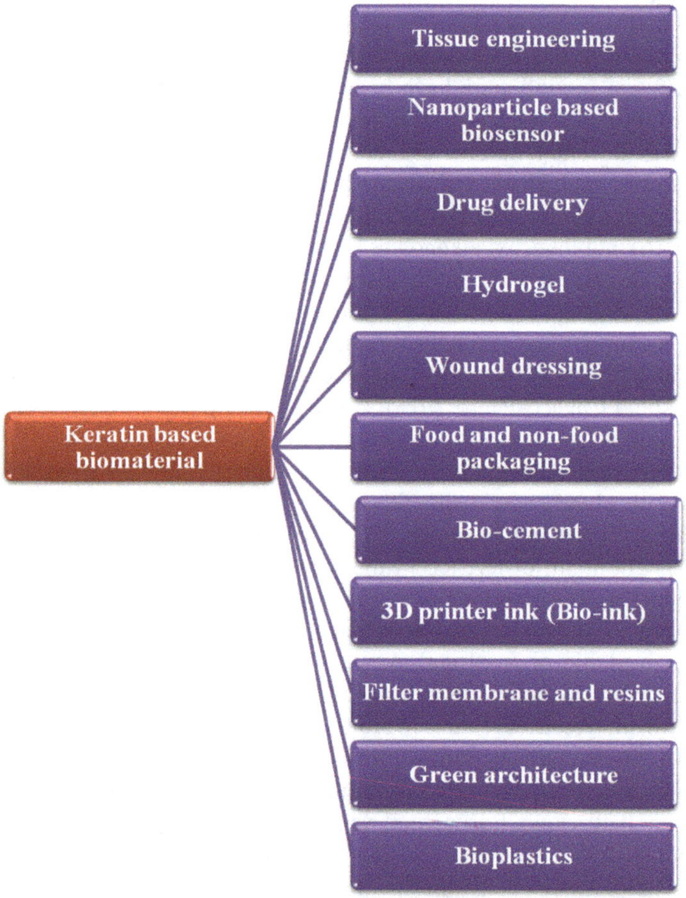

Fig. 1 Biotechnological applications of keratin-based biomaterial

2 Application of Keratin-Based Biomaterial

2.1 *Tissue Engineering and Biomedical Application*

Tissue engineering is a multidisciplinary area that concerns with the principles of engineering and biological science to develop appropriate alternatives that improve, maintain and/or restore tissue's biological function (O'Brien 2011), as shown in Fig. 2. Recently, the biodegradable polymers are playing a crucial role in tissue engineering field, because it facilitates removal of the implant without surgery. Thus, much attention has been given to the development of biomaterials from biopolymers. In medical applications, biopolymers have been applied successfully for the treatment of ligament injury, blood vessels damage, skin graft, drug delivery and wound treatment, etc. (Khan et al. 2015). It is believed that intrinsic properties of biopolymers can be improved by appropriate process engineering methods for optimum results such as cell attachment, elongation, and proliferation. Biocompatibility is the ability of a material to perform with an appropriate host response in a specific application (Williams 2011). The rationale of conducting biocompatibility study is to determine the suitability of a biomaterial particularly for human use and to confirm its noncytotoxic effects as well (Ciapetti et al. 1993). It is a key factor for designing any scaffold for tissue engineering application because scaffold continuously interacts with the subject's body during treatment. If this parameter is not taken into consideration, then the material used can infect application site.

Biocompatibility studies of a scaffold involve the in vitro and in vivo experiments in order to investigate its effects on cultured cells. In vitro cell culture method is more often used as compared to in vivo test to monitor the biocompatibility of any material (Spector 2007). Permanent cell lines are commonly engaged for such study like a rat primary culture of fibroblast, keratinocyte and osteoblasts, etc. (Restle et al. 2015). Considerable advancements in the area of implantable devices and the combination products such as artificial organs (Fine and Tobias 2007; Kizilel et al. 2005; O'Connor et al. 2008), tranquilize eluting stents (Moses et al. 2002; Kipshidze et al. 2005; Frazier and Jacob 2007; Belverud et al. 2008), catheters (Callahan and Natale 2008), biosensors (Koschwanez and Reichert 2007), platforms for tissue designing (Salgado et al. 2004; Abukawa et al. 2006; Matsumoto et al. 2007), and heart valves (Nozynski et al. 2001) have been reported. Nevertheless, the biocompatibility of implantable platforms remains a basic issue in restricting device lifespan and usefulness, especially on account of biosensors. The foreign body response to implantable materials likewise introduces a huge hazard for patients. However, various natural, semisynthetic, and synthetic materials are used in the creation of implantable material like chitosan (Khor and Lim 2003), HA (Vercruysse and Prestwich 1998), collagen (Geiger et al. 2003), alginate (de Vos et al. 2002), dextran (Draye et al. 1998; Cadée et al. 2001), poly(lactic-co-glycolic acid) (PLGA) and PLA (Royals et al. 1999; Lunsford et al. 2000), 2-hydroxyethyl methacrylate (Royals et al. 1999), polyethylene glycol (PEG) (Tsai et al. 2002; Dalsin et al. 2003), and PVA (Paradossi et al. 2003; Maruoka et al. 2006). In spite of the fact that materials produced using these

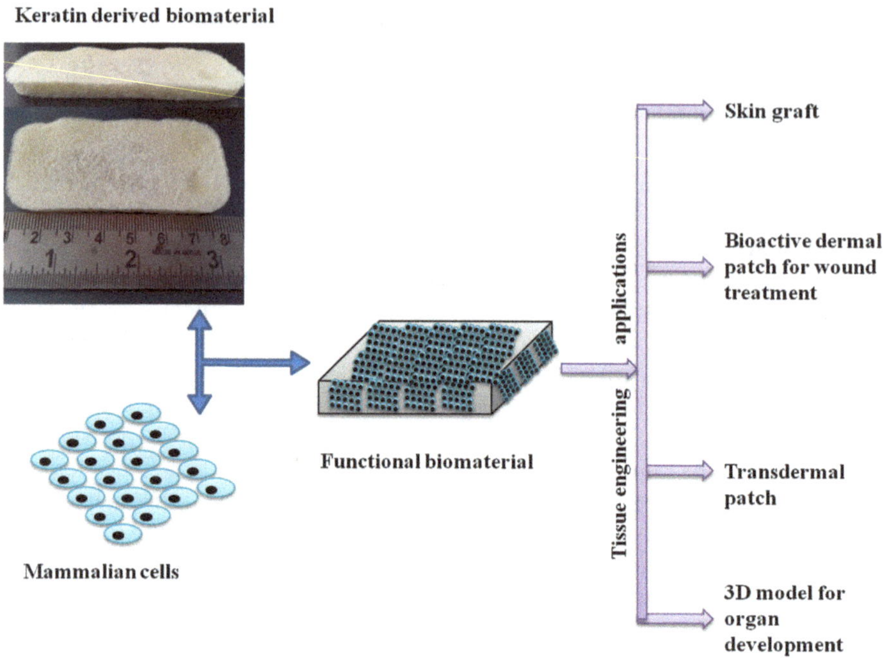

Fig. 2 Applications of keratin-based biomaterial on tissue engineering

polymers are generally utilized and thought to be moderately biocompatible, various investigations have demonstrated that frameworks produced have biocompatibility issues (Mendes et al. 2001). International Organization for Standardization-(c) 10993 (ISO) gives a progression of principles to assess the biocompatibility of a therapeutic device before clinical testing (Nayak and Gupta 2015). In vitro testing is secured by ISO 10993-573 and incorporates critical parts of the test systems such as negative or positive control materials and extraction of cell lines or cell media.

In recent times, numerous scaffolds have been developed and studied for then in vitro biocompatibility to investigate cell regenerative capability (Rosellini et al. 2009). For example, in vitro biocompatibility of chitosan matrix was investigated particularly for skin tissue engineering by Keong and Halim (2009) and in 2 weeks of study, matrix surface was found suitable for the attachment of fibroblast cells. Huang et al. (2013) reported the non-cytotoxicity of scaffold made from carboxymethyl chitosan/gelatin on cutaneous wound, which reveals the material efficacy to induce the formation of granulation tissue and new skin cells. The potential application of starch-based degradable polymers has been recently confirmed in several biomedical fields ranging from bone plates/screws to drug delivery carriers and skin tissue engineering scaffolds (Marques et al. 2002). Collagen platforms derived from a marine source and their biocompatibility was studied by Song et al. (2006). In this study, a novel type of corrosive solvent collagen was extracted from jellyfish with an aim to obtain alternative collagen source and in vitro study confirmed that jellyfish-collagen displayed higher cell suitability than other normally inferred biomaterials including cow-like collagen, gelatin, glucan, and HA. Alginate/keratin-based half-breed hydrogel was effectively developed by Silva et al. (2014) in which self-assembly properties of keratin and its capacity to mimic the ECM were combined with the property of alginate to create 2D and 3D crossbreed hydrogel. All outcomes showed that the creation of the hydrogel significantly affected their physical properties, which can be tuned to get material for appropriate applications. The cell-material cooperation was evaluated using human cells and found that the alginate/keratin half breed biomaterials upheld cell connection, spreading and multiplication. The results demonstrated that such novel crossbreed hydrogel may discover applications as biomaterial for delicate tissue engineering. Balaji et al. (2012) reported the formation of 3D frameworks with very much interconnected network of high porosity utilizing chitosan, keratin, and gelatin for skin tissue engineering applications. In this investigation, fibroblasts were seeded on the permeable scaffold and their bond, as well as multiplication efficiency, was examined. The result indicated enhanced cell connectivity on the inner pore surface of the scaffold and the cells were found encircled with ECM. This shows that the composition and structure of matrix enabled fibroblasts to connect on to the framework surface and to infiltrate into the permeable structure of the platform. Bhardwaj et al. (2015) investigated blends of keratin and silk fibroin as wound healing substrates that facilitate fibroblast cell attachment and expansion. Cytocompatibility tests demonstrated significant improvements in cell expansion due to hybrid properties of scaffold. This examination shows the properties of keratin and silk mixed biomaterial as an ideal dermal substitute for skin tissue engineering. Nayak and Gupta (2015) studied keratin-based biomaterials in vitro with mammalian myofibroblast cells (Fig. 3) to investigate the potential for skin regenerative capacity. It was observed that the proposed biomaterials are competent to support the cell growth and help in proliferation by providing a favorable cell attachment surface. The formation of connective arms between the cells was observed and found to migrate in the curved projectile, which proved attachment of the cells in 3D surface of scaffolds and confirm scaffold's potential to support cell proliferation.

2.2 *Wound Dressing*

Wound can be defined as a distraction in the layers of skin cells resultant from physical or thermal damage such as cuts or punctures (Dhivya et al. 2015). In classical wound healing process, interaction amongst various factors required for the complete wound healing such as reepithelialization, inflammation, granulation tissue formation, angiogenesis, tissue maturation and wound contraction (Epstein et al. 1999; Blakytny and Jude 2006). During the proliferative phase of wound healing keratinocytes, fibroblasts, smooth muscle cells and endothelial migrate through the wound and proliferate to reepithelialize the damage surface, deposit an ECM, contract the wound size and form new blood vessels (Santoro and Gaudino 2005). In

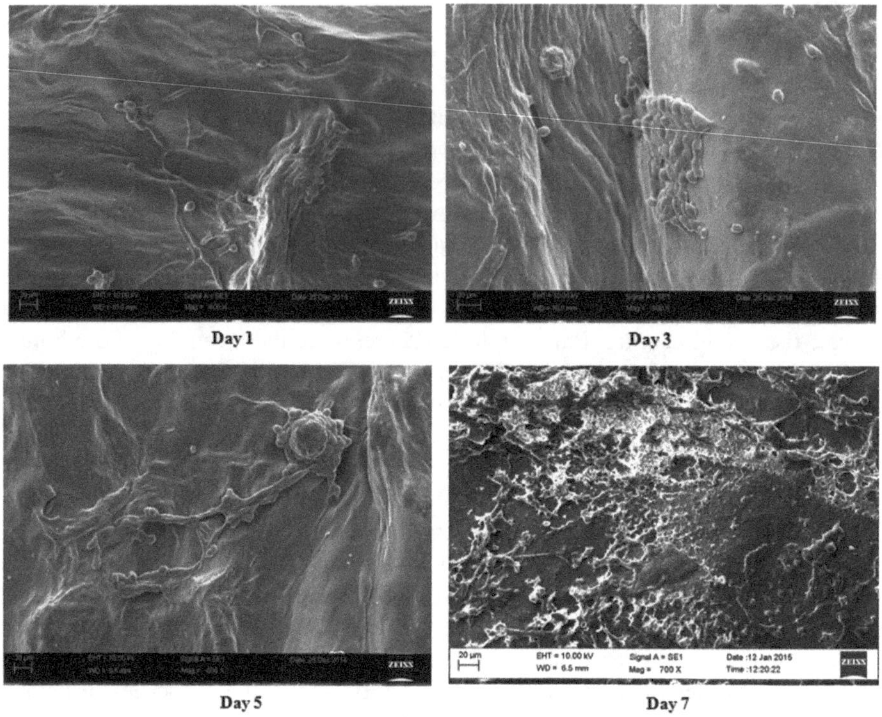

Fig. 3 The microscopic image illustrates the growth and adherence behavior of C_2C_{12} cell over a keratin/agar biomaterial, approximately above 60% of the scaffold surface was covered by the C_2C_{12} cell on the seventh day of culture. This result demonstrates the biocompatibility and skin regenerative potential of the keratin-based biomaterial (reproduced from Nayak and Gupta (2015), with permission from Elsevier)

the final stage, the recently shaped tissue is redesigned by the activity of ECM proteinases and its inhibitors, which adjusts the recovered dermis and strengthens the repaired tissue (Gurtner et al. 2008).

Wound dressing materials are the important tool in the medical field and accepted worldwide for the treatment of skin injuries. Traditional dressing materials such as lint, cotton wool gauzes and bandages, etc., have been used for the wound management (Dhivya et al. 2015). The primary goal of the wound management is good aeration, desertion of the wound exudates and prevention the microbial infection in the wound (Boateng et al. 2008). In the modern era, wound management is laid on the concept of producing an optimal atmosphere to permit the epithelial cell to grow unobstructed for faster treatment of wounds along with the therapeutic bioactive support of the dressing materials. The optimal environment for wound treatment includes a humid condition around the wound with good aeration for stimulating cell growth and low bacterial load (Bowler et al. 2001). Hence, viable injury management depends upon the understanding of various factors such as sort of wound being deal with, healing procedure, patient conditions (diabetic and nondiabetic) and the physicochemical properties of the available dressing materials (Boateng et al. 2008). Therefore, different dressings were assessed and tried as far as their physical properties and clinical execution for a given kind of wound and its stages, before being considered for routine application (Akita et al. 2013). Single dressing material is not available for the management of all kinds of wound like chronic, acute, dry, and exuding wounds. Further, the wound healing process has a few distinct stages (Boateng et al. 2008) that cannot be focused by a specific dressing type (Sarabahi 2012). Thus, it becomes important to develop a dressing that is attributed with the properties to manage all type of wounds at its preliminary or initial stage to prevent wound progression. Several advanced skin substrates are available to the consumers for clinical use, which is mostly made up by mammalian-derived matrix components or from human cells (El Ghalbzouri et al. 2004). In addition, skin recovery requires the synchronized action of various events in the extracellular wound environment overlapping phases of proliferation, inflammation and scaffold remodeling (Dickinson and Gerecht 2016). This process is extremely proficient during normal wound healing. However, chronic wounds are unable to improve via ordered and reparative wound healing process and fail to heal. To overcome this, scientific attention has to be focused on developing inventive biomaterial with more effective wound healing therapy.

There are a few wound care products that have been developed, which essentially provide cellular and structural components for wound healing. For example, Apligraf® (Organogenesis, Inc.) is made out of a cow-like type I collagen grid seeded with neonatal fibroblasts to create a neodermal layer. Upon this dermal component, a monolayer of human neonatal epidermal keratinocytes is added to approximate the epidermis and form a differentiated stratum corneum (Zaulyanov and Kirsner 2007). These result in a metabolically dynamic bilayered skin substitute, which gives both dermal and epidermal layer with living cells. In spite of the fact that the biofunctional activity of keratinocytes and fibroblasts in Apligraf® become dysfunctional after a month and a half once it applied into a patient (Hu et al. 2006). They are considered to be accountable for stimulating differentiation and proliferation via delivering of essential growth factors and cytokines (Falanga et al. 2002). Apligraf® was the main allogeneic cell-based item to be endorsed by the "food and drug administration" (FDA) in 1998 for the treatment of venous leg ulcers and diabetic foot ulcers (DFUs). As compared to conventional standard of care the clinical trials with Apligraf® resulted in a considerably higher rate of wound closure (Veves et al. 2001; Edmonds and European and Australian Apligraf Diabetic Foot Ulcer Study Group 2009). For the treatment of non-recuperating DFUs, Dermagraft® (Organogenesis, Inc.) was affirmed by the FDA in 2001 (Landsman et al. 2011). It contains neonatal dermal fibroblasts layer and used as a standard suture material (Capperauld 1989). In this material, cultured fibroblast cell synthesize proteoglycans, cytokines, collagens, and other key regulatory biomolecules, which convert this material in the form of bioactive matrix (Naughton et al. 1997). The rate of wound closure increased appreciably when Dermagraft® is applied in DFUs as compared to control (Marston et al. 2003). Another naturally dynamic human skin allograft is TheraSkin®, which is collected within 24 h after death and cryogenically prepared to preserve the viable keratinocytes, fibroblasts and ECM sequestered with essential growth factors and cytokines. TheraSkin® is powerful in the treatment of venous ulcers and DFUs (Landsman et al. 2011), because it contains more amount of key collagens which are important and critical to wound healing as compared to Apligraf®. The key regulatory proteins and cytokines delivered by using a living human dermal skin substitute that promoted fibroblast migration, angiogenesis and keratinocyte expansion to increase wound healing. It is reported, that the biomaterial like collagen, HA, alginate, and chitosan were used as skin graft for the treatment of burn wounds and its dermal patches are commercially available as well for wound management (Augst et al. 2006; Xu et al. 2007; Rath et al. 2015). Berglund et al. (2003) proposed the combination of polyhydroxybutyrate-co-hydroxyvalerate (PHBV) structures to induce skin regeneration in a full-thickness model. It is postulated that rapid degradation and faster regeneration of skin cells are a result of good integration of PHBV scaffolds with surrounding their tissues by allowing exudation and infiltration by inflammatory cells (Berglund et al. 2003). Collagen-based skin substitutes particularly in the form of hydrogel show superior stability and promising prospective (Zhang et al. 2011). Collagen gives strength and flexibility to connective tissues and allows physiological interaction of scaffold with cell and they eventually help in wound healing (Wang et al. 2013).

The proteins present in ECM matrix provide cell surface receptors for cell attachment, signaling factors reservoir to regulate cell proliferation and migration, tensile strength and structural support. This matrix has a complex 3D architecture and is made from combination of fibrous proteins, polysaccharides and proteoglycans, which play a significant role in wound healing (Badylak 2002; Tracy et al. 2016). In previous studies it was found that the keratin-based biomaterial supports the fibroblast cell proliferation, growth and attachment because of the material's characteristics such as porosity, degradability, hydrogel property as well as mechanical behavior, which confirm its suitability as wound dressing (Isarankura Na Ayutthaya et al. 2015; Nayak and Gupta 2015; Gupta and Nayak 2015a, b, 2016; Kumaran et al. 2017). Several keratin-based therapeutic dermal patches such as Keratin/Agar, Keratin/Alginate and Keratin/Gellan were developed by Nayak and Gupta (2017) and loaded specifically with basic fibroblast growth factor (bFGF) then engineered in the form of porous hydrogel dermal patches for wound treatment. In this study, the release rate of bFGF from dermal was conducted for 3 days and it was found that the patches were able to release bFGF up to 77.77–88.89%. The keratin-based dermal patches were able to release the bFGF near its vicinity and could accelerate the wound healing efficiency during practical application. The development strategies of therapeutic dermal patch from keratin-based biomaterial are shown in Fig. 4, opted from Nayak and Gupta (2017).

In expansion of the above work, three keratin-based scaffolds were studied to validate its in vitro skin regeneration potential. For this, mammalian fibroblast cells were seeded over developed keratin-based scaffold and examine by fluorescence and scanning electron microscope (SEM) technique (Fig. 5). The fluorescence images were taken at three stages 1st, 5th and 10th day of cell culture, presence of the

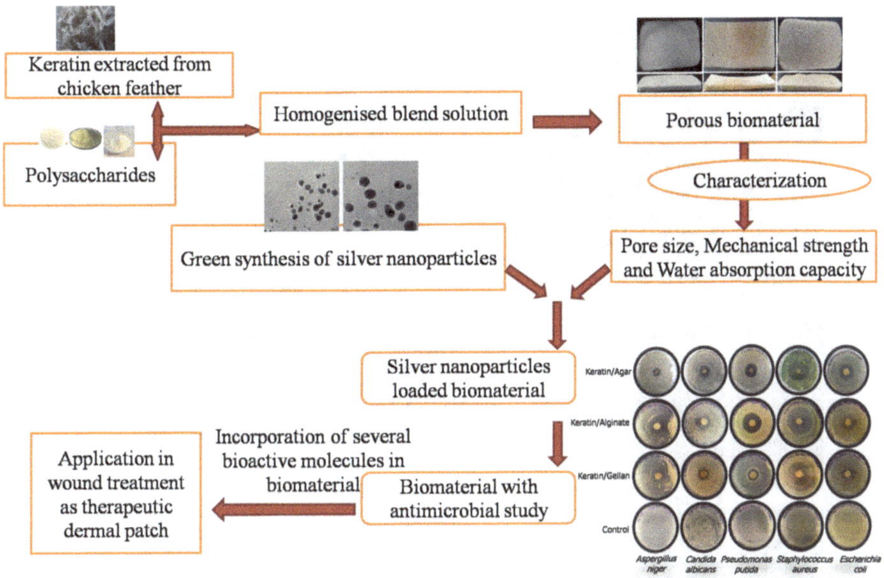

Fig. 4 Design and development strategies of therapeutic dermal patch from keratin-based biomaterial for several wound treatment

cells in scaffold's 3D environment was confirmed from the green dots while the cytoskeleton of the growing cells were confirmed with fluorescent red color (Fig. 5a). The alignment of the cultured cell observed in images reveals its migration in the 3D surface of scaffold, which confirms that the scaffold support cells proliferation and growth. In addition, cell's morphology and attachment on the scaffold surface is analyzed by SEM at 5th and 13th day of culture (Fig. 5b) and it was found to be in monolayer flat pattern, which is a characteristic of a fibroblast cell. The cytoskeleton of the cells was indicated by an arrow in the SEM images. The conclusion of this work is that the cells were able grown and adhered the surface of keratin-based scaffold and can spread their connective arm to form a cytoskeleton for tissue formation. Results disclose the utility of keratin-based scaffold in the field of skin tissue engineering and can replace the synthetic or semisynthetic biomaterials from extensively uses in skin tissue engineering.

In the similar approach, fibroblast and epidermal growth factors were loaded in HA-collagen (Choi et al. 2016) and chitosan sponges (Denkbas et al. 2003) dressing matrix for the treatment of wounds. Many natural polymers such as silk, alginate and chitosan are well defined biologically, that allows them to be used as viable materials for growth factors intake within the structures (Yun et al. 2010). Skin regeneration potential of keratin-based dermal patch motivated to develop the material as a wound dressing and assess its wound healing properties by in vitro method. It is also reported by several researchers that the antimicrobial activity of the keratin-based dressing was improved by loading silver nanoparticles (Rujitanaroj et al. 2008;

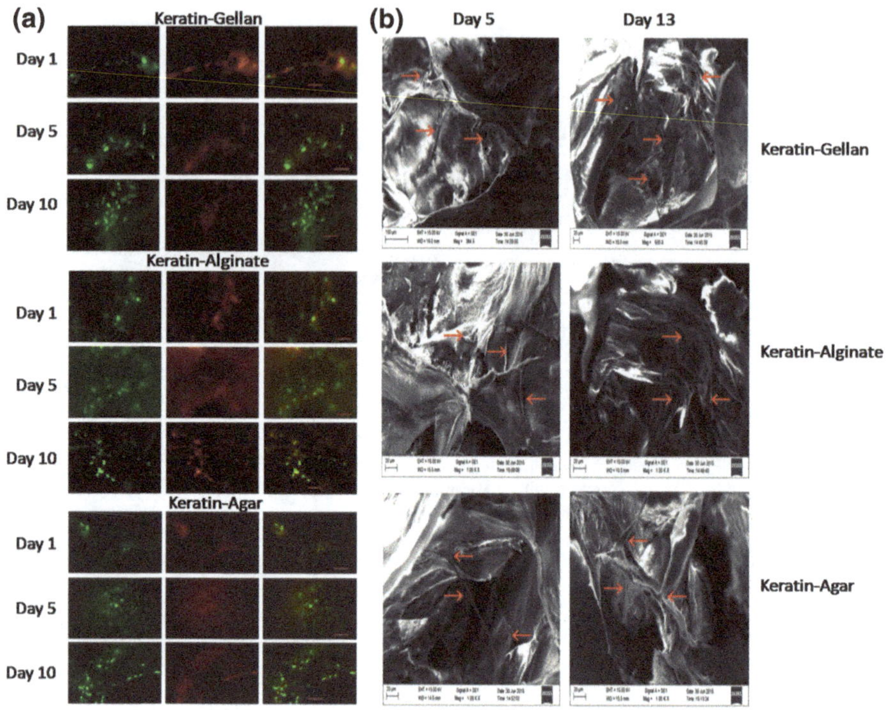

Fig. 5 Mammalian fibroblast cells attachment and migration study in keratin-based scaffold. In this figure, **a** represent the fluorescence images of cells and **b** illustrate the morphology of cells grown on the surface of keratin-based scaffold

Dickerson et al. 2013). Such dressings were found antimicrobial against the broad spectrum of pathogenic microorganisms, which is an essential property to a control wide range of microbial infection during wound healing. Skin regenerating potential of the keratin-based scaffold makes it a promising material further towards the development of a dressing material for wound treatment. Therapeutic wound dressings have been developed from keratin-based scaffolds, by incorporation of several bioactive components such as glucose oxidase, papain. and fibroblast growth factor into the biomaterial to control the glucose flux around the wound surface, for cell debridement and to promote cell proliferation, respectively (Nayak and Gupta 2017).

2.3 *Food and Nonfood Packaging Material*

The aim of biopolymer-based research is the preparation of a material that can imitate the function of synthetic polymer to some extent and replaced them from extensive use. Biocompatibility, complete degradation and low accumulation of the bio-originated polymers are the key points over synthetic polymers (Gupta and Nayak 2015a; Ramakrishnan et al. 2018), which attract investigator's attention. The origins of such biopolymers are from three kinds of renewable resources, viz., (a) plants such as starch, soy protein, and cellulose, (b) from animals such as chitosan, keratin and silk, (c) by microbial fermentation such as PHA and PHB (Gupta and Nayak 2015a). Nowadays, various polymer composites have been derived with novel characteristics due to advanced polymer engineering and technology and are explored significantly by various researchers (Nair and Laurencin 2007; Bandyopadhyay and Bose 2013). Protein is a strong candidate among the biopolymer and can be used to develop novel biocomposite material (Sharma et al. 2018). In biological sources, proteins are available either in globular or fibrous form, with the help of hydrogen, ionic, hydrophobic and covalent (disulfide) bonds. The physical and chemical behavior of the proteins depends on the fraction of amino acid residues and its placement along the polypeptide chain. The use of such amino acid sequence with other polymers can enrich polymer chemistry, which can be further investigated through protein engineering modifications to get a novel polymer design from proteins. Composites from protein like silk–chitosan (Bhardwaj and Kundu 2011), collagen–alginate (Petrulyte 2008), keratin–chitosan (Vazquez et al. 2015), keratin–polypropylene (Bertini et al. 2013) and keratin–polyethylene (Barone and Schmidt 2005) are studied by researchers and reported that the properties of these blend films (i.e., film strength, flexibility, and water vapor permeability, etc.) were improved when blended with protein polymer. Several globular proteins, including gluten, milk protein, soy protein, and whey protein have been investigated for their film properties (Pol et al. 2002). However, nonfood grade protein such as keratin and silk could be used in the development of packaging film for various material industries. The future application of protein-based biocomposite seems to be intense in the field of biomaterials, food and nonfood packaging and in coating industries (Gupta and Nayak 2015a). Protein-based edible films are attractive for food application because of their high nutritional quality, excellent sensory properties and good potential to adequately protect food products from their surrounding environment. Such films act as a carrier of antioxidant, flavor and bacteriostats can improve the quality of food products. Protein-based films have received considerable attention in recent years because of uses in edible and nonedible packaging materials (Gupta and Nayak 2015a). Because of the aforementioned nutrition properties the plant and animal proteins alike have been found to be appropriate for the production of edible film (Brandenburg et al. 1993; Cutter and Sumner 2002; Bourtoom 2009). After the study of various protein polymers like keratin, gluten, milk protein, zein, soy, silk, and jatropha protein, etc., scientific community concluded that the composites made up of proteins have a promising future in biotechnological application. Dou and colleagues, for biomedical and food packaging applications developed and characterized casting films based on feather keratin crosslinked by dialdehyde starch (Dou et al. 2015). The transparency value and moisture content of the film increased while the solubility decreased with the addition of dialdehyde starch. This indicated crosslinking between keratin and starch. Because of the increased water resistance, the keratin films found potential applications in food packaging. A transparent film for active packaging has been produced by mixing wool keratin and citric acid in water (Ramirez et al. 2017). This film can be used as an alternative to synthetic polymers as it has good biocidal effect, high elongation value (600%), and little loss of keratin after immersion in water. Barone and coworker (Barone et al. 2005) reported that the films prepared by mixing poultry feather keratin and glycerol were completely cohesive in contrast to being partially cohesive when prepared without glycerol. The films are "tough" and the mechanical properties show similarities to the properties of commercially available thermoplastics. The keratin films were produced in a few minutes without reducing or oxidizing agents. The keratin films could have potential environmental-friendly applications such as in food packaging and mulching films industries.

2.4 *Bio-cement for Hard Tissue Engineering*

Many of the filler agents are used in bone tissue engineering and dentistry to treat for the defect or injuries on bone and tooth (Hutmacher 2000; Kumar et al. 2013). The application of such filler agents started over 100 years ago with the implantation of calcium sulfate. The calcium sulfate was used initially by Dreesmann at Germany in 1892 to treat cavities in bone (Nilsson et al. 2013). This study was conducted over eight patients having large bone defects then grafting with b-calcium sulfate hemihydrates to treat bone defect and results showed complete regeneration of bone. In addition, it was also mentioned that calcium sulfate does not cause any complications even in infected cavities. In recent times, the majority of the studies were performed on calcium sulfate pellets (Parker et al. 2011) due to its outstanding biocompatibility and fast dissolution rate. But the single drawback of calcium sulfate is its availability for long term, which is needed for full bone regeneration. Hydroxyapatite is the most common calcium phosphate compound that has been widely used in hard tissue engineering due to its structural similarity and mineral composition like bone (Nilsson et al. 2013). It also expresses good biocompatibility, osteoconductive behavior, and bioactivity, which support the formation of chemical bonds between bones. Hydroxyapatite may remain at the site of implantation for years or even decades. This may be an advantage for certain applications, but is a limitation too for few applications, because in younger patients, resorbable implants should not present for longer time and are preferable to be replaced over time by bone tissue.

In clinical applications, bone substitutes from keratin derived material can be used for bone repair because of following properties (a) osteoconductivity, (b) injectability and (c) adequate strength and controllable resorption. A keratin–hydroxyapatite nanocomposite has been blended with PLLA to design a membrane by electrospinning (Li et al. 2013). The effect of this material on the growth of bone tissue was investigated in vitro and it is reported that the incorporation of keratin–hydroxyapatite into PLLA fibers improves bone formation with respect to control membrane of PLLA. Similarly, Arslan and colleagues studied the osteoinductive scaffold derived from keratin/jellyfish-collagen/eggshell-hydroxyapatite biocomposite for bone tissue engineering (Arslan et al. 2017). Moreover, reported that the developed biomaterial has the potential of being used in bone tissue engineering. In another study, keratin nanoparticles (nKer) were prepared and embedded with chitosan matrix to develop a scaffold for bone regeneration application (Saravanan et al. 2013). Against human osteoblastic cells, this scaffold was studied for its non-cytotoxicity behavior and the result proves the application of such material for bone tissue engineering applications.

In hard tissue engineering, another important aspect of keratin application is the preparation of dental cement (Meyers and Chen 2014). Dental pastes are fast setting mixtures formed by the blending of powder and liquid simultaneously. These may be prepared from acid–base cement or resin cement but should exhibit few characteristics as: (1) should not create irritation of gum and pulp, (2) shows enough chemical interaction with enamel and dentin, (3) does not dissolve easily, and (4) has chemical/thermal stability. Some of the common dental cements are zink phosphate, zink polyacrylate, glass ionomer cement, resin-based sealant, calcium sulfate, bioactive glass, calcium phosphate, and their composites (Meyers and Chen 2014). The biocompatibility and structural feature of keratin, similar to the above-stated material allow the scientific community to develop keratin-based dental cement as a value-added product for dentistry. Previously, keratinous materials were also reported as additives for the development of ceramics and concrete (Karthikeyan et al. 2007), hence keratin would be good choice for the development of bio-cement for various biotechnological applications.

2.5 *Bio-ink for 3D Printer*

3D printing is a process that refers to the fabrication of any three-dimensional object under computer controlled commands (Murphy and Atala 2014). In this process, a binder material (ink) deposits layer by layer onto a powder bed from head of the inkjet printer to form a 3D object. Fused filament fabrication or fused deposition modeling (FDM) is a technique, in which ink is melted or softened to produce the layers after deposition over the printing surface (Zhu et al. 2016). The specimen is also produced by spraying streams or small beads of material, which harden immediately to form layers. In addition, a filament of metal wire and thermoplastic material was used to develop 3D object, in this approach 3D printer extruder warm up metal and allow to flow on and off over the platform. After the deposition of the first layer, nozzle head moves in predefined direction over working platform for the addition of second layer. This process repeated until the formation of desire 3D shape of object. Other than FDM several techniques like selective laser sintering (SLS), stereolithography (SLA), and direct metal laser sintering (DMLS) are also available for 3D printing (An et al. 2015). These techniques differ only in terms of the ink that can be used and in the way the layers are deposited to create the object. SLS, SLA, and FDM are the most common methods in which biopolymers are used as ink. Hence, it is termed as bio-ink. Much attention on biopolymers for development of bio-ink has been given by scientific community due to similarity with extracellular matrix and ability to suspend cells in a gel at low temperatures (Li et al. 2014).

The biotechnological application of bio-ink in 3D printing has facilitated a new approach for the printing of cells and biomaterials to fabricate tissue analogous structures. For example, PHB a natural polymer that behaves like thermoplastic has attracted the interest of researchers for its biomedical applications because of its biocompatibility and biodegradability (Pereira et al. 2012). One of the advantages of PHB processing in 3D printing is non-utilization of plasticizers, unlike the traditional methods. Thermoplastic and biocompatible behavior of keratin favors its candidature to apply as bio-ink for the development of 3D scaffold, particularly for tissue engineering application. In addition, the ultimate goal of 3D printing technique is to provide a biomimetic structural atmosphere, which assists tissue construction along with host tissue integration like vascularization and cellular infiltration (Richards et al. 2013).

2.6 Other Applications

The application of keratin is not only limited to the construction of material for tissue engineering but also has utilized by several other industries like cosmetic, biofertilizer, cattle feedstock, leather, and textile processing industry for the development of value-added products (Sharma and Gupta 2016). Hydrolysate of keratin has various applications in cosmetic industries (creams, shampoos, hair conditioners) because of its hydrogel nature and biocompatibility (Karthikeyan et al. 2007). Such products have been reported to be useful for the treatment of human hair and skin. The blends of keratin with other natural polymers such as chitosan, collagen, and silk give value-added product for cosmetic industries (Sharma and Gupta 2016). Keratin is rich in nitrogen content and has supplemented with sulfur bond, due to which its other use would be the preparation of biofertilizer for the enhancement of plant metabolism and growth (Sharma and Gupta 2016). In addition, keratin extracted from natural sources by chemical or thermal hydrolysis is abundant in amino acid content and peptide along with 15–18% nitrogen, 2–5% sulfur, 3.20% mineral elements, and 1.27% fat (Karthikeyan et al. 2007). The composition of such hydrolysate was reported similar to the soybean protein and cottonseed protein hydrolysate, which is well known for diet source of farm animals. Hence, keratin hydrolysate from waste source would be a big raw material for the preparation of cattle feedstock. Karthikeyan et al., reported the potential application of keratin hydrolysate in leather tanning and retanning industry (Karthikeyan et al. 2007). They found that the keratin hydrolysate improves exhaustion of chromium in tanning bath up to 90%. They also reported that keratin-based material has been successfully utilized as retanning/filling agent in leather industry, which gives an opportunity to fill loose surface of leathers like flanks, belly and poor substance materials without contributing much load to tannery effluent.

3 Conclusions

The current trend of biomaterial development from keratin has motivated to fulfill the variety of applications that range from micro to macro level, i.e., pharmaceutical and health sector to cosmetic products. The chemical stability and biocompatibility of keratin greatly account for their utilization in biological devices and products. It is necessary to develop a biomaterial from waste keratin, which fulfills the required bioactivity, hydrogel nature, degradation, porosity, microstructure and mechanical properties for various biotechnological applications. Keratin-based biomaterials have advantages over conventional material due to biocompatibility, propensity for self-assembly and intrinsic cellular recognition. In addition, keratin's chemical structures and functional groups govern its morphology and other material properties. Specific modifications of the structure and functional groups within keratin allows the creation of desired molecular architectures having plethora of biomedical and tissue engineering applications. Scientific reports available till date revealed that keratin has a potential for the development of various value-added products. The demand of keratin as raw material at industrial scale would be fulfilled from the collection of waste biomass generated from animal industries. Thus, the utilization of such waste into the development of commercially important product can save earth's ecosystem and help in economy boost up of pharmaceutical and cosmetic industry. Keratin-derived biofertilizer is one of the prominent research areas, which is still unexplored on industrial scale and can be a good replacement of the chemical fertilizers.

References

Abukawa H, Papadaki M, Abulikemu M et al (2006) The engineering of craniofacial tissues in the laboratory: a review of biomaterials for scaffolds and implant coatings. Dent Clin North Am 50:205–216, viii. https://doi.org/10.1016/j.cden.2005.11.006

Ahmed S, Ikram S (2016) Chitosan based scaffolds and their applications in wound healing. Achiev Life Sci. https://doi.org/10.1016/j.als.2016.04.001

Akita S, Akino K, Hirano A (2013) Basic fibroblast growth factor in scarless wound healing. Adv Wound Care 2:44–49. https://doi.org/10.1089/wound.2011.0324

An J, Teoh JEM, Suntornnond R, Chua CK (2015) Design and 3D printing of scaffolds and tissues. Engineering 1:261–268. https://doi.org/10.15302/j-eng-2015061

An T, Zhang C, Han X et al (2016) Hyaluronic acid-coated poly(β-amino) ester nanoparticles as carrier of doxorubicin for overcoming drug resistance in breast cancer cells. RSC Adv 6:38624–38636. https://doi.org/10.1039/c6ra03997a

Arslan YE, Sezgin Arslan T, Derkus B et al (2017) Fabrication of human hair keratin/jellyfish collagen/eggshell-derived hydroxyapatite osteoinductive biocomposite scaffolds for bone tissue engineering: from waste to regenerative medicine products. Colloids SurfS B Biointerfaces 154:160–170. https://doi.org/10.1016/j.colsurfb.2017.03.034

Augst AD, Kong HJ, Mooney DJ (2006) Alginate hydrogels as biomaterials. Macromol Biosci 6:623–633. https://doi.org/10.1002/mabi.200600069

Badylak SF (2002) The extracellular matrix as a scaffold for tissue reconstruction. Semin Cell Dev Biol 13:377–383

Bainbridge P (2013) Keratin nanofibers as a biomaterial. J Wound Care 22:407–412. https://doi.org/10.12968/jowc.2013.22.8.407

Balaji S, Kumar R, Sripriya R et al (2012) Preparation and comparative characterization of keratin–chitosan and keratin–gelatin composite scaffolds for tissue engineering applications. Mater Sci Eng, C 32:975–982. https://doi.org/10.1016/j.msec.2012.02.023

Bandyopadhyay A, Bose S (2013) Characterization of biomaterials, 1st edn. Elsevier, The Netherlands

Barone JR, Schmidt WF (2005) Polyethylene reinforced with keratin fibers obtained from chicken feathers. Compos Sci Technol 65:173–181

Barone JR, Schmidt WF, Liebner CFE (2005) Thermally processed keratin films. J Appl Polym Sci 97:1644–1651. https://doi.org/10.1002/app.21901

Belverud S, Mogilner A, Schulder M (2008) Intrathecal pumps. Neurotherapeutics 5:114–122. https://doi.org/10.1016/j.nurt.2007.10.070

Berglund JD, Mohseni MM, Nerem RM, Sambanis A (2003) A biological hybrid model for collagen-based tissue engineered vascular constructs. Biomaterials 24:1241–1254. https://doi.org/10.1016/s0142-9612(02)00506-9

Bertini F, Canetti M, Patrucco A, Zoccola M (2013) Wool keratin–polypropylene composites: properties and thermal degradation. Polym Degrad Stab 98:980–987

Bhardwaj N, Kundu SC (2011) Silk fibroin protein and chitosan polyelectrolyte complex porous scaffolds for tissue engineering applications. Carbohydr Polym 85:325–333

Bhardwaj N, Sow WT, Devi D et al (2015) Silk fibroin–keratin based 3D scaffolds as a dermal substitute for skin tissue engineering. Integr Biol (Camb) 7:53–63. https://doi.org/10.1039/c4ib00208c

Blakytny R, Jude E (2006) The molecular biology of chronic wounds and delayed healing in diabetes. Diabet Med 23:594–608. https://doi.org/10.1111/j.1464-5491.2006.01773.x

Boateng JS, Matthews KH, Stevens HNE, Eccleston GM (2008) Wound healing dressings and drug delivery systems: a review. J Pharm Sci 97:2829–2923. https://doi.org/10.1002/jps

Bourtoom T (2009) Edible protein films: properties enhancement. Int Food Res J 16:1–9

Bowler PG, Duerden BI, Armstrong DG (2001) Wound microbiology and associated approaches to wound management. Clin Microbiol Rev 14:244–269. https://doi.org/10.1128/cmr.14.2.244-269.2001

Brandenburg AH, Weller CL, Testin RF (1993) Edible films and coatings from soy protein. J Food Sci 58:1086–1089

Bullions TA, Hoffman D, Price-O'Brien J, Loos AC (2003) Feather fiber/cellulose fiber/polypropylene composites manufactured via the wetlay papermaking process. In: International Nonwovens Technical Conference (INTC 2003), Balt, MD, USA

Bullions TA, Gillespie RA, Price-O'Brien J, Loos AC (2004) The effect of maleic anhydride modified polypropylene on the mechanical properties of feather fiber, kraft pulp, polypropylene composites. J Appl Polym Sci 92:3771–3783

Cadée JA, Brouwer LA, den Otter W et al (2001) A comparative biocompatibility study of microspheres based on crosslinked dextran or poly(lactic-co-glycolic)acid after subcutaneous injection in rats. J Biomed Mater Res 56:600–609

Calkins M (2009) Materials for sustainable sites: a complete guide to the evaluation, selection, and use of sustainable construction materials. Wiley, Hoboke

Callahan TD, Natale A (2008) Catheter ablation of atrial fibrillation. Med Clin North Am 92(179–201):xii. https://doi.org/10.1016/j.mcna.2007.09.001

Capperauld I (1989) Suture materials: a review. Clin Mater 4:3–12. https://doi.org/10.1016/0267-6605(89)90021-8

Choi SM, Ryu HA, Lee K-M et al (2016) Development of stabilized growth factor-loaded Hyaluronate–Collagen Dressing (HCD) matrix for impaired wound healing. Biomater Res 20:9. https://doi.org/10.1186/s40824-016-0056-4

Ciapetti G, Cenni E, Pratelli L, Pizzoferrato A (1993) In vitro evaluation of cell/biomaterial interaction by MTT assay. Biomaterials 14:359–364. https://doi.org/10.1016/0142-9612(93)90055-7

Ciardelli G, Chiono V, Vozzi G et al (2005) Blends of poly-(epsilon-caprolactone) and polysaccharides in tissue engineering applications. Biomacromolecules 6:1961–1976. https://doi.org/10.1021/bm0500805

Cutter CN, Sumner SS (2002) Application of edible coatings on muscle foods. In: Gennedios A (ed) Protein-based films and coatings. CRC, Boca Raton, FL, pp 467–484

Cziple FA, Marques AJV (2008) Biopolymers versus synthetic polymers. Eftimie Murgu University of Resita, Romania, pp 125–132

Dalsin JL, Hu B-H, Lee BP, Messersmith PB (2003) Mussel adhesive protein mimetic polymers for the preparation of nonfouling surfaces. J Am Chem Soc 125:4253–4258. https://doi.org/10.1021/ja0284963

de Vos P, Hoogmoed CG, Busscher HJ (2002) Chemistry and biocompatibility of alginate-PLL capsules for immunoprotection of mammalian cells. J Biomed Mater Res 60:252–259

Denkbas EB, Ozturk E, Ozdemi N et al (2003) EGF loaded chitosan sponges as wound dressing material. J Bioact Compat Polym 18:177–190. https://doi.org/10.1177/0883911503035388

Dhivya S, Padma VV, Santhini E (2015) Wound dressings—a review. BioMedicine 5:22. https://doi.org/10.7603/s40681-015-0022-9

Dickinson LE, Gerecht S (2016) Engineered biopolymeric scaffolds for chronic wound healing. Front Physiol 7:341. https://doi.org/10.3389/fphys.2016.00341

Dickerson MB, Sierra AA, Bedford NM et al (2013) Keratin-based antimicrobial textiles, films, and nanofibers. J Mater Chem B 1:5505–5514. https://doi.org/10.1039/c3tb20896f

Dou Y, Huang X, Zhang B et al (2015) Preparation and characterization of a dialdehyde starch crosslinked feather keratin film for food packaging application. RSC Adv 5:27168–27174. https://doi.org/10.1039/c4ra15469j

Draye JP, Delaey B, Van de Voorde A et al (1998) In vitro and in vivo biocompatibility of dextran dialdehyde cross-linked gelatin hydrogel films. Biomaterials 19:1677–1687

Dreifke MB, Ebraheim NA, Jayasuriya AC (2013) Investigation of potential injectable polymeric biomaterials for bone regeneration. J Biomed Mater Res A 101:2436–2447. https://doi.org/10.1002/jbm.a.34521

Edmonds M, European and Australian Apligraf Diabetic Foot Ulcer Study Group (2009) Apligraf in the treatment of neuropathic diabetic foot ulcers. Int J Low Extrem Wounds 8:11–18. https://doi.org/10.1177/1534734609331597

El Fray M, Prowans P, Puskas JE, Altstadt V (2006) Biocompatibility and fatigue properties of Polystyrene–Polyisobutylene–Polystyrene, an emerging thermoplastic elastomeric biomaterial. Biomacromolecules 7:844–850. https://doi.org/10.1021/bm050971c

El Ghalbzouri A, Hensbergen P, Gibbs S et al (2004) Fibroblasts facilitate re-epithelialization in wounded human skin equivalents. Lab Invest 84:102–112. https://doi.org/10.1038/sj.labinvest.3700014

Epstein FH, Singer AJ, Clark RAF (1999) Cutaneous wound healing. N Engl J Med 341:738–746. https://doi.org/10.1056/nejm199909023411006

Falanga V, Isaacs C, Paquette D et al (2002) Wounding of bioengineered skin: cellular and molecular aspects after injury. J Invest Dermatol 119:653–660. https://doi.org/10.1046/j.1523-1747.2002.01865.x

Feughelman M (1959) A two-phase structure for keratin fibers. Text Res J 29:223–228

Fine DM, Tobias AH (2007) Cardiovascular device infections in dogs: report of 8 cases and review of the literature. J Vet Intern Med 21:1265–1271

Frazier OH, Jacob LP (2007) Small pumps for ventricular assistance: progress in mechanical circulatory support. Cardiol Clin 25(553–64):vi. https://doi.org/10.1016/j.ccl.2007.09.001

Geiger M, Li RH, Friess W (2003) Collagen sponges for bone regeneration with rhBMP-2. Adv Drug Deliv Rev 55:1613–1629

Gupta P, Nayak KK (2015a) Characteristics of protein-based biopolymer and its application. Polym Eng Sci 55:485–498. https://doi.org/10.1002/pen.23928

Gupta P, Nayak KK (2015b) Compatibility study of alginate/keratin blend for biopolymer development. J Appl Biomater Funct Mater 13:e332–e339. https://doi.org/10.5301/jabfm.5000242

Gupta P, Nayak KK (2016) Optimization of keratin/alginate scaffold using RSM and its characterization for tissue engineering. Int J Biol Macromol 85:141–149. https://doi.org/10.1016/j.ijbiomac.2015.12.010

Gurtner GC, Werner S, Barrandon Y, Longaker MT (2008) Wound repair and regeneration. Nature 453:314–321. https://doi.org/10.1038/nature07039

Harrison J, Pattanawong S, Forsythe JS et al (2004) Colonization and maintenance of murine embryonic stem cells on poly(alpha-hydroxy esters). Biomaterials 25:4963–4970. https://doi.org/10.1016/j.biomaterials.2004.01.054

He Y, Lu F (2016) Development of synthetic and natural materials for tissue engineering applications using adipose stem cells. Stem Cells Int 2016:5786257. https://doi.org/10.1155/2016/5786257

Hemsri S, Asandei AD, Grieco K, Parnas RS (2011) Biopolymer composites of wheat gluten with silica and alumina. Compos Part A: Appl Sci Manuf 42:1764–1773. https://doi.org/10.1016/j.compositesa.2011.07.032

Hu S, Kirsner RS, Falanga V et al (2006) Evaluation of Apligraf® persistence and basement membrane restoration in donor site wounds: a pilot study. Wound Repair Regen 14:427–433. https://doi.org/10.1111/j.1743-6109.2006.00148.x

Huang X, Zhang Y, Zhang X et al (2013) Influence of radiation crosslinked carboxymethyl-chitosan/gelatin hydrogel on cutaneous wound healing. Mater Sci Eng, C 33:4816–4824. https://doi.org/10.1016/j.msec.2013.07.044

Hutmacher DW (2000) Scaffolds in tissue engineering bone and cartilage. Biomaterials 21:2529–2543. https://doi.org/10.1016/s0142-9612(00)00121-6

Isarankura Na Ayutthaya S, Tanpichai S, Wootthikanokkhan J (2015) Keratin extracted from chicken feather waste: extraction, preparation, and structural characterization of the keratin and keratin/biopolymer films and electrospuns. J Polym Environ 23:506–516. https://doi.org/10.1007/s10924-015-0725-8

Jung Y, Kim S-S, Kim YH et al (2005) A poly(lactic acid)/calcium metaphosphate composite for bone tissue engineering. Biomaterials 26:6314–6322. https://doi.org/10.1016/j.biomaterials.2005.04.007

Karthikeyan R, Balaji S, Sehgal PK (2007) Industrial applications of keratins—a review. J Sci Ind Res 66:710–715

Keong LC, Halim AS (2009) In vitro models in biocompatibility assessment for biomedical-grade chitosan derivatives in wound management. Int J Mol Sci 10:1300–1313. https://doi.org/10.3390/ijms10031300

Khan F, Tanaka M, Ahmad SR et al (2015) Fabrication of polymeric biomaterials: a strategy for tissue engineering and medical devices. J Mater Chem B 3:8224–8249. https://doi.org/10.1039/c5tb01370d

Khor E, Lim LY (2003) Implantable applications of chitin and chitosan. Biomaterials 24:2339–2349

Kipshidze NN, Tsapenko MV, Leon MB et al (2005) Update on drug-eluting coronary stents. Expert Rev Cardiovasc Ther 3:953–968. https://doi.org/10.1586/14779072.3.5.953

Kizilel S, Garfinkel M, Opara E (2005) The bioartificial pancreas: progress and challenges. Diabetes Technol Ther 7:968–985. https://doi.org/10.1089/dia.2005.7.968

Koschwanez HE, Reichert WM (2007) In vitro, in vivo and post explantation testing of glucose-detecting biosensors: current methods and recommendations. Biomaterials 28:3687–3703. https://doi.org/10.1016/j.biomaterials.2007.03.034

Kumar P, Vinitha B, Fathima G (2013) Bone grafts in dentistry. J Pharm Bioallied Sci 5:S125–S127. https://doi.org/10.4103/0975-7406.113312

Kumaran P, Gupta A, Sharma S (2017) Synthesis of wound-healing keratin hydrogels using chicken feathers proteins and its properties. Int J Pharm Pharm Sci 9:171. https://doi.org/10.22159/ijpps.2017v9i2.15620

Landsman AS, Cook J, Cook E et al (2011) A retrospective clinical study of 188 consecutive patients to examine the effectiveness of a biologically active cryopreserved human skin allograft (TheraSkin®) on the treatment of diabetic foot ulcers and venous leg ulcers. Foot Ankle Spec 4:29–41. https://doi.org/10.1177/1938640010387417

Levingstone TJ, Matsiko A, Dickson GR et al (2014) A biomimetic multi-layered collagen-based scaffold for osteochondral repair. Acta Biomater 10:1996–2004. https://doi.org/10.1016/j.actbio.2014.01.005

Li J-S, Li Y, Liu X et al (2013) Strategy to introduce an hydroxyapatite–keratin nanocomposite into a fibrous membrane for bone tissue engineering. J Mater Chem B 1:432–437. https://doi.org/10.1039/c2tb00460g

Li X, Cui R, Sun L et al (2014) 3D-printed biopolymers for tissue engineering application. Int J Polym Sci 2014:1–13. https://doi.org/10.1155/2014/829145

Liu H, Slamovich EB, Webster TJ (2006) Less harmful acidic degradation of poly(lacticco-glycolic acid) bone tissue engineering scaffolds through titania nanoparticle addition. Int J Nanomedicine 1:541–545

Lu L, Peter SJ, Lyman MD et al (2000) In vitro and in vivo degradation of porous poly(dl-lactic-co-glycolic acid) foams. Biomaterials 21:1837–1845. https://doi.org/10.1016/s0142-9612(00)00047-8

Lu T, Qiao Y, Liu X (2012) Surface modification of biomaterials using plasma immersion ion implantation and deposition. Interface Focus 2:325–336. https://doi.org/10.1098/rsfs.2012.0003

Lunsford L, McKeever U, Eckstein V, Hedley ML (2000) Tissue distribution and persistence in mice of plasmid DNA encapsulated in a PLGA-based microsphere delivery vehicle. J Drug Target 8:39–50. https://doi.org/10.3109/10611860009009208

Ma PX (2008) Biomimetic materials for tissue engineering. Adv Drug Deliv Rev 60:184–198. https://doi.org/10.1016/j.addr.2007.08.041

Maitz MF (2015) Applications of synthetic polymers in clinical medicine. Biosurface Biotribology 1:161–176. https://doi.org/10.1016/j.bsbt.2015.08.002

Marques AP, Reis RL, Hunt JA (2002) The biocompatibility of novel starch-based polymers and composites: in vitro studies. Biomaterials 23:1471–1478. https://doi.org/10.1016/s0142-9612(01)00272-1

Marston WA, Hanft J, Norwood P et al (2003) The efficacy and safety of Dermagraft in improving the healing of chronic diabetic foot ulcers: results of a prospective randomized trial. Diabetes Care 26:1701–1715

Maruoka S, Matsuura T, Kawasaki K et al (2006) Biocompatibility of polyvinylalcohol gel as a vitreous substitute. Curr Eye Res 31:599–606. https://doi.org/10.1080/02713680600813854

Matsumoto T, Okazaki M, Nakahira A et al (2007) Modification of apatite materials for bone tissue engineering and drug delivery carriers. Curr Med Chem 14:2726–2733

McKittrick J, Chen P-Y, Bodde SG et al (2012) The structure, functions, and mechanical properties of keratin. JOM 64:449–468. https://doi.org/10.1007/s11837-012-0302-8

Mendes SC, Reis RL, Bovell YP et al (2001) Biocompatibility testing of novel starch-based materials with potential application in orthopaedic surgery: a preliminary study. Biomaterials 22:2057–2064

Meyers MA, Chen P-Y (2014) Biological materials science: biological materials, bioinspired materials, and biomaterials. Cambridge University Press and the Materials Research Society, Cambridge

Moses JW, Kipshidze N, Leon MB (2002) Perspectives of drug-eluting stents: the next revolution. Am J Cardiovasc Drugs 2:163–172

Murphy SV, Atala A (2014) 3D bioprinting of tissues and organs. Nat Biotechnol 32:773–785. https://doi.org/10.1038/nbt.2958

Nair LS, Laurencin CT (2007) Biodegradable polymers as biomaterials. Prog Polym Sci 32:762–798. https://doi.org/10.1016/j.progpolymsci.2007.05.017

Naughton G, Mansbridge J, Gentzkow G (1997) A metabolically active human dermal replacement for the treatment of diabetic foot ulcers. Artif Organs 21:1203–1210

Nayak KK, Gupta P (2015) In vitro biocompatibility study of keratin/agar scaffold for tissue engineering. Int J Biol Macromol 81:1–10. https://doi.org/10.1016/j.ijbiomac.2015.07.025

Nayak KK, Gupta P (2017) Study of the keratin-based therapeutic dermal patches for the delivery of bioactive molecules for wound treatment. Mater Sci Eng, C 77:1088–1097. https://doi.org/10.1016/j.msec.2017.04.042

Nilsson M, Zheng MH, Tägil M (2013) The composite of hydroxyapatite and calcium sulphate: a review of preclinical evaluation and clinical applications. Expert Rev Med Devices 10:675–684. https://doi.org/10.1586/17434440.2013.827529

Nozynski JK, Religa Z, Wszołek J et al (2001) Biological heart valve an alternative to mechanical valve. Med Sci Monit 7:550–562

O'Brien FJ (2011) Biomaterials & scaffolds for tissue engineering. Mater Today 14:88–95. https://doi.org/10.1016/s1369-7021(11)70058-x

O'Connor RC, Lyon MB, Guralnick ML, Bales GT (2008) Long-term follow-up of single versus double cuff artificial urinary sphincter insertion for the treatment of severe postprostatectomy stress urinary incontinence. Urology 71:90–93. https://doi.org/10.1016/j.urology.2007.08.017

Oh S (2003) Fabrication and characterization of hydrophilic poly(lactic-co-glycolic acid)/poly(vinyl alcohol) blend cell scaffolds by melt-molding particulate-leaching method. Biomaterials 24:4011–4021. https://doi.org/10.1016/s0142-9612(03)00284-9

Panchal B, Bagdadi A, Roy I (2013) Polyhydroxyalkanoates: the natural polymers produced by bacterial fermentation. In: Thomas S, Visakh P, Mathew A (eds) Advances in natural polymers, vol 18. Springer, Berlin, pp 397–421

Paradossi G, Cavalieri F, Chiessi E et al (2003) Poly(vinyl alcohol) as versatile biomaterial for potential biomedical applications. J Mater Sci Mater Med 14:687–691

Parker AC, Smith JK, Courtney HS, Haggard WO (2011) Evaluation of two sources of calcium sulfate for a local drug delivery system: a pilot study. Clin Orthop Relat Res 469:3008–3015. https://doi.org/10.1007/s11999-011-1911-1

Pereira TF, Oliveira MF, Maia IA et al (2012) 3D printing of poly(3-hydroxybutyrate) porous structures using selective laser sintering. Macromol Symp 319:64–73. https://doi.org/10.1002/masy.201100237

Petrulyte S (2008) Advanced textile materials and biopolymers in wound management. Dan Med Bull 55:72–77

Pol H, Dawson P, Acton J, Ogale A (2002) Soy protein isolate/corn-zein laminated films: transport and mechanical properties. J Food Sci 67:212–217

Ramakrishnan N, Sharma S, Gupta A, Alashwal BY (2018) Keratin based bioplastic film from chicken feathers and its characterization. Int J Biol Macromol 111:352–358. https://doi.org/10.1016/j.ijbiomac.2018.01.037

Ramirez DOS, Carletto RA, Tonetti C et al (2017) Wool keratin film plasticized by citric acid for food packaging. Food Packag Shelf Life 12:100–106. https://doi.org/10.1016/J.FPSL.2017.04.004

Rath G, Hussain T, Chauhan G et al (2015) Collagen nanofiber containing silver nanoparticles for improved wound-healing applications. J Drug Target 1–10. https://doi.org/10.3109/1061186x.2015.1095922

Reichl S (2009) Films based on human hair keratin as substrates for cell culture and tissue engineering. Biomaterials 30:6854–6866. https://doi.org/10.1016/j.biomaterials.2009.08.051

Restle L, Costa-Silva D, Loureno ES et al (2015) A 3D osteoblast in vitro model for the evaluation of biomedical materials. Adv Mater Sci Eng 2015:1–8. https://doi.org/10.1155/2015/268930

Richards DJ, Tan Y, Jia J et al (2013) 3D printing for tissue engineering. Isr J Chem 53:805–814

Rosellini E, Cristallini C, Barbani N et al (2009) Preparation and characterization of alginate/gelatin blend films for cardiac tissue engineering. J Biomed Mater Res, Part A 91:447–453. https://doi.org/10.1002/jbm.a.32216

Rowlands AS, Lim SA, Martin D, Cooper-White JJ (2007) Polyurethane/poly(lactic-co-glycolic) acid composite scaffolds fabricated by thermally induced phase separation. Biomaterials 28:2109–2121. https://doi.org/10.1016/j.biomaterials.2006.12.032

Royals MA, Fujita SM, Yewey GL et al (1999) Biocompatibility of a biodegradable in situ forming implant system in rhesus monkeys. J Biomed Mater Res 45:231–239

Rujitanaroj P, Pimpha N, Supaphol P (2008) Wound-dressing materials with antibacterial activity from electrospun gelatin fiber mats containing silver nanoparticles. Polymer (Guildf) 49:4723–4732. https://doi.org/10.1016/j.polymer.2008.08.021

Salgado AJ, Coutinho OP, Reis RL (2004) Bone tissue engineering: state of the art and future trends. Macromol Biosci 4:743–765. https://doi.org/10.1002/mabi.200400026
Santoro M, Gaudino G (2005) Cellular and molecular facets of keratinocyte reepithelization during wound healing. Exp Cell Res 304:274–286. https://doi.org/10.1016/j.yexcr.2004.10.033
Sarabahi S (2012) Recent advances in topical wound care. Indian J Plast Surg 45:379–387. https://doi.org/10.4103/0970-0358.101321
Saravanan S, Sameera DK, Moorthi A, Selvamurugan N (2013) Chitosan scaffolds containing chicken feather keratin nanoparticles for bone tissue engineering. Int J Biol Macromol 62:481–486. https://doi.org/10.1016/j.ijbiomac.2013.09.034
Schuster J (2003) Polypropylene reinforced with chicken feathers. In: International conference on composite materials (ICCM-14), San Diego, CA
Sharma S, Gupta A (2016) Sustainable management of keratin waste biomass: applications and future perspectives. Braz Arch Biol Technol 59:1–14. https://doi.org/10.1590/1678-4324-2016150684
Sharma S, Gupta A, Kumar A et al (2018) An efficient conversion of waste feather keratin into ecofriendly bioplastic film. Clean Technol Environ Policy 1–11. https://doi.org/10.1007/s10098-018-1498-2
Shi W, Dumont MJ (2014) Review: bio-based films from zein, keratin, pea, and rapeseed protein feedstocks. J Mater Sci 49:1915–1930. https://doi.org/10.1007/s10853-013-7933-1
Silva R, Singh R, Sarker B et al (2014) Hybrid hydrogels based on keratin and alginate for tissue engineering. J Mater Chem B 2:5441–5451. https://doi.org/10.1039/c4tb00776j
Song E, Yeon Kim S, Chun T et al (2006) Collagen scaffolds derived from a marine source and their biocompatibility. Biomaterials 27:2951–2961. https://doi.org/10.1016/j.biomaterials.2006.01.015
Spector M (2007) Biomaterials-based tissue engineering and regenerative medicine solutions to musculoskeletal problems. Swiss Med Wkly 137(Suppl):157S–165S
Tanabe T, Okitsu N, Tachibana A, Yamauchi K (2002) Preparation and characterization of keratin–chitosan composite film. Biomaterials 23:817–825
Tracy LE, Minasian RA, Caterson EJ (2016) Extracellular matrix and dermal fibroblast function in the healing wound. Adv Wound Care 5:119–136. https://doi.org/10.1089/wound.2014.0561
Tsai WB, Grunkemeier JM, McFarland CD, Horbett TA (2002) Platelet adhesion to polystyrene-based surfaces preadsorbed with plasmas selectively depleted in fibrinogen, fibronectin, vitronectin, or von Willebrand's factor. J Biomed Mater Res 60:348–359
Vasconcelos A, Cavaco-Paulo A (2013) The use of keratin in biomedical applications. Curr Drug Targets 14:612–619
Vazquez N, Chacón M, Meana A et al (2015) Keratin–chitosan membranes as scaffold for tissue engineering of human cornea. Histol Histopathol 30:813–821. http://dx.doi.org/10.14670/HH-11-585
Vercruysse KP, Prestwich GD (1998) Hyaluronate derivatives in drug delivery. Crit Rev Ther Drug Carrier Syst 15:513–555
Veves A, Falanga V, Armstrong DG et al (2001) Graftskin, a human skin equivalent, is effective in the management of noninfected neuropathic diabetic foot ulcers: a prospective randomized multicenter clinical trial. Diabetes Care 24:290–295
Wang YX, Cao XJ (2012) Extracting keratin from chicken feathers by using a hydrophobic ionic liquid. Process Biochem 47(5): 896–899
Wang YC, Lin MC, Wang DM, Hsieh HJ (2003) Fabrication of a novel porous PGA-chitosan hybrid matrix for tissue engineering. Biomaterials 24:1047–1057
Wang HM, Chou YT, Wen ZH et al (2013) Novel biodegradable porous scaffold applied to skin regeneration. PLoS ONE 8:1–11. https://doi.org/10.1371/journal.pone.0056330
Williams DF (2011) The Williams dictionary of biomaterials. Liverpool University Press, Liverpool
Wrzesniewska-Tosik K, Adamiec J (2007) Biocomposites with a content of keratin from chicken feathers. Fibres Text East Eur 15:106–112
Xing ZC, Yuan J, Chae WP et al (2011) Keratin nanofibers as a biomaterial. In: 2010 international conference on nanotechnology and biosensors (IPCBEE), pp 120–124

Xiong H, Tang S, Tang H, Zou P (2008) The structure and properties of a starch-based biodegradable film. Carbohydr Polym 71:263–268. https://doi.org/10.1016/j.carbpol.2007.05.035
Xu H, Cai S, Xu L, Yang Y (2014) Water-stable three-dimensional ultrafine fibrous scaffolds from keratin for cartilage tissue engineering. Langmuir 30:8461–8470. https://doi.org/10.1021/la500768b
Xu H, Ma L, Shi H et al (2007) Chitosan–hyaluronic acid hybrid film as a novel wound dressing: in vitro and in vivo studies. Polym Adv Technol 18:869–875. https://doi.org/10.1002/pat
Yang J, Wang N, Chiu H (2014) Preparation and characterization of poly(vinyl alcohol)/sodium alginate blended membrane for alkaline solid polymer electrolytes membrane. J Memb Sci 457:139–148. https://doi.org/10.1016/j.memsci.2014.01.034
Yu X, Tang X, Gohil SV, Laurencin CT (2015) Biomaterials for bone regenerative engineering. Adv Healthc Mater 4:1268–1285. https://doi.org/10.1002/adhm.201400760
Yun YR, Won JE, Jeon E et al (2010) Fibroblast growth factors: biology, function, and application for tissue regeneration. J Tissue Eng 1:218142. https://doi.org/10.4061/2010/218142
Zaulyanov L, Kirsner RS (2007) A review of a bi-layered living cell treatment (Apligraf) in the treatment of venous leg ulcers and diabetic foot ulcers. Clin Interv Aging 2:93–98
Zhang L, Li K, Xiao W et al (2011) Preparation of collagen–chondroitin sulfate–hyaluronic acid hybrid hydrogel scaffolds and cell compatibility in vitro. Carbohydr Polym 84:118–125. https://doi.org/10.1016/j.carbpol.2010.11.009
Zhijiang C, Chengwei H, Guang Y (2012) Poly(3-hydroxubutyrate-co-4-hydroxubutyrate)/bacterial cellulose composite porous scaffold: preparation, characterization and biocompatibility evaluation. Carbohydr Polym 87:1073–1080. https://doi.org/10.1016/j.carbpol.2011.08.037
Zhu W, Ma X, Gou M et al (2016) 3D printing of functional biomaterials for tissue engineering. Curr Opin Biotechnol 40:103–112. https://doi.org/10.1016/J.COPBIO.2016.03.014

FSC
www.fsc.org
MIX
Papier | Fördert
gute Waldnutzung
FSC® C083411

Zeitfracht Medien GmbH
Ferdinand-Jühlke-Straße 7
99095 Erfurt, Deutschland
produktsicherheit@kolibri360.de